Advanced Materials for Water Handling: Composites and Thermoplastics

1st Edition

Advanced Materials for Water Handling: Composites and Thermoplastics

1st Edition

Derick Scott

ELSEVIER
ADVANCED
TECHNOLOGY

UK	Elsevier Science Ltd, The Boulevard, Langford Lane, Kidlington, Oxford OX5 1GB, UK
USA	Elsevier Science Inc, 665 Avenue of the Americas, New York, NY 10010, USA
JAPAN	Elsevier Science Japan, Tsunashima Building Annex, 3-20-12 Yushima, Bunkyo-ku, Tokyo 113, Japan

Copyright © 2000 Elsevier Science Ltd.

All rights reserved. No part of this publication may be reproduced, stored in a retrieval system or transmitted in any form or by any means: electronic, electrostatic, magnetic tape, mechanical, photocopying, recording or otherwise, without permission in writing from the publishers.

First edition 2000

Library of Congress Cataloging-in-Publication Data
Scott, Derick Vernon
  Advanced materials for water handling: composites and thermoplastics / Derick Scott.–1st ed.
    p. cm.
  ISBN 1-85617-350–X
  1. Thermoplastics. 2. Thermosetting plastics. 3. Water-pipes–Materials. I. Title.
TA455.P5 S39 2000
628.1–dc21   00-042976

British Library Cataloguing in Publication Data
A catalogue record for this title is available from the British Library.

ISBN 1 85617 350 X

No responsibility is assumed by the Publisher for any injury and/or damage to persons or property as a matter of products liability, negligence or otherwise, or from any use or operation of any methods, products, instructions or ideas contained in the material herein.

Published by
Elsevier Advanced Technology,
The Boulevard, Langford Lane, Kidlington, Oxford OX5 1GB, UK
Tel.: +44(0) 1865 843000
Fax: +44(0) 1865 843971

Typeset by Variorum Publishing Ltd, Rugby
Printed and bound in Great Britain by
Biddles Ltd, Guildford and King's Lynn

Preface

The word 'plastic' has tended to be used in a general and sometimes derogatory manner for many years. The author apologises for using this word frequently, but has found no alternative to describe both thermoplastic and thermosetting composite materials in one word.

Thermosetting and composite materials are considered as identical in this book, unless specific reference is made to a thermosetting resin, and in this case all information concerns the resin itself, rather than a composite material. FRP (fibre reinforced plastic) or GRP (glass fibre reinforced plastic) may also be used instead of composite, especially when used in relation to industrial manufacturers and some standards.

The author has more than 30 years of experience in the fields of composite and thermoplastic materials, and has been actively involved in the design, manufacture, sales and installation of a large range of equipment for the water, oil and chemical markets in Europe, the Middle East and North America. He has considerable experience with ISO 9000 quality systems in relationship to pipe systems and vessels manufactured from thermoplastic and composite materials, as well as ASME Section X for the design, manufacture and testing of composite high pressure vessels. He has previously published with Elsevier Advanced Technology a review of the European FRP Tank and Vessel market, and has worked as a consultant in the field of manufacture and sales of composite materials for the last five years.

Comments, requests for further clarification and/or further contributions from suppliers of raw material, finished or semi-finished products, end users or organisations concerned with the specification, design or working of plastic materials in contact with water are more than welcome and will be considered for inclusion in the first revision of this handbook.

Derick Scott

Please contact the author, Derick Scott, at ICCR (International Composites Consultancy and Representation) at:
12 Rue Abbé Vacandard, 76000 Rouen, France.
Tel: +33 2 35 89 48 43; Fax: +33 2 35 88 61 64
E-mail: dscott@wanadoo.fr

Acknowledgements

The Author and the Publisher would like to thank the following organisations for their kind permission to reproduce the illustrations listed below.

	Page Number	Company
Figure 1	14	Degrémont
Figure 2	19	ACWa Services
Table 1	25	DSM Composite Resins
Table 4	131	Smith Fiberglass Product Company
Figure 21	156	KWH Pipe Ltd.
Figure 22	158	Plasson Ltd.
Figure 23	159	Dyka (UK) Ltd.
Figure 24	160	Dyka (UK) Ltd.
Figure 25	161	Hydrodif
Figure 26	163	George Fischer
Figure 27	164	KWH Tech.
Figure 29	168	Fusion Group Manufacturing
Figure 30	172	Bauku (www.bauku.de)
Figure 31	177	Forbes
Figure 32	178	Forbes
Figure 33	190	Tankinetics Inc.
Figure 34	192	Dewey Waters
Figure 35	193	Brinar Plastic Fabrications Ltd.
Figure 36	208	FilmTec Corp.
Figure 37	208	FilmTec Corp.
Figures 38–45	211–215	Structural NV
Table 6	225	CertainTeed Corp.
Figures 47, 48	226, 227	USF Filtration and Separation
Figure 49	229	Iniziative Industriali SpA
Figures 50–56	235–238	Subterra
Figures 57, 58	240, 241	TROLINING GmbH

	Page Number	Company
Figures 59–63	249–252	Steuler Industriewerke GmbH-Surface protection systems
Figure 65	265	International Grating Inc.
Figures 66, 67	272, 273	F Z Fantoni SpA
Table 7	274	F Z Fantoni SpA
Figure 68	276	Bio-Plus Environmental Systems Ltd.
Table 8	276	Bio-Plus Environmental Systems Ltd.
Figure 69	277	WPL Ltd.
Table 9	277	WPL Ltd.
Figure 70	279	Feluwa Pumpen GmbH
Figure 71	282	2H Kunststoff GmbH
Figure 72	283	Munters Euroform GmbH
Figure 73	285	Zenon Environmental Systems Inc.
Figure 74	289	Marley Cooling Tower Co.
Figure 75	294	CertainTeed Corp.
Figures 76, 77	295, 296	Europelec/SFA
Figure 78	298	Mono Pumps Ltd.
Figure 79	300	Tankinetics Inc.
Figures 80, 81	302, 304	C P Composites
Figures 82, 83	306, 307	Redman Fisher Engineering Ltd.
Table 10	308	Redman Fisher Engineering Ltd.
Figure 84	311	EWWA Sarl
Figure 85	314	Aform Projects Ltd.
Figures 87, 88	319, 320	GPC
Figure 89	323	Johnston Pipes Ltd.
Figure 90	325	Smith Fiberglass Products Company
Figure 91	326	Aform Projects Ltd.
Figure 92	328	Ershigs Inc.
Figure 93	330	Smith Fiberglass Products Company
Figure 94	333	Flowtite Technology A/S
Figure 95	335	Subterra
Figures 96–99	337–340	KWH Pipe Ltd.
Figure 100	341	Pall Corp.

CONTENTS

Preface		v
Acknowledgements		vii

Chapter 1 Introduction
1.1 Historical Background of the Water and Plastic Industries — 2
1.2 Economical and Engineering Trends — 5

Chapter 2 The Water Industry
2.1 Water Treatment — 10
    2.1.1 Rainwater — 10
    2.1.2 Surface water — 10
        2.1.2.1 River water and lakes – upstream — 10
        2.1.2.2 River water and lakes – downstream — 11
    2.1.3 Underground sources — 11
    2.1.4 Sea or salty water — 11
    2.1.5 Miscellaneous water specifications — 11
2.2 Potable Water — 12
    2.2.1 Quality standards — 12
        2.2.1.1 Microbiological — 13
        2.2.1.2 Physical and chemical — 13
        2.2.1.3 An example — 13
    2.2.2 Treatment process: potable water — 14
        2.2.2.1 Screening — 15
        2.2.2.2 Pre-disinfection or oxidation — 15
        2.2.2.3 Micro-screening — 15
        2.2.2.4 Coagulation and flocculation — 15
        2.2.2.5 Decantation — 16
        2.2.2.6 Filtration — 16
        2.2.2.7 Disinfection — 16
        2.2.2.8 Fluoridation and mineralisation — 17

		2.2.2.9	Storage	17
		2.2.2.10	Distribution	17
	2.3	Wastewater and Effluents		19
		2.3.1	Treatment processes	19
		2.3.1.1	Recuperation	19
		2.3.1.2	Screening and grit removal	20
		2.3.1.3	Grease and oil removal	20
		2.3.1.4	Flocculation, settling and flotation	20
		2.3.1.5	Aerobic and anaerobic biological treatment	21
		2.3.1.6	Dewatering and sludge thickening	21
		2.3.1.7	Odour control	21

Chapter 3 Materials

	3.1	Thermosets/Composite Materials		24
		3.1.1	Thermosetting resins	24
		3.1.2	Vinyl ester resins	28
		3.1.3	Bisphenol fumerate resins	30
		3.1.4	Liquid epoxy resins	31
		3.1.4.1	Anhydride cured epoxy resins	32
		3.1.4.2	Polyamine cured epoxy resins	33
		3.1.5	Solid epoxy resins	33
		3.1.6	Polyurethane resins	34
		3.1.7	Other thermosetting resins	34
		3.1.7.1	Furane resins	34
		3.1.7.2	Phenolic resins	34
		3.1.7.3	HET resins	34
		3.1.8	Curing and hardening agents	36
		3.1.9	Miscellaneous fillers/additives	36
		3.1.10	Reinforcing fibres	37
		3.1.10.1	Glass manufacture	37
		3.1.10.2	Forming	38
		3.1.10.3	Sizing	38
		3.1.10.4	Types of glass fibre	38
		3.1.10.5	Forms of glass fibre reinforcement	39
		3.1.11	Moulding compounds	42
	3.2	Thermoplastics		44
		3.2.1	ABS — acrylonitrile butadiene styrene	44
		3.2.2	ECTFE — ethylene-chlorotrifluoroethylene	44
		3.2.3	FEP — fluorinated ethylene propylene	45
		3.2.4	PA — polyamide	45
		3.2.5	PBT — polybutylene terephtalate	45

		3.2.6	PC — polycarbonate	45
		3.2.7	Pe — polyethylene	45
		3.2.8	PET — polyethylene terephtalate	45
		3.2.9	PFA — perfluoro-alkoxy	45
		3.2.10	Polyether ketone	46
		3.2.11	PMMA — polymethylmethacrylate	46
		3.2.12	POM — polyacetal	46
		3.2.13	PP — polypropylene	46
		3.2.14	PPE — polyphenylene ether (see PPO)	46
		3.2.15	PPO — modified polyphenylene oxide	46
		3.2.16	PPS — polyphenylene sulphone	46
		3.2.17	PS — polystyrene	47
		3.2.18	PTFE — polytetrafluoroethylene	47
		3.2.19	PVC — polyvinyl chloride (PVC-U, PVC-C)	47
		3.2.20	PVDF — polyvinylidene fluoride	47
	3.3	Thermoplastic Materials for the Construction of Dual Laminates		48
	3.4	Summary		50

Chapter 4 Production Technology

	4.1	Thermosets/Composite Materials		54
		4.1.1	Contact moulding	58
		4.1.2	Spray up moulding	59
		4.1.3	Filament winding	60
		4.1.4	Centrifugation or centrifugal casting	62
		4.1.5	Resin transfer moulding	63
		4.1.6	Compression moulding	64
		4.1.7	Pultrusion	65
	4.2	Thermoplastics		67
		4.2.1	Injection, compression and transfer moulding	67
		4.2.2	Extrusion	68
		4.2.3	Rotational moulding	69
		4.2.4	Thermoforming	70
		4.2.5	High-frequency welding	70
		4.2.6	Boiler work (welded tanks and other vessels)	71
		4.2.7	Machining	71
		4.2.8	Dip moulding or coating	72

Chapter 5 Design, Production and Testing Standards

	5.1	Composite Pipe Systems		75
		5.1.1	North American standards	75
			5.1.1.1 Specifying and testing of composite pipe systems	76

			5.1.1.2	Recommended practices	78
			5.1.1.3	Test methods	78
			5.1.1.4	Fire or flame resistance	83
		5.1.2	Individual national standards — Europe		83
			5.1.2.1	France	83
			5.1.2.2	Germany	84
			5.1.2.3	UK	85
		5.1.3	European and cross-border standards (EN)		86
		5.1.4	International standards (ISO)		87
		5.1.5	Japanese standards		89
	5.2	Composite Tanks and Vessels			91
		5.2.1	North America		91
		5.2.2	Europe		93
			5.2.2.1	France	94
			5.2.2.2	Germany	94
			5.2.2.3	UK	94
		5.2.3	Japan		94
	5.3	Thermoplastic Pipe Systems			95
		5.3.1	North America		95
		5.3.2	Europe		96
			5.3.2.1	UK national standards	96
			5.3.2.2	German national standards	98
			5.3.2.3	French national standards	99
		5.3.3	Japanese standards		100
		5.3.4	Cross-border standards		101
			5.3.4.1	International standards (ISO)	101
			5.3.4.2	European standards (EN)	105
	5.4	Thermoplastic Tanks and Vessels			108
	5.5	Composite Profiles and Associated Constructions			109

Chapter 6 **Using non-metallic Materials with Potable Water**

	6.1	Laws and Regulations		112
	6.2	Comparison of Standards in Europe and North America		114
		6.2.1	France	114
		6.2.2	Germany	115
			6.2.2.1 KTW recommendations	115
			6.2.2.2 DVGW W270 testing	115
		6.2.3	The Netherlands	115
		6.2.4	UK	116
		6.2.5	European legislation and directives	117
		6.2.6	North America	117

xiii

		6.2.6.1	FDA	117
		6.2.6.2	NSF Standard 61	118
	6.3	Testing Organisations or Official Bodies where Relevant Information can be Obtained		121
		6.3.1	Austria	121
		6.3.2	Belgium	121
		6.3.3	Denmark	121
		6.3.4	Finland	121
		6.3.5	France	121
		6.3.6	Germany	122
		6.3.7	Italy	122
		6.3.8	The Netherlands	122
		6.3.9	Norway	122
		6.3.10	Spain	122
		6.3.11	Sweden	122
		6.3.12	UK	122
		6.3.13	North America	123

Chapter 7 Products and Applications — Composite and Thermoplastic Pipe Systems

	7.1	Composite Materials — Filament Wound Epoxy Pipes and Fittings		126
		7.1.1	General information	126
		7.1.2	Design basis	127
		7.1.3	Product range	127
		7.1.4	Jointing systems	128
		7.1.5	Installation	130
		7.1.6	Supporting documentation	132
		7.1.7	Typical suppliers, products, designation, sizes and pressure/temperature ratings	132
		7.1.8	Recent developments	135
	7.2	Composite Materials — Large Diameter Filament Wound and Centrifugally Cast Polyester and Vinyl Ester Pipes and Fittings		138
		7.2.1	General information	138
		7.2.2	Design basis	139
		7.2.3	Product range	139
			7.2.3.1 Continuous filament wound pipe systems, trademark: Flowtite	139
			7.2.3.2 Reciprocal filament wound pipe	140
			7.2.3.3 Centrifugal cast pipe—Hobas and licensed companies operating the Hobas license	140

	7.2.4	Installation	141
	7.2.5	Supporting documation	141
7.3	Composite Materials — Small to Medium Diameter Filament Wound Polyester and Vinyl Ester Pipes and Fittings		142
	7.3.1	General information	142
	7.3.2	Design basis	142
	7.3.3	Product range	144
7.4	Composite Materials — Small to Medium Diameter Centrifugally Cast Epoxy and Vinyl Ester Pipes and Fittings		145
7.5	Composite Materials — Filament Wound Epoxy and Vinyl Ester Pipe Fittings		146
7.6	Composite Materials — Resin Transfer and Compression Moulded Epoxy, Vinyl Ester and Polyester Fittings		147
7.7	Thermoplastic Pipes		148
	7.7.1	Extruded pipe	149
		7.7.1.1 PVC	149
		7.7.1.2 PVC-C or chlorinated PVC	149
		7.7.1.3 Bi-oriented PVC or otherwise strengthened PVC pipe	150
		7.7.1.4 Polyethylene	151
		7.7.1.5 Cross linked HdPe or PE-X	153
		7.7.1.6 PB	154
		7.7.1.7 PP	154
		7.7.1.8 PVDF	154
		7.7.1.9 Dual laminate pipe systems	154
		7.7.1.10 Thermoplastic lined steel pipe system	155
	7.7.2	Spiral wound pipe	155
	7.7.3	Assembly of thermoplastic pipe systems	157
		7.7.3.1 Solvent cement bonding	157
		7.7.3.2 Mechanical joints	159
		7.7.3.3 Compression joints	159
		7.7.3.4 Spigot/socket with elastomeric seals	159
		7.7.3.5 Flanged joints	160
		7.7.3.6 Threaded joints	161
		7.7.3.7 Welding	161
		7.7.3.7.1 Socket fusion welding	162
		7.7.3.7.2 Butt welded polyfusion	163
		7.7.3.7.3 Electrically heated socket fusion welding	166

		7.7.3.8	Qualification of welders	168
	7.7.4	Thermoplastic pipe fittings		169
	7.7.5	Design of thermoplastic pipe systems		169
		7.7.5.1	Product design	169
		7.7.5.2	System design	169
	7.7.6	Pipe installation and testing		170
	7.7.7	Large-diameter fittings and access chambers for water and effluent mains		172

Chapter 8 Products and Applications — Composite and Thermoplastic Tanks, Silos and Other Vessels

8.1	Cylindrical Composite Tanks and Other Vessels		176
	8.1.1	Design	179
	8.1.2	Applications	183
8.2	Rectangular or Square Composite Storage Tanks		191
8.3	Thermoplastic and Dual Laminate Tanks and Vessels		195
	8.3.1	Rotationally moulded thermoplastic tanks	195
	8.3.2	Spiral wound and welded shell construction	196
	8.3.3	Thermoplastic welded tank and vessel construction	197
8.4	Thermoplastic Welding Technology		198
	8.4.1	Fusion welding	198
	8.4.2	Hot gas welding	199
	8.4.3	Extrusion welding	200
8.5	Dual Laminate Construction		201
8.6	High-Pressure Filters		203
8.7	Small Composite Low Pressure Filters, Containers and Storage Tanks		210

Chapter 9 Products and Applications — Miscellaneous Equipment

9.1	Tank Covers and Odour Control Equipment in Wastewater Treatment plants			218
9.2	Well and Borehole Equipment			223
	9.2.1	Thermoplastic well casings		223
	9.2.2	Composite tubing and well casing/screens		228
9.3	Coating and Linings			231
	9.3.1	Initial protection of new equipment		231
		9.3.1.1	Pipe systems	231
		9.3.1.2	Tanks, vessels and other constructions	231
	9.3.2	In-situ lining of existing equipment		231
		9.3.2.1	Pipe systems	231
		9.3.2.2	Protection of other constructions for fluid containment	244

9.4	District Heating and Cooling Processes	253
9.5	Industrial Process, Demineralised and Deionised Water	255
9.6	Ultra-Pure Water and Similar Applications	256
9.7	Composite Structures and Gratings	257
9.8	Water Treatment Associated with Air Pollution Control	260
9.9	Miscellaneous Equipment for Water and Effluent Treatment	261
9.10	Onshore and Offshore Oilfield Applications	262
9.11	Buried Fire Mains	264
9.12	Jetties and Locks	265
9.13	Irrigation	268

Chapter 10 Novel Applications

10.1	Composite Modular Settling Tanks	272
10.2	Package Sewage Treatment Plants	275
10.3	Prefabricated Pumping Stations	279
10.4	Modular Biomedia	281
10.5	Advanced Polymeric Membranc Systems for Potable and Wastewater Treatment	285
10.6	Cooling Towers and Cooling Tower Components	287
10.7	Restrained Joint PVC Pipe	293
10.8	Aeration and Oxidation — Aerators and Fine Bubble Air Diffusers	295
10.9	Dynamic Screening with the Discreen Family	298
10.10	Oblated FRP Tanks	300
10.11	Gravity Filter Floors in Composite Sandwich Construction	302
10.12	Articulated Lamellae Filters for Easy Cleaning	304
10.13	FlowGRiP Composite Planks	306

Chapter 11 Case Histories

11.1	Nanofiltration Plant at Mery sur Oise, Paris, France	310
11.2	Water Treatment Plant — Odour Control at Howdon, UK	313
11.3	Geothermal Sourced District Heating	315
11.3.1	Jonzac, France	315
11.3.2	Le Mée sur Seine, France	317
11.3.3	Down hole tubing and casing	318
11.4	Vertical Composite Sewer Pipe — Ipswich, UK	322
11.5	A 2100 Metre FRP Pipeline Insertion — Portland, USA	324
11.6	FRP Folding Covers — Farmoor, UK	326
11.7	FRP Storage Tanks — Massachusetts, USA	328

	11.8	FRP Pipe in Wastewater Treatment Plant — Arizona, USA	330
	11.9	FRP Pressure Pipe in Sewage Transfer Scheme in Grimmen, Germany	333
	11.10	Polyethylene Relining of 5500 Metres of 30 and 36 Inch Potable Water Pressure Mains	335
	11.11	Renovation of 33 000 Meters of Sewer Pipe with 160–225 HdPe	337
	11.12	HdPe Submarine Pipeline in Malaysia	339
	11.13	Reverse Osmosis Disc Tube (DT) and Landfill Leachate Treatment	341

Chapter 12 Quality Systems and a Summary of Considerations Specific to the use of Thermoplastic and Composite Materials

12.1	A Guide to the Design and Quality Control of Composite Vessels	346
	12.1.1 Vents and overflows	349
	12.1.2 Fitting design	349
	12.1.3 Handling and installation	350
	12.1.4 Inspection	350
	12.1.5 External inspection	351
	12.1.6 Internal inspection	351
	12.1.7 Conclusion	352

Chapter 13 Raw Materials

13.1	Trade and Registered Names of Raw Materials and the Relevant Producers	356
13.2	Suppliers of Raw Materials (for Composite Products)	362
	13.2.1 Polyester/vinyl ester/bisphenol/HET resins	362
	13.2.2 Epoxy resins	363
	13.2.3 Phenolic and/or furane resins	364
	13.2.4 Glass fibre reinforcements	364
	13.2.5 Glass fibre and synthetic surfacing veils	365
	13.2.6 Coating materials and formulations	366
	13.2.7 Thermoplastic sheet and pipe designed specifically for incorporation into dual laminates	368

Chapter 14 Trademarks, Equipment Suppliers and Manufacturers

14.1	Pipe Systems	370
	14.1.1 Filament wound epoxy pipes and fittings	370
	14.1.2 Filament wound polyester and vinyl ester pipes and fittings	371

	14.1.3	Centrifugal cast and filament wound epoxy and vinyl ester pipes and fittings	373
	14.1.4	Centrifugal cast polyester pipes and fittings	374
	14.1.5	RTM epoxy, vinyl ester and polyester fittings	374
	14.1.6	High-pressure filament wound pipe, tubing and casing	374
	14.1.7	Low-pressure filament wound tubing and casing	375
	14.1.8	Dual laminate pipe systems	375
	14.1.9	Thermoplastic lined steel pipe	376
14.2	Thermoplastic Pipe Systems		377
	14.2.1	Extruded thermoplastic pipes	377
	14.2.2	Spiral wound thermoplastic pipes	381
	14.2.3	Thermoplastic pipe fittings	381
	14.2.4	Large diameter fabricated fittings and access chambers for water and effluent mains	383
	14.2.5	Thermoplastic well casings and screens	383
	14.2.6	Thermoplastic valves	384
14.3	Composite Vessels		385
	14.3.1	Standard composite storage tanks, vessels for storage and treatment of water and silos	385
	14.3.2	Custom fabricated vessels	388
	14.3.3	Standard high- and low-pressure composite storage tanks/filter bodies	390
	14.3.4	Modular composite storage tanks	390
	14.3.5	Very large filament wound composite storage tanks (on-site manufacture, diameter above 8 m/25 ft)	392
14.4	Thermoplastic Vessels		393
	14.4.1	Rotationally moulded thermoplastic storage tanks	393
	14.4.2	Other standard thermoplastic storage tanks	393
14.5	Other Equipment		394
	14.5.1	High-pressure filters	394
	14.5.2	Reverse osmosis pressure vessels	394
	14.5.3	Spiral wound membranes	395
		14.5.3.1 Other membrane filtration technologies referred to in this handbook	396
	14.5.4	Tank covers and odour control equipment	396
	14.5.5	Biological carrier media	397
	14.5.6	Cooling towers (and components)	398
14.6	Composite Structures and Gratings		400

		14.6.1	Manufacturers of composite profiles and gratings	400
		14.6.2	Companies specialised in the design and fabrication of composite structures from pultruded sections	403
	14.7	Coating and Linings		405
		14.7.1	Pipe linings	405
		14.7.2	In-situ tank and vessel linings — applicators	406
	14.8	Mini/modular Wastewater Treatment Plants		407
	14.9	Thermoplastic Pipe and Sheet Welding Equipment		408
	14.10	Specialised Water Treatment Equipment		410
	14.11	Engineering and Contracting Compaies		411
	14.12	Consultants		412

Chapter 15 Governing Bodies, Authorities, and Associations of Relevance to the Scope of this Handbook
    15.1 The Field of Water Supply and Treatment   414
    15.2 The Field of Composites and Thermoplastics, and Products Manufactured from these Materials   416

Chapter 16 Glossary of Abbreviations/Technical Terms; Bibliography
    16.1 Summary of Abbreviations and Technical Terms Frequently Used in the Water or Plastic Industries   420
    16.2 Bibliography   424
          16.2.1 The water industry   424
          16.2.2 The composite and plastics industry   424

Editorial Index   427

CHAPTER 1

Introduction

1.1 Historical Background of the Water and Plastic Industries

Water is of course the one essential material required for all forms of life. Water covers 75% of the surface of our planet, and makes up 80% of the weight of an average human body. However, only 3% of water on the planet is potable and only one third of this water is readily available for use by people. More than 25% of the earth's population has no access to water which could be described as drinkable, and every day 25 000 people die because of the poor quality of their drinking water. With the ever growing world population and the associated increase in all forms of agricultural and economic activities, our limited water resources will require even more careful and innovative management over the coming years. The most recent forecasts predict that if no significant changes are made in how we access, use and dispose of water, in the year 2025 more than 50% of the population of the world will not have sufficient access to potable water.

Handling water has been a major part of the activity of all civilisations, and water has been collected, treated, stored, transported and distributed in practically all-imaginable materials, ranging from animal skins, to rock, wood, glass and ceramics, and practically all metals. The development of cities and whole civilisations can be traced back to the availability and good management of adequate water resources, the Roman civilisation being probably the best example in historic terms. The Romans constructed gigantic systems of aqueducts, and separated their supply into two or three different systems depending upon the quality of the water, and the use to which it was destined. Water treatment on an industrial scale as we appreciate it today has its origins in western Europe in the mid-sixteenth century, and from this period onwards numerous references can be found as to the requirement to treat both drinking water and sewerage. We have tended to accept that pollution of rivers is essentially a problem of the twentieth century, but many extreme cases of chronic pollution of major rivers such as the Thames can seen in the eighteenth and nineteenth centuries, probably amplified by the technological advance of the WC or water closet at the beginning of the nineteenth century, which entailed drainage systems discharging directly into the nearest river. A law was enacted in London in 1852 which required that all drinking water should be filtered, and on the effluent side various type of sewage farms were in place by the end of the nineteenth century. In an attempt to reduce the level of bacteria in water, the UK added chlorine to drinking water as from 1904, and was followed by the USA in 1919.

The actual combination of rarefaction of water resources and increased usage will only increase this dependability of civilisation on abundant and economic water supplies in the future. Fortunately new materials and technologies have already or are rapidly becoming available which will allow in many cases to reduce loss and wastage of water, recuperate and re-use water and effluent, or recover water from sources which have not been practical or economic to date. Composite and thermoplastic materials can play a major role in satisfying this increased demand, under extremely economic conditions when compared to technology considered as revolutionary and very expensive as little as 20 years ago.

The second half of the twentieth century has already witnessed the introduction and development of new products and technologies for handling, storing or treating water in all its forms, based on a wide range of thermoplastic and composite materials. However, we should not ignore the continual development of metallic, cement based or cement-lined and clay products which will continue to play major roles in many sectors of the water industry.

Probably the earliest significant development with synthetic plastic materials was the introduction of PVC pipe in the 1950s, and it is interesting to note the development of the product over the last 50 years, including recent developments in bi-oriented or otherwise strengthened tubular products as well as the PVC-C family of high-temperature pipe systems. At the same time one can trace the development of polyethylene-based systems which have already taken over from PVC in many fields. The introduction and development of composite pipe systems has further developed the use of plastic materials for the transport of large volumes of water over long distances. The composite business is relatively young, with less than 50 years of experience since the original industrial applications in North America, and whilst many companies have emerged and subsequently disappeared over the last 50 years, it is worthwhile noting that the original pioneers are still, in the main, the leaders in the industry today.

These and all the other developments will be explored throughout the following chapters of this book.

The object of this handbook is essentially to demonstrate to the user, engineer or constructor of water handling equipment in all forms and phases of industrial, private and municipal installations that the world of plastic materials offers a wide range of possible applications. These applications range from low cost, basic everyday products such as PVC drainage lines up to highly engineered, high-performance thermoplastic and composite materials such as high-pressure vessels tested to 420 bar, down hole tubing and casing up to 2000 m depth, drawing on one essential property of the plastics, that is the resistance to the corrosive effects of most forms of water, the enemy of many competitive metallic materials. Whilst many references will be made to applications of plastics in water industries, this book does not pretend to be in any way a guide on water treatment procedures or the design and construction of materials and equipment specifically for this industry, although this is a new tendency which must be followed through via a co-operation between the water industry and the composite and thermoplastic fabricators, using the wide range of properties of the different materials to solve both material and process problems.

Apart from the water industry as such, that is the collection, preparation and distribution of potable water, followed by the evacuation, treatment and disposal of waste water and effluent, other applications and usage of water are of importance to the economy of the world today. Applications of thermoplastic and composite materials in a wide range of industries will be discussed. These include such applications as industrial process water, collection and treatment of industrial effluent, seawater cooling of power and other industrial plants, handling of geothermal and geothermic water for district heating, injection of water at high-pressure into partially depleted oil fields, production and storage of demincralised and deionised water, and water for the food and drinks industries.

1.2 Economical and Engineering Trends

Corrosion and/or oxidation of metallic materials as well as erosion of metallic or non-metallic materials in the presence of various forms of water are well-known phenomena, and are well documented. The design of water handling equipment with conventional materials is often based on an accepted loss of wall thickness via oxidation and/or erosion over a specific period of time. The necessity to reline, coat or paint the equipment is built into the cost of the project, and in many cases the consideration of initial cost of purchase and installation plus long-term maintenance costs, or even replacement, over an extended service life, forms the real overall cost of the project. In some cases, plant down-time required to maintain or replace corroded material can be a significant consideration when comparing overall budgets of conventional and more advanced materials.

Composites and thermoplastics exist side by side, in many applications with water, with a wide range of conventional metallic materials such as steel, stainless steel, cast and ductile iron, galvanised steel, aluminium, brass, copper–zinc alloys, copper and copper–nickel alloys, and in older installations, lead. Concrete and cement-based mortar linings also enter into consideration for medium- to large-diameter pipe systems and storage tanks and reservoirs. The study of the chemical and abrasion resistance of all these materials in the presence of the wide range of types or qualities of water to be handled is extremely complex, and can require the consideration of factors which have no concern for the thermoplastic and composite alternatives.

Electrochemical corrosion of steel, with or without the presence of oxygen is always a basic consideration within the selective process of a construction material, and spontaneous or induced formation of protective layers or films and passivation can be used to combat to some extent this phenomenon. This complex study may call for even more complex solutions, such as involving the use of corrosion inhibitors, maintaining a sufficient level of alkaline or oxidising media to obtain passivation of the material surfaces, in some cases increasing the pH level, or for buried pipelines, cathodic protection. In many cases, these solutions will add to either the initial cost of investment or to the running costs of the installation, or both, and many possible anti-corrosion solutions may have such an influence on the quality of the water that they may not be employed.

The effects of the type and level of the mineralisation of the water will also have to be taken into account, with both small levels of chlorides and sulphates being capable of causing corrosion of both mild and stainless steels. Further areas of consideration may include:

- The effect of temperature on the corrosive activity of oxygenated waters, especially at temperatures above 60°C.
- The corrosive effect of microorganisms.
- The effects of the velocity of the fluid on corrosion of the pipe.
- The effects of the pH level in oxygenated water.

Moving up market, so to speak, in the metallic materials, the wide range of ferritic, austenitic and austeno-ferritic (duplex and super duplex) stainless steels offer considerable opportunities to resolve corrosion problems, usually associated with high material and welding costs. High-temperature applications excepted, composite materials can offer significant initial and long-term advantages over many grades of stainless steel. The chloride resistance of the more economic grades of stainless steel is very limited, and exotic materials such as 254SMO or super-duplex materials are usually required when handling saline or seawater, whereas on the corrosion side a basic and very economic PVC or isophthalic based polyester resin will perform extremely well.

Concrete, or the cement basis of concrete can be attacked by an association of mechanical and chemical factors. Basic mechanical or structural damage may be caused by:

- Excessive permeability of the concrete due to low levels of cement dosage.
- Presence of cracks and crevices due to the low ductility of the concrete, caused by incorrect ratios of water/cement or the lack of plasticizers.
- Erosion due to high water flow velocities or excessive heat.

Chemical factors causing the deterioration of concrete or cement based linings may include:

- Carbonic attack initiated by carbon dioxide in the water.
- Attack by any acidic components within the fluids being handled, especially phosphoric, sulphuric, hydrochloric and nitric acid based contaminants.
- Attack from ammonia contained within effluent via the development of an acidic nitrification.
- Bacterial corrosion within municipal effluent, with an initial formation and release of H_2S, and a subsequent formation of sulphurous and sulphuric acids.

The expected service life of plastic-based materials is often used in design codes and standards, alongside the usual factors concerning temperature, pressure and other mechanical loading. Many short- and medium-term test procedures have been developed and which forecast the loss of mechanical properties over extended periods, which can vary between 10 and 50 years or more, depending upon the application under consideration. Many major manufacturers of basic equipment such as pipe systems, tanks and filters have applications in service for 30 years or more, and are able to offer some considerable proof as to the validity of what was, at the start, an empirical and relatively insecure design basis.

Current design and economical trends are somewhat complicated. We live in a society which requires low prices to obtain contracts, high return on investment for shareholders, and long-term guarantees on materials and performances. Thermoplastic and composite materials are not 'wonder products', they do not offer immediate and cost efficient solutions to all corrosion problems, but they do offer significant advantages in many applications, sometimes associated with lower installed costs than many basic conventional materials. On the one hand, subject to careful material selection they do offer a high level of chemical resistance to all forms of water and solutions based upon water, up to working temperatures depending upon the specific material used. Based upon this criterion alone, they can be shown to have an almost unlimited service life, in other words an investor's dream. On the other hand, and in complete opposition to the major characteristics of conventional materials, they show significant loss of mechanical properties over a period of time, which may be more or less long depending upon such factors as temperature or pressure or other mechanical loading. This phenomenon is not necessarily a handicap, as long as the design engineer takes it into account, and that the subsequent equation of cost/performance is not detrimental to the long-term viability of the product or project (see Section 5.1.1.3 for details of the hydrostatic design basis of a composite pipe system and Section 8.1.1 for information concerning the design of composite vessels).

The quality and ultimate performance of plastic products are heavily dependent upon the production process used and the associated quality control procedures. One of the great drawbacks for the industrial composite market has been the extremely poor performance of many small or not so small companies who have taken advantage of the fact that even relatively large and/or complex structures may be built with little or no investment in production, control and testing equipment. Significant failures have occurred in all markets, and unfortunately, due to their very nature, if composite materials fail, then they tend to do so in a catastrophic manner. The author cannot stress too much the importance of purchasing equipment from reputable sources which are able to demonstrate a high and constant level of quality, via their production and testing equipment, allied to an in-depth knowledge and application of relevant standards and codes. Mass produced equipment such as pipes, standard tanks and pressure vessels manufactured by computer-controlled filament winding or centrifugal casting machines would tend to fall into this category, whereas one-off or custom designed equipment needs a much higher level of control during design and manufacture.

Most thermoplastic materials require a combination of high manufacturing volumes to justify relatively high production machinery and tooling costs. This requirement for a high level of investment combined with in depth market research and associated mechanical design preclude 'low cost' companies from entering this market sector, and the catastrophic failures experienced with some composite materials can be avoided. This does not mean that expensive failures have not been experienced with thermoplastic equipment, but on analysis one can usually trace back the origins of the problem to poor design or erroneous material and performance specifications.

In spite of the financial trends indicated above, an ever-increasing number of engineers and buyers are specifying composite materials for applications in the water industry, often taking the precaution to specify a manufacturer or type of equipment which has performed well in the past. Long-term guarantees as to the mechanical performance of materials have been requested and obtained, with one to three years being standard, five years being possible, and ten years having been awarded in some cases, depending upon the type of product and the service conditions.

Engineering includes not only the selection of materials and production procedures to manufacture specific items, but increasingly so, for all applications handling potable water, a raw material or semi-finished product which will not pollute or change in any way the quality of the water. For many years the suitability of materials to handle potable water has been overlooked, that is their tendency to transfer taste or harmful elements to the water. An obvious case is the intensive use of lead pipes in domestic and municipal water systems until very recently. Actual analysis of potable water in countries such as France show areas having levels of lead above limits of European legislation, and well above levels which must be attained within the next few years. The ever-increasing requirement for materials to conform to national and international standards in this domain must not be neglected, and the position for plastic (or non-metallic materials) is examined in Chapter 6.

CHAPTER 2

The Water Industry

As indicated previously this handbook is not in any way intended to instruct the reader in the procedures and technologies used to treat water and effluent. A brief introduction and overview of the water industry as a whole has however been included, and may help to define the main areas of interest for plastic materials. For those having developed a desire to obtain an introduction into the mechanical, chemical and biological aspects of water treatment in many forms, from all types of industrial to municipal applications, the author would strongly recommend the reader to acquire a copy of the *Water Treatment Handbook* and other associated documents published by Degrémont (in English and French, see bibliography) which offers the most comprehensive information generally available, used by both the Degrémont organisation as well as some of their competitors!

2.1 Water Treatment

Water is available, as a raw material, either by collection directly as it falls as rain, or by collection from holes or depressions in the earth's surface, as river, lake or marsh water, of very variable quality depending upon any natural filtration or mineralisation, or as salt or brackish water collected in general from or close to the sea. The treatment of the various types of water depends on the application to which it is destined, together with a huge range of other factors such as volume, temperature, type of pollution if any, environmental considerations and cost.

A description of the types of water available for treatment is described below.

2.1.1 Rainwater

Rainwater can be collected in most areas of the world, except for those subject to heavy industrialisation where it can be contaminated by dust and gas in the atmosphere. However, the problems of captivation and storage on any significant scale generally override any advantage of availability.

2.1.2 Surface water

This includes all stored or flowing water on the surface of the earth, and can be broken down into the following sub-sections, each of which will present different chemical compositions. These compositions will depend upon the type of terrain through which the water has flowed, and will include dissolved gases, suspended solids, organic matter and plankton. In any given river or lake the actual chemical make-up may vary immensely from day to day, or season to season depending upon ambient temperature, amount of sunlight and levels of precipitation, for example. Change may be slow and progressive as one moves from one season to another, or dramatic in the case of storms, which cause sudden flooding.

2.1.2.1 River water and lakes — upstream

Defined as generally hilly or mountainous countryside, with low-density population and no significant industrial activity. In stormy conditions, the water could contain high levels of solids, but the general level of bacteria would be very low. Water temperatures tend to be low.

2.1.2.2 River water and lakes — downstream

Generally situated in flat countryside with a high-density population and high levels of industry and/or agricultural activities. The water would tend to contain high levels of bacteria, together with high levels of organic and non-organic materials coming from municipal wastewater treatment plants, industrial effluent discharge or treatment plants and run-off water from cultivated fields and animal farms.

2.1.3 Underground sources

This type of water tends to have extremely low bacteria content and no organic materials, due to the long-term filtration process through the different geological structures. This type of water requires little treatment to render it both potable and to conform to local legislation.

2.1.4 Sea or salty water

Seawater or brackish water is used to produce drinking water or process water for large industrial plants and power stations when fresh or soft water is not available. The actual salinity of seawater varies immensely, with concentrations from as high as 43 000 mg/l in the Red Sea, down to 33 000 mg/l in the Pacific Ocean, and as low as 17 000 mg/l in the Baltic Sea. Seawater characteristics will also vary as far as contaminants are concerned. Close to the shore the seawater may have slightly higher salt contents, and may also contain suspended solids and organic materials coming from industrial plants or municipal wastewater facilities. Further off shore plankton build up may be of some concern, whereas in river estuaries the mixing of fresh and salt water causes significant variations in salt contents and levels of suspended solids.

2.1.5 Miscellaneous water specifications

Although the water classifications indicated above cover the visible side of the water handling market, this book will also look at various categories of 'process water' such as demineralised and deionised water, as well as ultra-pure water, geothermal water, and the highly important industrial and domestic effluent and sewerage market.

2.2 Potable Water

Potable, or drinking water, is, by definition, water which has been obtained from one of the sources of water described above, and has, by some form of treatment, been made suitable for drinking by humans. In developed countries the consumption of potable water may be as high 50 to 200 litres per day per person; however, considerable wastage of potable (treated) water is evident from the fact that actual water consumed by drinking or in cooking may be as low as 2 to 5 litres per day. This situation has often been evoked when considering parallel distribution of potable and non-potable water, which would result in lower costs and more economic usage of natural resources, but would increase considerably initial investment costs of constructing two separate distribution systems.

It is also worth pointing out the considerable levels of wastage of potable water within distribution pipe systems. The level of wastage is defined as the difference between the amount of treated water pumped out of a potable water plant and the total amount of water consumed by the clients of the plant. Recently, within Western Europe, wastage levels of as high as 50% have not been exceptional, and even after considerable levels of investment waste levels of 20 to 25% are still common. This situation is of considerable significance when considering the construction and jointing materials used to build new water mains, as discussed at a later stage of this book.

2.2.1 Quality standards

Quality standards and associated testing requirements concerning the actual quality of potable water vary from one country to another, depending upon the local, national and international legislation which is applied. Within the European community, a previous somewhat confused situation existed due to the differences between the legislation of each member state. Some considerable results in the standardisation have been achieved over the last 20 years, and the European Directive 'Drinking Water' dated 3 November 1998 reinforces the quality standards applied to potable water, and once transposed into the legislation of each member state around the end of the year 2000 should show a high level of uniformity in the standards imposed.

At the same time various directives are pushing the European countries towards standardised testing programmes for both the quality of water as well as the requirements for the materials used to handle potable water.

We have moved on from previous unique reliance on the somewhat subjective criteria such as clarity, lack of smell and a good taste to measurements of specific properties, although the more subjective tests still have a significant role to play. Standards and testing procedures vary from one country to another, but have a generally similar basis as outlined below.

2.2.1.1 Microbiological

Each sample of water should not contain any coliform faecal bacteria, no other organisms of faecal origin, and no pathogenic or parasitic organisms. Various sub-conditions generally apply allowing for the presence of very small concentrations of bacteria over an extended period of time without the necessity to classify the water as non-potable.

2.2.1.2 Physical and chemical

Legislation will also contain instructions as to the measurement of colour and opacity, as well as maximum permissible contents of a wide range of chemicals which may be introduced into the water by completely natural events, by long-term sustained industrial, agricultural or domestic pollution, or by accidental or deliberate discharge of chemical substances into, for example, drain systems and rivers.

2.2.1.3 An example

One can consider the current European legislation. The Council Directive 98/83/EC of 3 November 1998 on the quality of water intended for human consumption lists the parameters to be analysed by a process of check monitoring in order to provide information on the organoleptic and microbiological quality of potable water.

The obligatory criteria to be checked are listed below, although the directive also allows member states to add other parameters if deemed appropriate:

- Aluminium (only if used as a flocculant).
- Ammonium.
- Colour.
- Conductivity.
- *Clostridium perfringens* (including spores) but only if the water originates from or is influenced by surface water.
- *Escherichia coli* (E. coli) — ISO 9308-1.
- Hydrogen iron concentration.

- Iron (only if used as a flocculant).
- Nitrite (only when chloramination is used as a disinfectant).
- Odour.
- *Pseudomonas aeruginosa* (when water is offered for sale in bottles or containers) — prEN ISO 12780.
- Taste.
- Enumeration of culturable microorganisms, colony count at 22°C and 37°C — prEN ISO 6222.
- Coliform bacteria — ISO 9308-1.
- Turbidity.

The directive includes some instructions on test procedures to follow, as indicated above together with guidance on frequencies of testing. Similar testing procedures can be identified in most other countries.

No methods of analysis are specified for the following parameters: colour, odour, total organic carbon and turbidity.

2.2.2 Treatment process: potable water

In order to obtain potable water from one of the sources mentioned above it is necessary to treat the raw water in some way. Many similarities can be found in the process treatment of both potable and wastewater, but both will be reviewed separately.

Figure 1. Potable water treatment plant. Mont Valerien (Paris). Courtesy of Degrémont.

The typical treatment procedures from the original point of collection of raw water up to the actual storage and distribution of potable water can be summarised as follows.

2.2.2.1 Screening

An initial retention of large sizes of solid matter can be obtained by passing the raw water through a calibrated screen. This operation is designed to protect the more sophisticated treatment facilities further into the treatment sequence. Various screening configurations, operating and cleaning methods are available, but few of them call for thermoplastic or composite materials at this stage. The screen calibration can vary from as small as 10 mm for fine screening, up to 40 or 50 mm for initial coarse screening. In many cases two- or three-stage screening is used to obtain maximum efficiency and protection of subsequent plant, whilst at the same time facilitating cleaning of the screens.

2.2.2.2 Pre-disinfection or oxidation

A reduction in the concentration of microorganisms and an oxidation of organic matter can be obtained via the injection of ozone, chlorine or chlorine dioxide.

2.2.2.3 Micro-screening

Fine calibrated screening is done to remove very small particles in suspension. It is generally used to treat relatively non-contaminated surface water with low turbidity and coloration. It is not used if the treatment requires coagulation, flocculation and decantation.

2.2.2.4 Coagulation and flocculation

The turbidity and colour of raw water are caused in the main by small particles in suspension, called colloidal matter, which are difficult to filter out and which will only settle out with considerable difficulty.

Coagulation followed by flocculation is the main process used to remove this type of particle.

The principle of coagulation is to destabilise the particles and to facilitate their coagulation into larger blocks of material. This is normally carried out by the injection and dispersion of specific chemical products within a coagulation chamber, via a static mixer or within a specifically designed reactor.

Flocculation promotes, via a slow mixing, the contact of the larger blocks, and these particles consolidate to form a floc, which can then be eliminated by decantation or filtration. Flocculation is carried out within specifically designed and dimensioned tanks, called flocculators, equipped with paddle- or propeller-type mixers.

Some usage has been made of composite and thermoplastic materials in the field of reactors, flocculation tanks and mixers, but due to the relatively low levels of corrosion these applications have found more success in municipal, and more especially industrial, wastewater treatment plants.

2.2.2.5 Decantation

Decantation is used in practically all water treatment plants in order to separate out all solids in suspension which have a density higher than that of water. These solids are in the main the larger agglomerates of materials provoked by previous treatment procedures. The solids accumulate at the bottom of the decanter from which they are removed either continuously or periodically depending upon the type of decanter being used by means of one or more scrapers. Settling tanks may vary in dimension from a few metres to 40 or 50 m in diameter, or in some cases can be rectangular in form.

The equipment supplied in and on typical decanters is often made of thermoplastic or composite materials (lamella plates, covers, access walkways, ladders and platforms, all of which are discussed in the following chapters).

2.2.2.6 Filtration

An almost infinite number of filter processes and configurations exist, many of them developed alongside a proprietary process. A basic distinction can be made between those using granular filtration media and those using candle or frame filters.

Granular filters may be closed and pressurised or open to the air, but all involve a floor or supporting structure for the filtration media and nozzles for distribution of the fluid to be cleaned as well as backwash cleaning. Typical filtration media can vary from garnet (0.35 mm) to sand (2 mm) up to anthracite (5 mm) depending upon the type and quality of water used and produced.

Candle filters are of closed design, consisting of a sealed cylindrical casing into which are placed a certain number of filtering bodies (candles) attached to a base plate.

Composite and thermoplastic materials have been used to build filtration floors and nozzles as well as pressurised vessels for both granular and candle filtration.

2.2.2.7 Disinfection

Four main procedures exist for final disinfection of potable water prior to storage or distribution, using chlorine, chlorine dioxide, ozone or ultraviolet radiation.

Thermoplastic and composite materials offer excellent corrosion resistance in the field of manufacture, storage and distribution of chlorine and chlorine dioxide, and can be used in the fabrication of pipe systems, generators, scrubbers and storage units. The same materials are, however, of no interest in installations using ozone or ultraviolet radiation.

Chlorine is used extensively in the treatment of potable water and in cooling systems. Chlorine dioxide has much fewer applications in the municipal sector but is preferred in some industrial processes, and can also be of benefit when treating water containing traces of phenols which by a combination of chlorine with the phenol could result in the formation of chlorophenols, giving an unpleasant taste to the water.

Ozone can improve the taste and colour of drinking water. It is also used for the removal of bacteria, and the oxidation of organic matter and micropollutants, and is also used to improve the visual aspect of water in bathing pools.

Ultraviolet radiation is generally reserved to small, compact distribution systems, as the effect of the radiation is not long lasting.

2.2.2.8 Fluoridation and mineralisation

Fluoridation of potable water, a process approved by, for example, the WHO and the US and UK governments but is contested by medical authorities in other parts of the world, involves the addition of fluoride in the form of sodium hexafluorosilicate, hexafluorosilicic or sodium fluoride, with the intent of promoting the formation of dental enamel, thereby protecting the teeth of the population.

Mineralisation involves the replacement and addition of minerals to water from which they may have been removed, essentially during filtration procedures such as reverse osmosis or nanofiltration. This replacement of minerals is required to remove the corrosive effect of demineralised water, and also to recover an acceptable taste.

2.2.2.9 Storage

Potable water is generally stored in large concrete structures, with or without internal linings, large metallic structures, usually with a protective inner lining, or more recently in self-standing composite storage tanks or large, modular FRP constructions which may or may not rely on a metallic supporting or reinforcing structure.

2.2.2.10 Distribution

The distribution of potable water is achieved by networks of high-pressure mains, branching off into smaller and smaller diameter distribution lines up to the delivery point of each house, block of flats, office building, industrial plant, and so on. Various thermoplastic and composite materials offer economical solutions to the problems of distribution of potable water, essentially via buried pipelines, which because of the fact that they are buried, often under roads, require the long-term guarantees of both physical and corrosion resistant properties which these materials can offer.

2.3 Wastewater and Effluents

As mentioned above, there are many similarities when examining the chain of events in the preparation of potable water and the treatment of effluent, both municipal and industrial. By their very nature, effluent and municipal wastewater are, however, much more corrosive materials than potable water, and the field of intervention for thermoplastic and composite materials is much wider and varied.

2.3.1 Treatment processes

The following review covers the most significant stages of treatment, in a normal sequence of events, although each particular application will vary as to content and sequence.

2.3.1.1 Recuperation

Essentially, a buried pipe system collects municipal or industrial effluent from the source and transfers it to the treatment plant. Industrial effluent is often of

Figure 2. Brewery Effluent Treatment System, Riddles Brewery Ltd, Oakham. Courtesy of ACWa Services.

a very corrosive nature and is a natural application for thermoplastic and composite materials. Municipal effluents vary in their level of aggressiveness, with the greatest problems being generated in countries with higher ambient temperatures, although all sewage lines are subject to the permanent effects of the waste being handled and long-term deterioration of non-corrosion resistant materials can be expected.

2.3.1.2 Screening and grit removal

Screening of effluent follows the same basic process as indicated above for potable water, although the nature and dimensions of solids may be somewhat different. In some cases the actual screen may be replaced with an equipment comprising a number of indented and rotating discs which rather than retaining large batches of solid effluent, will shred it into smaller elements which facilitates subsequent collection or further treatment.

Although grit may be present in water fed into potable water treatment plants, especially when collecting surface water from lakes and rivers and must be removed in order to avoid build-up of deposits in pipes and pumps, the problem of grit is generally more apparent in industrial and wastewater treatment plants, and may be more difficult to remove as it will be mixed up with the more important organic solids.

Grit may be removed using natural hydraulic forces developed by passing the water through specially contoured vessels, or in aeration chambers, whereby the turbulence induced by the diffusion of air promotes the separation of organic matter bound to the grit.

2.3.1.3 Grease and oil removal

Grease, generally a solid material of animal or vegetable origin, may be separated through a natural or engineered flotation effect, and may be present in municipal and industrial wastewater in relatively large quantities. Oil, however, will only be present in very small quantities, coming essentially from rain water run-off in urban and industrial areas.

Both oil and grease separators have become standard units for many fabricators of FRP environmental control equipment, operating on the basis of a grease or oil trap, providing sufficient retention time and volume to recuperate more than 80% of the pollutant.

FRP systems have also been developed to collect oil spillage and pollution within areas such as oil refineries and petrochemical plants.

2.3.1.4 Flocculation, settling and flotation

This applies in the same manner as for potable water, although of course the size and amount of solid materials requires a completely different approach in dimensioning the equipment.

2.3.1.5 Aerobic and anaerobic biological treatment

This treatment, specific to wastewater treatment plants can be divided into two different procedures, activated sludge and fixed growth.

The purification of sludge in an activated sludge plant is achieved via the absorption and oxidation of the polluting elements by the bacterial mass (or biological floc) and this floc is then separated from the associated purified water. The aeration system in an activated sludge tank has a dual purpose, that is to provide the aerobic microorganisms with the oxygen they need, generally taken from the air, and at the same time to cause sufficient homogenisation and mixing so as to ensure close contact between the live medium, the polluting elements and the water thus oxygenated.

Both surface aerators and air diffusion equipment in composite and thermoplastic materials are standard equipment in many wastewater treatment plants, and some considerable experience has also been obtained with associated composite pipe systems on the bed of aeration tanks.

Separation of the bacterial floc is usually obtained by clarification in a large settling tank.

The attached growth process may be built using trickling filters with either traditional or thermoplastic fill. The thermoplastic fill offers high void ratios which reduces considerably the risks of clogging, and as the weight of the fill is much less than mineral fill, it is possible to design taller plants, thus reducing the land surface occupied.

2.3.1.6 Dewatering and sludge thickening

Filtration is the predominant method of dewatering sludge, and methods of filtration may vary from simple drainage through sand beds up to sophisticated and highly automated equipment.

Typical procedures are:

- Drying beds, which require large surfaces, high levels of manpower, and are subject to climatic conditions.
- Vacuum filtration, as a rotary drum, or as a high-pressure filter press, often using thermoplastic filter plates.
- A wide range of belt filters.
- Centrifugal separation.

Conventional cylindro-conical continuous centrifuges are also used to thicken sludge.

2.3.1.7 Odour control

A wide range of odours emanate from all wastewater and industrial effluent treatment plants, and vary considerably from plant to plant, or even within the same plant depending upon atmospheric conditions and throughput of the station. Hydrogen sulphide and mercaptans (thiols) are generally the most obvious components of these emissions and also provoke the greatest reaction from populations living close to and down wind from these plants. Odour control can be approached in two ways, either the reduction of odour emission via specific engineering and design procedures, sometimes associated with the dosage of effluent with specific chemicals, or the containment and subsequent elimination of odours via the usage of covers on tanks and scrubbing or neutralisation of the trapped gases. This second procedure relies heavily on thermoplastic and composite materials for the fabrication of covers, ducts, fans, absorbers and gas scrubbers.

CHAPTER 3

Materials

This study is based upon two main families of plastic materials, thermosetting resins and thermoplastics. Thermosetting resins, such as epoxies, generally require heat to transform them into an irreversible three-dimensional molecular chain, whereas thermoplastic materials, such as polyethylene, become malleable as the temperature is increased, and this transformation is reversible. Both resin types can be used with fibre reinforcement or fillers to improve their mechanical properties or to facilitate a particular moulding process.

Plastic materials are generally composed principally of carbon and hydrogen atoms, associated with a wide range of other elements such as chlorine, oxygen and nitrogen.

The origins of modern thermoplastic materials can be traced back to the USA, and the discovery in 1869 of celluloid, the result of research into the replacement of ivory in the manufacture of billiard balls. The first material recognised as a real thermosetting resin was discovered by the chemist Léo Baekeland in 1907, and to which he gave the name Bakelite.

A relatively detailed description of thermosetting resins is given due to the strong link between the choice of a specific resin and the properties of the finished composite material. This more detailed description is further facilitated by the comparatively small number of major manufacturers of thermosetting resins for the composite business. However, the complexity and diversity of thermoplastic resins is such that only a brief overview by resin type can be made. The manufacturer of the moulded equipment can supply any detailed information that the end user may require concerning the resin formulations used, or the overall properties of the finished goods.

3.1 Thermosets/Composite Materials

Composite materials as far as this book is concerned are essentially combinations of a thermosetting resin and glass fibre reinforcements, although some reference is made to synthetic materials used as reinforcement of corrosion resistant liners. The water industry does not actually require high-performance fibre reinforcement such as carbon or aramid, although current examination of mechanical handling devices and other structures which require continuous or intermittent movement may lead to new developments in the very near future. The different procedures to combine the resin matrix with fibre reinforcement are covered in a following chapter. The different resin types and forms of reinforcement employed for handling water are described below.

Mechanical, thermal and electrical properties of laminates are dependent not only on the type of resin, but also on the type and structure of the reinforcing material, glass/resin ratios and degrees of cure for example. To illustrate the variety and range of properties which can be obtained with a specific resin and a range of reinforcing materials we include as an example a document established for the Palatal A 410-01 isophthalic resin from DSM Composite Resins (see Table 1).

Thermoplastic materials such as PVC, polyethylene and polypropylene, discussed in detail in following chapters, are in some cases combined with a thermoset composite, taking advantage of the chemical and abrasion resistance of the thermoplastic material, and the mechanical strength and resilience of the composite. This combination of materials, designated 'dual laminate', can be used to fabricate pipe systems, tanks and hoods for example.

3.1.1 Thermosetting resins

The thermosetting resin family includes the wide range of unsaturated polyesters, generally identified as UP resins, the epoxies, under the general designation EP, and the various forms of polyurethanes, identified generally as Pu. All three are extremely important within the range of applications under consideration. Many other thermosetting resin families exist, such as phenoplasts, aminoplasts and polyimides, but these materials have little or no interest within the context of this handbook.

Table 1. Properties of cured Palatal A410-01. Mass fraction of glass fibre[1]. Courtesy of DSM Composite Resins

Property[2]		30–35% Mat (S.B.)[1]	40–45% Mat (I.B.)[1]	52–55% Mat (I.B.)	50–55% Mat (S.B.) +roving fabric, alternating	50–80% Roving fabric	60–70% Filament fabric	70–75% Unidirectional roving fabric	Unit	Test method
Mass density at 20°C		1.48	1.52	1.56	1.70	2.00	2.00	2.25	g/cm^3	DIN 53479
Tensile strength		130	150	170	280	300	340	630	MPa	DIN 53455
Ultimate elongation (50 mm between gauge marks)		≤2	≤2	≤2	≤2	≤2	≤2	≤2	%	DIN 53455
Modulus of elasticity in tension		10 200	12 000	14 250	17 270	20 120	24 350	30 000	MPa	DIN 53457
Flexural strength[3]		210	260	280	360	320	420	420	MPa	DIN 53452
Modulus of elasticity in flexure[3]		8200	9500	11 800	13 600	17 800	22 000	28 500	MPa	DIN 53457
Compressive strength		150	150	160	170	200	290	230	MPa	DIN 53454
Notched impact strength[4]		250/220	280/240	332/218	342/326	327/323	257/200	340/340	kJ/m^2	DIN 53453
Thermal conductivity[5]		0.17	0.19	0.22	0.23	0.29	0.28	0.21	W/m·K	DIN 51612
Specific heat[5]		1.1	1.1	1.0	1.0	1.0	1.0	0.9	kJ/kg·K	DIN 40685 part 2
Thermal coefficient of linear expansion		28×10^{-6}	25×10^{-6}	20×10^{-6}	20×10^{-6}	15×10^{-6}	12×10^{-6}	8×10^{-6}	K^{-1}	VDE 0304
Dielectric constant at 50 Hz/1 kHz	dry	3.8/3.7	3.8/3.7	3.8/3.8	3.9/3.9	4.4/4.2	3.8/3.8	–/4.0		DIN 53483
	wet[6]	10/6.3		4.2/4.1	6.8/5.3		5.2/4.9			
Dielectric factor at 50 Hz/1 kHz	dry	0.002/0.004	0.002/0.004	0.001/0.003	0.011/0.087	0.025/0.016	0.002/0.002	0.006/0.006		DIN 53483
	wet[6]	0.5/0.2		0.02/–	0.2/0.1		0.08/0.06			
Volume resistivity	dry	10^{16}	10^{16}	10^{16}	10^{16}	10^{15}	10^{15}	10^{15}	Ω cm	DIN 53482
	wet[6]	10^{10}	10^{11}	10^{13}	10^{10}	10^{8}	10^{13}	10^{7}		
Surface resistivity		10^{13}	10^{13}	10^{13}	10^{13}		10^{13}		Ω	DIN 53482
Dielectric strength		30	30	30			30		kV/mm	DIN 53481
Tracking resistance		KA 3c	KA 3c	KA 3c	KA 3c	KA 3c	KA 3c		stage	DIN 53480

[1](S.B. = soluble binder) (I.B. = insoluble binder).
[2]Standard reference atmosphere 23°C/50%-2 according to DIN 50014, unless otherwise stated.
[3]Beam supported at three points.
[4]Measured on a standard bar specimen.
[5]Measured between 0 and 60°C.
[6]Specimens immersed in drinking water for 24 hours.

Initially identified in 1941, polyester resins, due to their relatively low cost and relatively good mechanical and chemical resistance, are the backbone resins for many applications with water. They can be split into two different sub-types, corresponding to different levels of price and performance.

Orthophthalic polyester resins often used in the reinforcing structure on dual laminates or for applications requiring limited resistance to temperature or chemicals are the cheapest thermosetting resin systems available.

Isophthalic polyester resins and terephthalic polyesters offer an improved all round performance compared to the basic orthophthalic, and may also offer improved moulding properties.

Before continuing with information on even higher performance resins, it is perhaps worthwhile, in order to reach an understanding of the diversity and relative performance of the different resins, to compare the resin systems with comparable metallic materials.

One could compare (using comparative price levels and corrosion resistance) the different resins, from basic orthophthalic polyesters, through bisphenols, vinyl esters and epoxies, to the newer, high-performance vinyl ester resins and fluoropolymer-based dual laminates, with the full range of metallic materials. A parallel comparison would start with basic mild steels, progress through the different ranges of stainless steel, up to special grades of alloys and materials such as titanium. There exists an in-depth knowledge and long-term proven performance of metallic materials and material engineers are able to select the lowest cost material for a specific application. In the composite market, there is a distinct tendency to over-specify and to use a resin system which will cost more than the lowest-cost system which will do the job. This is especially true in the water treatment industry where low-cost polyesters will perform as well as more expensive bisphenols or vinyl ester resins.

The user of most metallic materials has the advantage of being able to specify and purchase pipe, fittings, plates or profiles to national or international standards, thereby ensuring that material purchased to the same or equivalent standards in Scandinavia, Germany or the USA, for example, will have an identical material composition, chemical and temperature resistance, as well as identical physical properties. This is unfortunately not the case with resins and composite systems generally. Each resin manufacturer has their own standards and specifications, and although an individual resin from a specific manufacturer will always conform to the resin specification, within tolerances (which should not be overlooked), it cannot be confirmed that similar resins from different manufacturers will behave in an identical manner.

To return to the isophthalic resin family mentioned above, and which is probably the dominant resin type used in the manufacture of composite material destined for the water market, it is useful to use this family to demonstrate some of the basic characteristics of the polyester/vinyl ester and bisphenol resins, as compared to epoxy resins which rely on a fundamentally different chemistry and curing system.

Isophthalic resins may be slightly more expensive than orthophthalic resins, perhaps by between 10 and 20%, but they are far cheaper than basic vinyl esters, perhaps as little as 50% of the cost of the latter. It can be seen that a limited extra investment in the isophthalic resin will bring enhanced properties and probably extended performance, whereas the more important extra cost required to build with vinyl esters may only be justified in more critical applications.

Isophthalic polyester resins are produced via a chemical reaction between isophthalic acid, maleic anhydride and a glycol diluted within a styrene monomer. By controlling the proportions of raw materials and reaction times, the resin manufacturer can vary the properties of the finished resin.

The most important properties of the resin as a raw material for manufacture of a composite material are:

- the viscosity, which will determine the suitability of the resin for a specific laminating technique; and
- the resin reactivity to specific curing agents and resultant cure times and exothermic reactions.

The main consideration of the properties of the cured resin include:

- the chemical resistance to the fluids being handled;
- the heat distortion temperature which will indicate the maximum service temperature to which the resin may be exposed; and
- mechanical data important for the design of the equipment to be manufactured, such as the flexural and tensile strength and modulus.

Other basic considerations concerning the exposure of the resin to any form of water would include:

- the resistance to hydrolysis;
- impact strength; and
- fire resistance

all depending upon the requirements of the application in hand. Isophthalic polyester resins offer the optimum performance/cost ratio when considering the above-mentioned criteria and the requirements identified within municipal potable and wastewater treatment plants, as well as many other low pressure handling and storage systems for seawater, brackish water and mildly corrosive industrial effluent.

Typical orthophthalic resins, with their basic properties, are:

- Crystic 196 from Scott Bader, for laminates with low taint and good resistance to acidic conditions in operating temperatures up to 55°C.
- Crystic 198 from Scott Bader, for operating temperatures up to 100°C.
- Polylite 33053-10 from Reichhold for laminates with thin to medium cross-sections.
- Polylite 33131.00 from Reichhold, developed for the manufacture of laminates with thick-wall sections.

Typical isophthalic resins are:

- Aropol 7240 from Ashland Specialty Chemical Co, for good corrosion resistance and general purpose corrosion-resistant equipment.
- Crystic 272 from Scott Bader, a low-viscosity resin, suitable for filament winding processes, and for temperatures up to 70°C.
- Crystic 199 from Scott Bader, for operating temperatures up to 120°C.
- Crystic 392 and 397PA from Scott Bader, for operating temperatures up to 75°C and 90°C, respectively. Both resins give good adhesion to PVC for the construction of dual laminates.
- Dion 6631T from Reichhold, offering good corrosion resistance and excellent physical properties.
- Dion 6334 from Reichhold, a resilient resin developed for filament winding and centrifugal cast pipe intended for ambient sewage and wastewater.
- Hetron 99P from Ashland Speciality Chemical Co, flame resistant with good corrosion resistance.

Other manufacturers provide a wide range of isophthalic polyester resins, including:

- DSM Composite Resins Structural Resins under the Palatal and Synolite trademarks;
- AOC under the trade mark Vipel;
- Cray Valley Resins;
- Neste Polyester; and
- Reichhold offer the wide range of Norpol polyesters from the former Jotun Polymer organisation recently acquired by Reichhold.

3.1.2 Vinyl ester resins

The first vinyl ester resin was developed in North America in the early 1960s, and was introduced into Europe in the 1970s. This resin group has evolved considerably over the last few years, and has become the dominant resin system for mild to severe chemical applications, or where considerations of temperature or mechanical loading preclude the use of polyesters. They are not a dominant resin system for handling water, but should be considered, along with the bisphenol resins, for all applications handling municipal or industrial effluent, for which the corrosion resistance of polyester or epoxy resins is considered as insufficient. On large volume cost-sensitive effluent pipe schemes, favourable cost and performance levels have been obtained with dual-resin systems, using a relatively flexible but resilient vinyl ester resin for the internal corrosion-resistant layer, and an isophthalic polyester for the supporting structural laminate.

The vinyl ester resin family offers typically very high mechanical properties allied to an excellent corrosion resistance. The resins are derived from an epoxy base, which is modified in such a way that they can be cured by the same methods as traditional unsaturated polyester (UP) resins, i.e. via a free radical mechanism with styrene as a co-curing monomer. An epoxy-based vinyl ester can be produced from almost any basic epoxy, depending upon the properties required. For example, a novolac-based epoxy will give higher thermal resistance, tetrabromo-bisphenol A will improve the fire resistance due to the bromine content, or even a rubber-modified epoxy can be used to give high impact strength. The bisphenol A-based grades have good all round performance, balancing excellent mechanical properties with good thermal resistance.

Most manufacturers of vinyl ester resins now market a wide range of materials, which can be defined as

- Standard vinyl ester such as the Derakane 411 series from Dow, or Dion 9100NP from Reichhold which have a very high resistance to a wide range of acids, alkalis, bleaches and organic solvents.
- Enhanced performance vinyl ester such as Derakane 441, Atlac 580 from Reichhold or DSM Composite Resins, all of which offer higher temperature resistance, improved corrosion resistance and a greater elongation.
- High-temperature vinyl esters such as the Derakane 470 series from Dow, or Dion 9480NP from Reichhold, with high resistance to strong acids, oxidising media and organic solvents, allowing fluid temperatures up to $120°C$, depending upon the nature of the fluid being handled.
- Fire retardant vinyl ester such as the Derakane 510 series from Dow, or Dion 9300 NP from Reichhold, which offer high fire-retardant properties with and without the addition of fire-retardant compounds, associated with outstanding corrosion resistance.
- High-performance vinyl ester such as Derakane XZ 92435 from Dow, which under certain conditions can extend the service capability of FRP constructions handling hot, corrosive gases up to $220°C$.

Alongside these basic formulations other more specialised types exist, of which Derakane 8084, a rubber-modified vinyl ester is used for forming the basis of many lining formulations, due essentially to an increased adhesive strength, and an improved resistance to abrasion and severe mechanical strength.

Dow has recently introduced an enhanced parallel version of their basic Derakane range under the trademark Derakane Momentum. According to Dow the Momentum range of products, identified as

- Derakane Momentum 411.350;
- Derakane Momentum 441.400;
- Derakane Momentum 470.300; and
- Derakane Momentum 510C.350

offers considerable technology improvements in the fabrication of FRP applications.

The Derakane Momentum resins have the same chemical backbone and corrosion resistance as standard Derakane resins but offer higher reactivity, easier working, reduced gel times, as well as fewer air bubbles, reduced stress cracking and a lighter colour for the finished product. This family of resins require reduced and simplified catalytic formulations, but improve on gel times by as much as 20%. The higher reactivity and lower catalyst content also reduces air entrapment and foaming, allowing higher productivity combined with a lower void content. Lower exotherm will also allow thicker single-cure stage laminates or reduced staging for multi-cure-stage laminates, with once again improved levels of productivity. Finally, the lower level of air entrapment combined with the lighter clear colour will improve laminate quality and facilitate visual inspection of the finished laminate.

One other highly significant property of this new family of resins is the extremely long shelf life, 10 to 12 months, which will greatly facilitate fabricators in areas with limited distribution circuits or on site installations in countries having no local resin suppliers.

The following companies also supply a range of vinyl ester resins:

- Ashland Specialty Chemical Co under the HETRON and Norpol trademarks;
- Scott Bader under the designation Crystic; and
- DSM Composite Resins under the Atlac trademark.

3.1.3 Bisphenol fumerate resins

For many years, bisphenol resins were the specified resin system for many corrosion-resistant applications, but we have in recent years seen their replacement in many applications by vinyl esters.

Bisphenol resins provide excellent corrosion resistance to strong acid and have the highest resistance of all thermosetting resins to alkali at elevated temperatures. They are derived from bisphenol-A and modified hydrogenated bisphenol-A with fumerate reactive groups to ensure the highest possible thermal stability. These resins are characteristically rigid with high heat distortion temperatures.

Typical resins are the Atlac range supplied by Reichhold and DSM Composite Resins as follows:

- Atlac 382.05A — Formulated for contact moulding. Provides a wide range of corrosion resistance, particularly with alkali.
- Atlac 4010A — Formulated for filament winding applications, and recommended for tanks and piping.
- Atlac 711.05AS — A flame-retardant brominated bisphenol fumerate polyester providing good high temperature performance in corrosive environments.
- Reichhold also market Dion 6694 — Described as a premium grade modified bisphenol fumerate resin, especially useful in chloralkali applications and chlorine handling equipment.

3.1.4 Liquid epoxy resins

First developed in 1946, these are now widely used in the manufacture of adhesives, laminates and coatings. In the field of water treatment and handling of water in all its forms, liquid epoxy resins are especially useful in the manufacture of high-performance pipe systems and all forms of composite pressure vessels.

Basic epoxy formulations can be modified via the use of a wide range of different curing systems, diluents, modifiers and fillers, and thus offer an extremely wide range of product characteristics.

The main points of interest in using epoxy resins are:

- Because they show minimal shrinkage and exotherm during curing, they maintain a good overall dimensional stability.
- The chemical resistance is good, although lower than with vinyl ester resins, but is more than satisfactory for most applications with water.
- No volatile fumes are given off during the manufacturing and curing process.
- The finished product shows extremely high mechanical properties, retention of significant mechanical properties at high temperatures and a good resistance to impact.
- The bonding of the resin to the glass fiber reinforcement is better than with any other resin system.

Typical epoxy resins offered by the three principal suppliers are as follows. From Ciba Specialty Chemicals, under the trademark Araldite:

- Araldite LY 556, a bisphenol A epoxy resin often used with a Ciba anhydride hardener (HY 917), and a heterocyclic accelerator (DY 070) in the production of filament wound pipe.
- Araldite LY 554, with HY 916 and an amine accelerator (HY 960) offers a system with a reduced viscosity and a higher reactivity.
- Araldite LMB 5446, with a range of possible hardeners and accelerators will offer good fire retardant properties.

From Shell, under the trademark Epikote:

- Epikote 828, a standard unmodified liquid epoxy resin produced from epichlorohydrin and disphenylolpropane (DDP-bisphenol A).
- Epikote 826, a blend of two resins produced respectively from epichlorohydrin and DPP, and epichlorohydrin and diphenylolmethane (DPM-bisphenol F).

From The Dow Chemical Co under the trademark D.E.R.:

- D.E.R. 330 is a liquid reaction product of epichlorohydrin and bisphenol A, forming a low molecular weight liquid resin with low viscosity, ideally suited to the filament winding production process.
- D.E.R. 331 and 336, a basic D.E.R. 330 with modified a viscosity which suit different winding and impregnation technologies used by the different manufacturers of composite materials.

3.1.4.1. Anhydride cured epoxy resins

Epoxy-based composite products, especially pipe systems, manufactured with an anhydride curing system, have long been criticised for poor performance, with limited service temperature and poor resistance to water. These problems have largely been resolved with the introduction of new and improved systems, although of course the service temperatures which can be envisaged remain well below those attainable with amine-cured systems. In applications which do not require a service temperature above 80°C, anhydride-cured systems are generally more than adequate for handling water, and do allow significant financial advantages.

Typical anhydride curing agents are hexahydrophthalic anhydride (HHPA) and phthalic anhydride (PA), and a few of the major pipe manufacturers have developed their own in house systems and procedures which allow extremely rapid curing cycles, sometimes as low as 20 minutes.

3.1.4.2 Polyamine cured epoxy resins

Polyamine-cured epoxy systems probably account for more than 80% of all epoxy pipe systems manufactured in the world, although the improved anhydride systems become generally more available.

Aromatic amines, such as metaphenylene diamine (MPDA) and diamino diphenyl sulphone (DDS) have been the mainstay of the industry for many years, but recent and forecast legislation is causing a progressive replacement of aromatic amines by aliphatic polyamines, which are much less aggressive products and facilitate material handling during production. The disadvantage of aliphatic polyamine based curing systems are their higher cost, which again will assist the market penetration of anhydrides, and the fact that the finished laminate is opaque rather than translucent, thereby precluding or at least hindering visual inspection of thicker structural laminates. This aspect is probably not very significant, as many products now include pigments, and more sophisticated quality control techniques have now been developed.

Typical amine curing agents are from Dow, as aliphatic amine compounds:

- D.E.H. 20 — diethylene triamine (DETA);
- D.E.H. 24 — triethylene tetramine (TETA); and
- D.E.H. 26 — tetraethylene tetramine (TEPA).

From Shell, under the trade name Epikure:

- Epikure DX 6052 — a cycloaliphatic polyamine;
- Epikure DX 6509 — an aromatic diamine;
- Epikure DX 6510 — a cycloaliphatic polyamine; and
- Epikure DX 6512 — an aromatic diamine.

To demonstrate the different curing cycles possible, an Epikote 828 resin, using Epikure agents detailed above, can present a gel time, at 60°C, of between less than half an hour and more than two days, and with corresponding T_g (glass transition temperature) ratings, dependant also on heat cure schedules, from 148 to 185°C.

Table 2 attempts to represent the main qualities and performance of the resin systems described above. All figures are approximate average values for the particular family of resins, and it should be noted that significant variations may be found within each resin in each specific family of resins, depending upon the moulding process or market area for which the resin is destined.

3.1.5 Solid epoxy resins

Solid epoxy resins are of interest for their use in paints and powder coatings and further reference to these materials will be found in Chapter 9 concerning coatings and linings.

3.1.6 Polyurethane resins

Polyurethane resins are often described as thermoplastic as they are often incorporated in thermoplastic moulding compounds. They are, however, true thermosetting resins, and because of their high impact resistance, toughness, abrasion resistance and excellent bonding properties, they often form the basis of a wide range of coatings or liners, discussed in further detail in Chapter 9.

3.1.7 Other thermosetting resins

3.1.7.1 Furane resins

As well as phenolic resins referred to below, these are actually of little interest within the context of this book, but we include a few words in order to complete the listing of the principal thermosetting resins used in the industrial composite market today. Both systems rely heavily on acidic catalysts which call for special materials and handling during the production of composite equipment.

Furane resins have an excellent resistance to alkaline solutions and acids containing chlorinated solvents, but these advantages are often offset by the handling problems referred to above, and relatively poor mechanical properties.

A typical furane resin, available from Ashland Specialty Chemical Co, is Hetron 800, described as a furane resin with excellent resistance to organic solvents.

3.1.7.2 Phenolic resins

These were the original thermosetting resins. This type of resin offers high heat resistance and an excellent level of non-flammability with very limited generation of smoke or toxic fumes in a fire condition. Physical properties are better than with furane resins, and may in some conditions even approach those of basic orthophthalic polyesters.

Typical resins, available from Blagden Cellobond are:

- Cellobond J 2042 L, a phenolic resin typically used for contact moulding with chopped strand mat or woven roving reinforcement.
- Cellobond J 2027 L, a phenolic resin normally used for filament winding with roving reinforcement, or band winding with stitched or woven tape.

3.1.7.3 HET resins

HET resins, or chlorendic resins are unsaturated halogenated polyester resins based on chlorendic anhydride, particularly well suited to high operating temperatures and in highly oxidising environments.

Table 2. Typical average properties and characteristics of thermosetting resins typically used in filament winding and contact moulding processes

Resin type	HDT	Specific gravity	Clear, non-reinforced resin				54° filament wound composite	
			Tensile strength	Tensile elongation	Flexural strength	Flexural modulus	Axial tensile strength at rupture	Hoop tensile strength at rupture
	°C		MPa	%	MPa	GPa	MPa	MPa
Orthophthalic polyester	75	1.10	70	2.50	125	3.8	45–65	200–240
Terephthalic polyester	85	1.10	70	2.50	125	3.8	50–70	200–260
Isophthalic polyester	110	1.10	70	2.50	130	3.9	50–70	240–300
Vinyl ester	105	1.11	80	5.00	130	3.2	60–80	240–300
Novolac vinyl ester	145	1.16	70	3.00	130	3.7	80–100	260–320
Bisphenol A fumerate	125	1.05	65	2.10	115	3.4	60–80	240–300
HET	115	1.18	56	2.00	82	3.8	60–80	240–300
Phenolic	250	1.25	40	2.00	70	4.0	40–50	200–300
Furane	105	1.10	36	1.00	72	4.1	40–50	160–180
Epoxy + amine cure	140	1.20	85	5.00	150	3.5	80–100	240–300
Epoxy + anhydride cure	100	1.20	80	4.00	135	3.8	80–100	300–380

Note: all mechanical properties are at ambient temperature and may vary considerably at higher or lower temperatures; Tensile strength based upon hydrostatic burst test to ASTM D1599 or equivalent; HDT: heat distortion temperature.

Typical resins, available from Ashland Specialty Chemical Co are:

- Hetron 92, a chlorendic polyester with high flame resistance often used for handling corrosive fumes.
- Hetron 197, a chlorendic polyester with high flame resistance often used for handling corrosive fumes.

3.1.8 Curing and hardening agents

The curing agents and procedures for epoxy resin have been described above. Furane and phenolic resins require very specific acidic catalysts which will not be covered in this book. All forms of polyester, vinyl ester, bisphenol and HET resins can be cured with similar systems.

Polyester, vinyl ester and similar resin systems can be, and often are, cured at ambient temperature, that is they are not placed inside an oven or other heating device. The chemical reaction provoked by the catalysts and accelerators create an exothermic reaction, and will take the temperature of the laminate being cured up to temperatures as high as 180 to 200°C. Control of this temperature is critical to achieve optimum cure, and at the same time avoid delamination caused by stress and shrinkage generated by excessive temperatures. Many systems may require a second-stage post-cure in order that the laminate achieves optimum physical, thermal and corrosion-resistant properties, which may include placing a complete large vessel inside a post-curing oven, or if this is not feasible, hot air or steam may be circulated through the vessel for several hours. This is of significant importance in the case of styrene-based resins to be used in equipment requiring certification for contact with potable water or foodstuffs.

Typical catalytic systems include the selected catalyst, usually a ketone peroxide (methyl ethyl ketone peroxide, MEKP) or a benzoyl peroxide (BPO), associated with accelerators and promoters such as dimethylaniline (DMA), cobalt solutions such as cobalt naphtenate at an activity rate of 6% (Cob-6).

No matter what the resin system retained for a particular laminate or application, the final result and overall level of performance is completely dependant upon the curing system retained, that is the additives used and the thermal cycle achieved, with or without a separate post-cure. This notion is of significant importance, and is often overlooked when a particular resin is specified for a particular application.

3.1.9 Miscellaneous fillers/additives

Many different materials may be introduced into a resin or laminate, for example, to reduce cost, to improve particular mechanical or other properties, to improve appearance or to facilitate or retard curing. It should be noted that the incorporation of a filler to enhance a specific property of the final laminate may often require a trade-off with other properties, some of which may decline as others are enhanced.

3.1.10 Reinforcing fibres

Within the context of this handbook our comments are restricted in the main to glass fiber reinforced plastics, and the most commonly used glass fibre materials are defined as:

- **Surface veils**, either in E, C or ECR glass to reinforce the inner resin layer of an anti-corrosion barrier or liner inside a pipe, tank or other vessel. Surface veils are also available in a wide range of synthetic materials, such as polyester, nylon either to reinforce the inner liner as indicated above, or to protect and reinforce an outer resin-rich finish on the outside of a pipe or vessel. The addition of a veil on the outside of a composite pipe or vessel shows a distinct improvement in the resistance to ultraviolet attack of the resin structure.
- **Chopped strand mat or chopped fibres**, used either to reinforce and increase the anti-corrosion inner layer of a laminate, or as part of the structural laminate made either entirely from chopped strand mat, or as a bonding layer between layers of continuous or woven rovings, where a laminate structure made entirely with rovings is not required or allowed by a particular specification or standard.
- **Continuous rovings** form the structural component of all filament wound structures.
- **Woven rovings** are used to build up the structural laminate in contact moulded or band wound structures, increasing the glass-to-resin ratio and the strength of the laminate.
- **Fabrics** are used in a similar manner as woven rovings, but may offer superior handling properties or strength.
- **Flakes or spheres** of glass are used to reinforce resin-based coatings and linings.

For the uninitiated it is perhaps interesting to outline the manufacturing technology used to produce glass fibre reinforcements. The production of glass fibres or strands calls upon three main technological areas: glass forming, textiles and chemical treatment.

3.1.10.1 Glass manufacture

The manufacture of glass is carried out in a special furnace at about 1550°C (E-glass) using finely ground raw materials from carefully selected quarries. The glass leaving the furnace at a very high temperature is used to feed bushings (blocks pierced with hundreds of holes) of platinum alloy.

The majority of products are based on E-glass. Other glasses are also used, for example, R-glass (high mechanical performance), D-glass (high dielectric performance), AR-glass (alkali-resistant glass), and ECR-glass (corrosion-resistant glass).

3.1.10.2 Forming

The basic strand is made by forming. Forming is achieved by drawing at high speed the molten glass flowing from the holes of the bushing. This gives rise to between 50 and several thousand filaments. These filaments are defined by their diameters, from 5 to 24 microns (1 µm=1/1000 mm).

Assembled, they give form to the basic strand defined by its linear mass expressed in tex (g/km). Fibres are available from 100 through 4800 tex, with 1200, 2400 and 4800 being the most common for the industries, production technologies and applications covered by this handbook.

3.1.10.3 Sizing

The individual filaments of reinforcements receive a coating of size at the forming stage, immediately after they leave the bushing. This size, made up of organic products dispersed in water, is designed to give the glass strand certain characteristics necessary for final processing. Each size is specially designed for a moulding or compounding process and for a different matrix type. Normally it contains a 'coupling agent' most often of the silane type.

Final products are either manufactured immediately after they leave the bushing (direct roving, continuous filament mat, chopped strands) or made up of basic strands subjected to finishing operations which are more or less complicated (assembly for the rovings, chopping of chopped strands, chopping and forming of the chopped strand mats).

3.1.10.4 Types of glass fibre

Traditionally the anti-corrosion composite market has used three of the five basic types of glass fibre reinforcement available, but the recent introduction of a new formulation, Advantex Glass Fiber, by one of the major suppliers, Owens Corning has somewhat changed the situation. The five basic products are described below, the first three being of interest for the anti-corrosion market:

- C-glass, or chemical glass, has a high degree of chemical resistance, and is used in areas where the fluid being handled may come into contact with the reinforcing materials. For this reason C-glass is one of the materials used for surface veils.

- E-glass (electrical), the most commonly used type of glass in the industry, is less expensive and has good mechanical properties. This type of glass is used throughout all structural laminates.
- ECR-glass (electric corrosion resistant) offers improved chemical resistance over E-glass in many applications, is more expensive than E-glass and may be specified throughout the total laminate for some type of applications, usually in very severe conditions.
- D-glass (dielectric) is, because of particularly good dielectric characteristics, used mainly in the electronics industry.
- R- (Europe) and S- (North America) glass is stronger than E-glass, but is substantially more expensive, and finds little or no usage in the anti-corrosion market.

Owens Corning claims that Advantex Glass Fiber combines the electrical and mechanical properties of traditional E-glasses with the acid corrosion resistance of ECR glass at a price level generally in line with that of E-glass. This new material, a boron-free formulation that brings environmental advantages to the manufacturing process, also offers possibilities for applications at higher temperatures than those obtainable with E-glass.

Owens Corning now markets a full range of products in the new formulation, including rovings for filament winding, pultrusion and weaving, choppable rovings for chop/hoop winding, surfacing veil for surface enhancement, as well as chopped strand mat.

3.1.10.5 Forms of glass fibre reinforcement

Glass fibre reinforcements are available in a wide range of forms, designed for a particular process market.

Surface veils are available in various thicknesses, dimensions and finishes, depending upon the resin system and type of application, for example.

Chopped strand mat is available in a large variety of weights, widths and finishes, the weight being specified as weight per surface area, for example 300 g/m^2.

Single end rovings, made from many individual filaments wound into a single strand, or direct wound rovings, a collection of parallel filaments wound together in a continuous strand, are both suitable for most filament winding properties.

Woven rovings, as described below, and glass fibre fabrics are used extensively throughout the industry.

Most suppliers of glass fibre reinforcement offer a large range of materials; themselves manufactured to suit the different resin systems and manufacturing technologies used by the different manufacturers.

The strength of the laminate is also dependant upon the direction of the glass fibres, and the use of continuous rovings and woven uni-, bi- or multidirectional rovings can all be used to tailor the mechanical properties of a laminate to suit each individual construction.

Using a major glass fibre supplier such as Vetrotex International, it is possible to define the different forms of glass fibre reinforcement in current use, with reference to specific examples.

- Chopped strand mat, Ref. M5, recommended for hand lay up, compatible with polyester and phenolic resins, using a high-solubility emulsion binder.
- Chopped strand mat, Ref. M125, recommended for hand lay up, compatible with polyester and vinyl ester resins, using a high-solubility powder binder, specifically recommended for the reinforcement of anti-corrosion resins and for parts which are difficult to mould.
- Chopped strand mat, Ref. M113, similar to Ref. M125, but using fine strands and available with a surface density of $100-900$ g/m^2.
- Direct roving, specifically for filament winding processes, Ref. R099-P103 and R099-P122, compatible with epoxy, polyester, phenolic and vinyl ester resins. Ref. R099-P139, compatible with epoxy, phenolic and vinyl ester resins.

Taking direct rovings as raw material, many forms of woven rovings can be manufactured. Woven rovings are interesting for contact moulding, band winding technology often used to manufacture pipe fittings, and as part of the laminate in many hoop winding operations in order to obtain sufficient axial strength.

Again, using Vetrotex products as an example, we can define four types of woven roving:

- Standard woven rovings, with 50% of the fibres in each direction, giving equal bi-directional properties. This type of material is available in plain twill, or satin weaves, in weights of 270 to 1000 g/m^2.
- Unidirectional woven rovings, with up to a 98/2% split in the direction of the rovings (with a majority of rovings in the weft or warp direction to suit the type of application. This material, again available in the different weaves mentioned above, is available in 250 to 1450 g/m^2.
- Woven rovings are also available using 'spun rovings', which increase the resistance of delamination of laminates using successive layers of woven roving. This type of material is, however, seldom used for corrosion-resistant FRP due to product specifications requiring alternative layers of chopped strand mat and woven rovings.
- A recent development from Vetrotex, amongst others, is the appearance of 'combination fabrics', an assembly of chopped strand mat and woven rovings, sewed together, and available in various different weights. The variation of width, weight, weaves, directional proportions, and the

various finishes gives the fabricator a wide choice of reinforcements to fit all configurations and manufacturing technologies.

A similar range of products is available from all of the major manufacturers, for example, PPG offers:

- A full range of choppable rovings under the trademark Hybon, compatible with polyester, vinyl ester, epoxy and polyurethane resins.
- Filament winding and pultrusion rovings for all the above-mentioned resin systems, as well as for phenolic resins.
- Chopped strand mat and woven rovings for polyester, vinyl ester and epoxy resins.

Another significant material development is the appearance of three-dimensional glass fabrics which allows the fabricator to build up an integral sandwich structure without the need to include conventional sandwich material such as polyurethane foam or balsa. During manufacture of the laminate, the three-dimensional glass fibre, in which the core and the two surfaces are already combined via a specific weaving process, the resin impregnation process compresses the fabric, but once the pressure is removed, the fabric springs back to its original height, resulting in a lightweight sandwich laminate up to 24 mm thick.

A typical supplier such as Parabeam 3D Glass Fabrics claims significant advantages over traditional sandwich or solid materials, such as:

- weight saving;
- high strength and stiffness;
- fast and easy processing;
- sandwich construction limited to glass and resin, with no bonding problems to a third material;
- integral sandwich laminates from 3 to 24 mm thick;
- good drapeability which allows the material to be used on many different forms and shapes;
- no core delamination in case of impact; and
- the possibility to use the interstitial space for insulation or for monitoring of leakage in double-wall vessel production.

Surfacing veils are available in a wide range of materials, based upon both synthetic and glass materials, the choice being dependent upon the properties required.

Whereas the resin system determines both the chemical and temperature resistance of a laminate, the reinforcement, with one exception, is responsible for the mechanical characteristics. This exception is the chemical resistance of the surfacing veils, which as their name implies, are relatively thin veils applied on or immediately behind the inner (and sometimes outer) surface of the pipe or vessel.

Veils can be obtained in glass (E-, C- or ECR-glass), in synthetic materials (e.g. polyester, nylon) and in carbon. Because of the contact of the surface veil with the fluid being handled, it is essential that the material has the chemical resistance required.

When handling water or effluent, in most cases a veil of C-glass is used but synthetic veils may be used dependent upon the required chemical and temperature resistance, or for other reasons, such as visual aspects and handling properties, for example.

PFG (Precision Fabrics Group) supplies a wide range of polyester surfacing veils, under the trademark Nexus. The veils have been shown to extend service life, improve corrosion resistance, improve abrasion resistance and reduce thermal shock cracking when compared to the more conventional veils.

Nexus is available in various grades:

- Style 100 — 10: an apertured, 1.3 oz veil for filament winding and contact moulding.
- Style 039 — 10: an apertured, resinated 1.0 oz stiffer product for filament winding.
- Style 111 — 05: a non-apertured, 1.6 oz low-elongation product for filament winding.
- Style 100 — 00: a non-apertured, 1.2 oz product for filament winding.

Conductive Nexus, having been treated with special conductive coating, has recently been introduced. This product offers the potential for static elimination, grounding, heating and electrostatic finishing, without having recourse to expensive carbon fibres or embedded wire screens. Conductive Nexus has also been specified for a system which allows the user of FRP tanks or pipes to control performance whilst in service.

Carbon fibre veils are used within surface layers to achieve an electrically conductive laminate and single carbon fibre rovings may be included within the structural laminate of filament wound pipes and vessels for the same reason. Carbon fibre veils are also used in applications with the new Turbo vinyl ester resins from Dow, in order to distribute surface heat as evenly as possible in very high temperature applications.

3.1.11 Moulding compounds

Most, if not all of the thermosetting resin systems mentioned above can be incorporated into a moulding compound, that is a mixture of resin, curing additives, fillers, pigments and fibre reinforcement, prepared for a particular and usually highly automated moulding process. Two main families of moulding compounds are of interest.

SMC or sheet moulding compound, which is basically a compound as described above, in continuous sheet form, which can be unrolled and cut and then used in some form of press moulding.

Bulk moulding compound (BMC) is, as its name suggests, a compound supplied in the form of a dough or sometimes as a thick continuous rope. This type of compound, with its many variations, can be used in compression, transfer or injection moulding.

3.2 Thermoplastics

The full range of thermoplastic materials available today is much too wide and complex to be detailed in this book. Many of the materials still considered as new and innovative today were in actual fact discovered in their original form more than 50 years ago, and derivations and other developments are still ongoing. Many of today's applications involve a modified polymer, or even a copolymer material obtained by mixing two polymers together, often with fillers or additives, either to facilitate moulding or machining processes, or to modify some of the properties of the base or main polymer.

In order to assist the inexperienced, and as an introduction to these materials, we include the designation, the usual abbreviation in alphabetic order, and a few notes on the major and basic thermoplastics materials available today and having some application within the context of this book.

3.2.1 ABS — acrylonitrile butadiene styrene

Discovered in 1946, this is a rigid product which shows high impact strength and durability, and is suitable for applications over a wide temperature range (−40 up to 80°C), and is especially useful at low and sub-zero temperatures due to its comparatively high level of impact resistance. Significant applications are in pipe systems for fluids and compressed air. Other styrene polymers exist, such as styrene acrylonitrile (SAN) and acrylonitrile styrene acrylic (ASA).

3.2.2 ECTFE — ethylene-chlorotrifluoroethylene

One of the family of fluoropolymers, offers extremely high chemical resistance, working temperatures of up to 190°C, good welding properties, and unique levels of purity for many ultra-pure pipe system applications. ECTFE and other fluoropolymer materials indicated below and used in ultra-pure applications are manufactured under extremely closely controlled conditions in purpose-built plants, using only carefully selected high-grade materials, in order that leaching of any impurities from the product to the fluid being handled is impossible. Important applications in pipe systems, tank linings and valves are to be found in the electronic and pharmaceutical industries.

3.2.3 FEP — fluorinated ethylene propylene

A copolymer of tetrafluoroethylene and hexafluoropropylene, it has a similar corrosion resistance to that of PTFE, with a slightly reduced maximum service temperature, but offering all thermoforming and welding properties of typical thermoplastics. FEP is often used in pipe systems, valves and pipe and tank linings.

3.2.4 PA — polyamides

Originally discovered in 1938, they exist in many different forms such as PA 6,6 (Nylon), PA 11 (Rilsan). Often used as an anti-corrosion coating on for example, steel flanges and supports.

3.2.5 PBT — polybutylene terephtalate

This is an opaque material offering similar properties to PET, with improved impact and scratch resistance.

3.2.6 PC — polycarbonate

Discovered in 1957, it offers the combination of transparency with an exceptional resistance to impact damage.

3.2.7 Pe — polyethylene

Originally discovered in 1937, this has now developed in a number of directions, as follows:

- HdPe — high density polyethylene;
- MdPe — medium density polyethylene;
- LdPe — low density polyethylene; and
- various high-performance grades of polyethylene.

3.2.8 PET — polyethylene terephthalate

This is a transparent rigid material with a high tensile strength, having good abrasion resistance, often used in bottles and water/drink containers.

3.2.9 PFA — perfluoro-alkoxy

This has a similar chemical resistance to that of PTFE, and can handle a wide range of fluids up to a maximum temperature of 260°C. PFA is often used to line pipe systems, valves, pumps and tanks.

3.2.10 Polyether ketone

This is a tough, high-temperature, highly corrosion resistant material, offering exceptional performances but at a very high cost. Used essentially within the aerospace, nuclear and oil industries.

3.2.11 PMMA — polymethylmethacrylate

Often defined by one of the registered trademarks such as Altuglas or Plexiglas, it was discovered as early as 1927, is more transparent than glass, and has a good resistance to ultraviolet radiation.

3.2.12 POM — polyacetal

Discovered in 1953, it is often used for machined or moulded components requiring high mechanical properties, such as end caps for high-pressure vessels, and moving parts such as cog wheels.

3.2.13 PP — polypropylene

Polypropylene, discovered in 1957, is light in weight and offers high impact strength and high service temperatures. These properties combined with the ease of assembly by polyfusion are the prime reasons for its importance in the manufacture of pipe systems, tanks and other vessels. Various modified forms of both homo- and copolymer polypropylene are available, with, for example, improved low-temperature performance, or, in a form designated PPs, as a flame-retardant material.

3.2.14 PPE — polyphenylene ether (see PPO)

3.2.15 PPO — modified polyphenylene oxide

PPO and PPE are similar in composition and in performance, both having high strength, high modulus, excellent dimensional stability, very good impact resistance and high thermal distortion resistance. Low water absorption and dimensional stability has given these materials an important place in close tolerance moulding of, for example, components for pumps, valves and spiral wound membranes.

3.2.16 PPS — polyphenylene sulphone

This is a flame-resistant, high-temperature-resistant material (up to 260°C), with good dimensional stability and solvent resistance. Most applications involve high-precision moulding of components.

3.2.17 PS — polystyrene

Discovered in its initial form in 1930, it led to other forms such as PSE — foamed polystyrene — in 1951 and impact-resistant polystyrene (IPS).

3.2.18 PTFE — polytetrafluoroethylene

Discovered in 1941, the original fluorocarbon resin shows exceptional chemical resistance, but is handicapped by limited transformation possibilities when compared to the more recent fluorocarbon resins described in this chapter.

3.2.19 PVC — polyvinyl chloride (PVC-U — unplasticised PVC)

This material was initially identified as far back as 1838, although the development of the principles of an industrial manufacturing process took another 70 years. Large-scale industrial production actually started in 1938.

Unplasticised polyvinyl chloride is one of the most used thermoplastics for handling water due to its rigidity and versatility, and can be used for service temperatures from 0 to 60°C, although care must be taken when handling the material at low temperatures.

PVC-C (chlorinated PVC)

A modified PVC due to the incorporation of higher levels of chlorine, which offers a higher service temperature of close to 100°C. Both PVC and PVC-C are used extensively for pipes, fittings and valves.

3.2.20 PVDF — polyvinylidene fluoride

Offers the weldability of polypropylene, an extended range of operating temperatures, from −40 to 140°C, and a chemical resistance which may approach that of more exotic materials such as PTFE for example. Generally used in pipe systems, or in sheet form for the construction of small vessels and tanks. Some manufacturers supply a high purity grade material for applications with ultra-pure water.

3.3 Thermoplastic Materials for the Construction of Dual Laminates

PVC-U and PVC-C are used as corrosion resistant inner liners for both pipe systems and vessels, using standard pipe and sheet in both materials. A chemical bonding between the PVC liner and the structural laminate is achieved by using an appropriate polyester resin such as Palatal A 410-01 from DSM Composite Resins. This resin is in fact an unsaturated polyester resin based on isophthalic acid and neopentyl glycol, dissolved in styrene. Palatal A 410-01 adheres well to rigid PVC pipe or sheet as long as a mat or woven or non-woven roving fabric is placed on top of the degreased PVC and the resin is applied about 15 minutes prior to gelification. Direct winding of rovings, via a filament winding process, directly onto even cleaned and/or roughened PVC will not achieve a satisfactory bond. Both thermal and mechanical properties of the bond can be achieved by using a 50/50 resin mix of A 410-01 and Palatal E 200-01 (an unsaturated polyester based on neopentyl glycol, orthophthalic acid and adipic acid, together with a curing formulation based on a cyclohexanone peroxide (CHP).

PVDF liner pipe is supplied by companies such as Simona AG and Georg Fischer GmbH with an outer surface prepared for basically mechanical adhesion to a composite laminate.

Polypropylene (PP) based dual laminate systems rely on the fabricator winding a woven or stitched fibre glass roving fabric onto the heated and softened outer surface of standard, relatively thin polypropylene pipe. The structural laminate is then built up on this glass fibre surface after cooling.

Polypropylene, PVDF, ECTFE, FEP and PFA sheets with a glass or synthetic fibre partially impregnated into one face of the sheet are available from Simona AG, Georg Fischer GmbH or Symalit AG, and allow the fabricator to thermoform and weld sheets together to form a tank or vessel skin or inner liner, onto which a structural laminate is applied by either hand lay up or filament winding.

Various standards define the test methods and values concerning the bond strength achieved between the thermoplastic liner and the structural composite reinforcement. Typical values and methods to obtain these values in both shear and peel configuration are given in the standards BS 4994 and NFT 57-900.

3.4 Summary

The physical properties of the most common thermoplastic materials supplied in finished or semi-finished form are summarised in Table 3.

Table 3. Typical properties of thermoplastic materials

	Test standard	Units	PVC-U	PVC-C	HdPe	PP	PVDF	FEP	PFA	ECTFE
Mechanical properties										
Density	DIN 53479	g/mm³	1.42	1.55	0.95	0.92	1.78	2.15	2.15	1.7
Tensile properties										
Yield stress	DIN 53455	N/mm²	45–55	57	22	33	40–60	21–28	24–30	44
Elongation at yield	DIN 53455	%		3	9	8	6			
Elongation at rupture	DIN 53455	%	20–30	15	300	70	20	240–350	300	200
Tensile modulus	DIN 53457	N/mm²	1800–3500	3000	800–1100	1200	2400	350–500	280	900–1300
Module in flexion	ASTM D790	N/mm²	3200	3000	800		2000	660	650	1300
Flexural Stress	DIN 53452	N/mm²	70–80	90	30–40	47	74	18	18	12
Impact properties										
Impact strength		kJ/m²	No rupture	No rupture	No rupture	No rupture	No rupture			
Notched impact strength		kJ/m²	4	8	12	>4	12			
Thermal properties										
Vicat softening point	DIN 53460	°C	78	105						
Average coefficient of thermal expansion	DIN 53752	k⁻¹(°C⁻¹)	0.8×10^{-4}	0.6×10^{-4}	1.8×10^{-4}	1.6×10^{-4}	1.3×10^{-4}	1.0×10^{-4}	1.4×10^{-4}	1.0×10^{-4}
Thermal conductivity	DIN 52612	W/m/k	0.12	0.14	0.38	0.22	0.14	0.25	0.19	0.24
Other properties										
Inflammability	DIN 4102	Class	B1	B1	B2	B2	B1			
Water absorption	DIN 53495	%/24 hr	<0.1	<0.2	<0.01	<0.01				
Abrasion resistance	DIN 53754	mm³/1000U								
Maximum continuous service temperature		°C	60	100	85	100	150	205	260	160
Minimum continuous service temperature		°C	–15	–15	–50	–15	–40	–190	–190	–70

CHAPTER 4

Production Technology

Production techniques and technologies for moulding or forming thermoplastic and composite materials are too numerous and complex to be discussed in detail in this book. We have limited the information to a brief presentation of the main technologies used within the context of the water industry. For more detailed information the author recommends the *Reinforced Plastics Handbook*, published by Elsevier Advanced Technology.

4.1 Thermosets/Composite Materials

The different moulding procedures described below are all designed to combine a thermosetting resin with a glass fibre reinforcement. Except for moulding compounds such as BMC and SMC, for example, the glass fibre reinforcement is always applied in successive layers, and the characteristics of the finished laminate will depend upon the relative proportions of resin and glass, the type of glass reinforcement used, and the direction in which the fibres are laid, the number of layers of each type of material, the characteristics of the resin used, and the type and duration of the curing process. For example, moulding compounds are made with cut fibres which are placed in a random pattern, and the resulting physical properties are the same in all directions. Polyester and vinyl ester resins are cured by the addition of a variety of catalysts, promoters, accelerators and inhibitors, all of which may be mixed into the resin in small accurately weighed proportions, generally in the range of 0.5 to 2.0% by weight. Small variations in prescribed amounts of additives will have little effect on the overall cure, either delaying or advancing slightly the time to gel and achievement of a full cure. Substantial variations from prescribed amounts could have serious consequences, with either extremely rapid gel times, associated high exothermic reaction, with eventually self-induced combustion of the material under cure, or if too little additives are used, the gel and cure time may be of such duration that a full cure is never achieved.

Epoxy resins are much more sensitive to the amount of hardener used, especially when using amine curing agents, and in order to obtain the optimum properties of the cured resin the precise ratio of hardener to resin must be calculated, which, depending upon the charactcristics of both materials, may vary between 10 and 30%.

To summarise, the strength of a composite laminate will vary, at a given temperature, as a function of the following main points:

- laminate thickness;
- type of resin and associated curing agents;
- the use or not of heat cure, with or without a post-curing operation;
- type and weight per fixed unit of the glass fibre reinforcement;
- the orientation of the reinforcing fibres;

- the ratio of resin to reinforcement;
- the type and percentage used of any fillers or additives; and
- the production process used.

Other points may also require consideration, such as the inclusion of air bubbles and the subsequent void content of the laminate, the uniform distribution of the reinforcement, especially when using spray up techniques, the relative humidity and ambient temperature of the workshop or job site during the lay up of the laminate. The original properties of the laminate will continue to be modified by any significant change in ambient temperature or any change due to the temperature of the fluid in contact with the laminate, and, depending upon the stress loading onto the finished equipment, will generally deteriorate over a period of time, but usually within the limits of the design code used to define the equipment itself.

It can be understood that the characteristics of the finished laminate can vary immensely as a function of all the basic characteristics indicated above, and that the specification of a particular laminate can be a complex problem. These characteristics of the laminate, will also depend upon the type of production process used, as described below.

The very nature of the laminate requires some consideration, in order that not only the physical properties are well defined and controlled, but that also the required level of corrosion resistance is obtained.

Using the production processes described below and grouping them into three sections according to their main production technologies, it is possible to reach a general understanding of the different types of laminate and their levels of corrosion resistance:

- Group 1 includes contact moulding, spray up, filament winding and centrifugal casting;
- Group 2 combines compression moulding and resin transfer moulding; and
- Group 3 includes pultrusion.

The production processes in Group 1 are those generally used to manufacture pipes, many fittings, tanks and other vessels as they offer the greatest degree of corrosion resistance.

These anti-corrosion laminates are generally composed of:

- a liner or inner layer of material which will not corrode in the presence of the fluids being handled, and which is as non-porous as possible;
- a structural laminate which ensures the structural integrity of the equipment in question; and
- an external coating which provides resistance to the external environment, either in a corrosive atmosphere, or against general or specific weather conditions.

We can examine in some detail each of these three layers or components of a typical laminate.

The liner or inner layer can be made with the same resin as the structural laminate, a different resin, or a different material, such as a thermoplastic sheet or pipe in case of a 'dual laminate' construction. The role of this layer is to first of all ensure the correct level of chemical resistance so that the equipment does not corrode, and to maintain a separation between the fluid and the structural laminate. It may also be required to add further specific properties to the laminate which are not possessed by the structural laminate, such as abrasion or impact resistance, electrical conductivity in order to avoid an accumulation of static electricity, approval for contact with potable water or foodstuffs, or resistance to fire, for example.

Thermoplastic liners are generally used because of a specific chemical resistance, abrasion resistance or impact resistance, or to reduce the overall cost of a corrosion-resistant pipe or vessel which requires a fluoropolymer material, by using a relatively economic composite laminate for the structural component rather than solid fluoropolymers or other systems. Typical materials are PVC, PVC-C, PP, PVDF, ECTFE, FEP or PFA. All of these materials can receive special treatment, either on the outer surface of a pipe, or on one face of a sheet, in order that a good peel and shear strength between the two materials is achieved. Wall thickness is generally between 2 and 4 mm, although the liner thickness of dual laminate pipe may be considerably thicker, as dependent upon standard thermoplastic pipe for most applications with PVC, PVC-C and polypropylene.

A thermosetting resin is the most economic material for the liner, and accounts for the great majority of composite pipes and vessels. The thickness of the liner may vary between 0.2 and 5 mm or more, depending upon the corrosive nature of the fluid being handled. In the case of water, including most industrial effluent, the maximum liner thickness would be 2.5 mm.

A liner of between 0.2 and 0.5 mm would generally be constructed from one or two layers of surface veil, heavily impregnated with resin, and showing a resin content of at least 90% by weight, although thermosetting resin liners without any fibre reinforcement can be made. The reinforcing veil is usually essential in avoiding cracking of the liner.

Manufacturing a pipe or vessel body by filament winding requires that the making of the liner is the first operation in the production cycle, as is the case in most applications using contact or spray up technologies. The opposite is true for centrifugal casting, when the resin rich liner is applied once the structure is in place, and in some cases, essentially when required to obtain a laminate approved for contact with potable water, it may even be applied in a secondary operation once the structural laminate has been cured.

Should the aggression of the fluid being handled require a thicker liner, the initial inner liner is reinforced with a laminate up to 2 mm thick, composed of layers of low-weight chopped strand mat (300 or 450 g/m^2) or chopped fibres should spray up be used. Typical resin content will be approximately 70% in order to maintain a high level of chemical resistance and resistance to weeping. No matter what the design code, the wall thickness of the liner should not be taken into account when defining the basic physical properties of the laminate.

In most cases, a structural laminate will be relatively porous, such that any high-pressure testing or sustained hydrostatic pressure will result in premature weeping of the laminate. It is obviously preferred to have a structural laminate which shows the lowest possible level of porosity, essentially to ensure satisfactory behaviour both in the long term and also should the inner liner receive some form of superficial damage. Some major manufacturers of pipe systems, particularly Smith Fiberglass Products Company, have, however, developed filament winding technologies, essentially for small to medium diameter anhydride- and amine-cured epoxy and vinyl ester pipes, which rely totally on the quality of the structural laminate, as these products have no inner liner whatsoever. Typical working pressures can be as high as 200 bar for diameters up to six inches (150 mm), and 16 bar for diameters up to 24 inches (600 mm). These products, under the trade names of Red Thread and Blue Streak are essentially used for applications with water or aqueous mildly corrosive fluids.

The structural layer behind the inner corrosion-resistant liner needs to be as thin as possible, in order to keep the overall cost of the equipment competitive with other materials. Whether manufactured by filament winding, contact moulding or centrifugation, the concept requires a glass fibre content of around 70% by weight. Glass contents over 75% will tend to show premature signs of delamination, due to the poor bonding of glass and resin, whereas glass contents less than 70% will decrease physical properties and increase wall thickness and cost. This type of laminate is obtained by filament winding, or in the case of centrifugation and hand lay up, via the use of laminate sequences relying heavily on woven rovings and fabrics. Spray up technology is limited to small components or non-load-bearing structures, as glass contents over 40% are difficult to obtain.

The outer layer of this type of laminate may vary from virtual non existence, in that the final structural layer is left with a resin-rich finish, or, because the pipe or vessel may be exposed to a severe and aggressive environment, a real corrosion-resistant layer of perhaps 1 mm thick may be built up in a manner similar to the internal liner. Generally, however, the external finish is an unreinforced resin-rich layer, including possibly pigments or other additives to reduce ultraviolet light attacking the surface resin. The use of specific surfacing veils will generally avoid any significant surface deterioration due to general exposure to various weather conditions and at the same time provide an aesthetically acceptable appearance.

Going on to examine compression and resin transfer moulding we meet with a completely different laminate build up and composition. Both moulding technologies, and many similar and derived technologies, result in a homogeneous structure, which has no special characteristics on either surface or central structure. The moulding is in fact a compromise between corrosion resistance and mechanical properties, both of which are lower than any of the laminates described above. The technologies call for short-cut fibres, mixed with resins, fillers and various other additives depending upon tooling, production process and required specific performance values. Corrosion resistance is suitable for such applications as flanges and nozzles, as well as low-pressure pipefittings and other components, which will be exposed to water and mildly aggressive aqueous effluent.

Pultrusion also produces a composite which is generally uniform and homogeneous throughout the total structure, being essentially manufactured from continuous rovings with only a 50% resin content in premium grade materials. The outer surface can be more or less resin rich depending upon the requirements of the product, or include a surfacing mat for optimum weathering and wear resistance. Woven rovings and chopped strand mat can be included within the structure should it be necessary to improve transverse properties.

4.1.1 Contact moulding

Contact or hand lay up moulding is a relatively simple, highly labour intensive operation whereby successive layers of fibre reinforcement are impregnated with a resin inside or on a mould. The process begins with the application of a thin layer of resin, reinforced with one or more surface veils, applied directly onto the surface of the prepared mould. In the case of a pipe, tank or other vessel this layer will form the beginning or all of the corrosion-resistant inner liner. This initial layer may then be reinforced with one or more layers of chopped strand mat, or in the case of spray up (see below), by chopped fibres. The total thickness of the corrosion-resistant layer can vary between 0.5 and 5 mm or more, depending upon the application, and on an average thickness of 2 to 2.5 mm, will contain approximately 30% glass by weight.

A structural laminate will then be added to this liner, either by hand, continuing the contact moulding process, or, in the case of pipe or tank production, this structure may be made by filament winding (see below). In the case of a structural laminate made by hand lay up, this laminate may be composed of further layers of chopped strand mat of variable weight, a variable sequence of chopped strand mat and woven rovings, or, in some applications such as small diameter pipe fittings, the laminate may be built up entirely of woven rovings applied in the form of a band. The glass content will vary between 40 and 70% by weight, depending upon the different forms of reinforcement used.

Contact or hand lay up laminates should in no way be considered automatically as being of low quality, as in many applications, and in

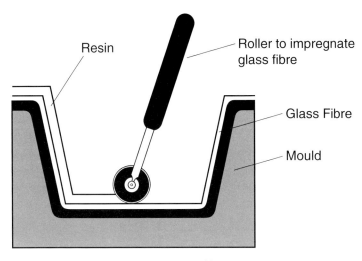

Figure 3. Contact moulding.

conformity with major national standards such as BS 4994, the high chemical resistance of this type of laminate when correctly cured cannot be surpassed. However, the extremely low level of investment required to produce composite equipment with this type of moulding can lead to companies having insufficient levels of training and control and producing large, complicated structures which may not perform to the level expected by the end user. Comparative price levels for contact moulding may reflect the level of competence and control of the different companies offering supposedly identical products.

Contact moulded items are essential to practically all composite systems supplied to the water industry, including many pipe fittings, flat bottoms and dished ends for storage tanks and other vessels, and the assembly of parts or components into finished equipment.

4.1.2 Spray up moulding

Spray up production technology can be seen as an extension of contact moulding. This process involves feeding chopped fibres into a stream of resin fed from a spray gun onto the surface of the mould. Whilst this process is faster, and therefore more economic than conventional hand lay up, the main problems are the inclusion of air bubbles within the laminate, and the difficulty in controlling an even deposit of material in order to achieve a minimum but uniform wall thickness of the finished laminate. Both of these negative points can be overcome by using high-performance and computer-controlled equipment, and this may be the case for high-volume production runs of standard parts.

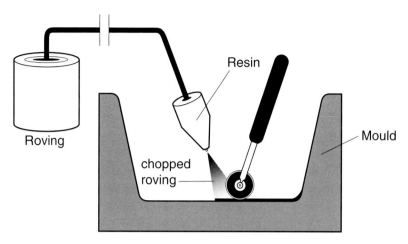

Figure 4. Spray up moulding.

Spray up may also be used in combination with filament winding, either to apply extra reinforcement in a particular area or to increase the resin/glass ratio of a filament wound structure.

4.1.3 Filament winding

Filament winding is based upon a simple process of winding resin impregnated fibres around a revolving mould or mandrel, generally a cylinder, but other forms such as ovoid or conical are also commonly wound structures. The horizontal translation of the head of the machine, that is the dispenser of the fibres, in relation to the diameter and speed of rotation of the mandrel defines the winding angle, and, by consequence, the axial and radial mechanical properties of the finished product. Most machines have two axes control, whereas more sophisticated machines mat have a third or fourth axes control, essentially to control the proximity of the head of the machine to the mandrel. Small to medium diameter pipes and tanks, for example, are usually manufactured using a reciprocal winding machine, on which the glass fibre is wound onto the mandrel at a balanced angle of $\pm 54°$. This particular angle produces the ideal relationship (2:1) for radial/axial properties. This angle is modified to suit particular specifications. For example, some high-pressure pipe and down hole tubing and casing may have reciprocal winding angles as flat as $70°$ in order to improve the axial characteristics of the product, or, on the contrary, large-diameter pipe for buried use at low pressure may have essentially a winding angle as close to $90°$ as possible in order that the pipe structure shall be as rigid as possible, with a greatly reduced axial strength.

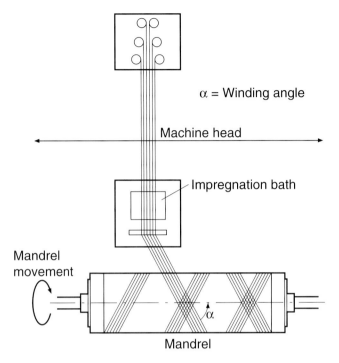

Figure 5. Filament winding.

An alternative to reciprocal winding, which still maintains a balance of winding angles in both directions, is to combine radial winding, that is as close as possible to 90°, with the interspersion of axial reinforcement, 0°, in the form of unidirectional woven rovings. This technique is used in the manufacture of large-diameter pipes and vessels, and especially for very large diameters where helical winding is impracticable. A similar construction can be obtained by using a combination of radial winding and simultaneous spray up of chopped fibres in order to obtain the relatively low axial strength required for storage tanks and other similar vessels. This form of filament winding can be transferred to an on-site production unit, for the manufacture of large tanks and gas scrubbers of 25 m in diameter or more.

Filament winding machine design varies from one supplier to another, and many manufacturers of composite equipment actually build their own winding machines. Depending upon the diameter of the pipe or tank shell being wound, the machines may be fitted with one, two, four or even more mandrels. The machine may be winding on two mandrels at one time, whilst the other mandrels are in the cure or demoulding process.

The methods described above are all considered as discontinuous moulding operations, in that the mandrel has to be removed from the machine for demoulding after each shell construction has been made. Although some of them are highly mechanised, they still require a relatively high labour input to complete the full production cycle from preparation of the raw material up to the cutting and machining of a finished pipe or vessel shell. Two other forms of manufacture exist, which between them allow the production of pipe or cylindrical tank shells up to 3500 mm in diameter, namely continuous filament winding or centrifugal casting, as discussed below.

Continuous filament winding as developed and used by Owens Corning Pipe Technology and its licensees involves a machine containing a continuous steel band supported by beams which form a cylindrically shaped mandrel. As the mandrel moves, fine-graded filler, glass rovings, resin and surface materials are metered onto it in precise amounts under the direction of a programmable logic controller (PLC) and a small computer (PC). The PLC–PC modules provide integrated process control based on pre-programmed recipes. Only basic pipe data such as diameter, pressure and stiffness class needs to be entered and the computer calculates all the machine settings.

Material consumption, as well as pipe thickness are continuously monitored and logged. The logged data are accumulated and reports printed as needed. The temperature of the laminate is measured at 20 different locations in the curing zone, and these temperatures are shown graphically on the PC monitor. Curing of the laminate is accomplished with a combination of induction heating through the steel band and infrared elements directly heating the laminate.

The saw unit is synchronised with the continuous longitudinal movement of the laminate, which ensures a clean perpendicular cut of the FRP pipe or tank shell. Cutting lengths may vary between 300 mm and 18 m.

After passing the cutting station the section is moved on through automatic lifting tables and conveyors up to the on-line chamfering and calibration unit, and then on to the hydrostatic test bench.

4.1.4 Centrifugation or centrifugal casting

Centrifugal casting as used in the HOBAS process is another highly automated production process. Pipe is manufactured by building up the pipe structure from its external surface by feeding the raw materials into a rotating mould. The manufacturing process can be briefly described as follows:

- While the mould is revolving at a relatively low speed, the resins, catalysts and aggregates are evenly fed into and along it at a controlled rate by a feeder arm which moves backwards and forwards along its length.
- The resins for both the liner and the structural laminate are stored in tanks. The structural resin is mixed with fillers and additives at the mixing station. The mixture is transferred to the feeder which pumps it

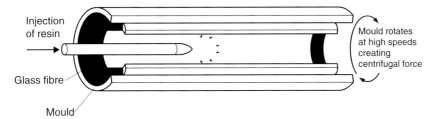

Figure 6. Centrifugal moulding.

as required to the end of the feeder arm where it is combined with catalyst immediately prior to being deposited in the mould. Liner resin is mixed separately and is pumped to the mould at the required moment.
- Dry aggregates are stored in bulk silos, and the material is transferred to a holding bin from which it is pumped to the hopper on the feeder arm. According to the required manufacturing programme material is fed at controlled rates to the end of the feeder arm and discharged into the mould.
- Filler is delivered in bulk and stored in a silo from where it is transferred to the batching silo. The filler is batch weighed prior to mixing with the resin.
- The HOBAS process uses E-glass rovings which are received in the form of wound packages or spools. The feeder carries a number of these spools from which the rovings are drawn to the end of the feeder arm where they are chopped into selected lengths by the glass cutter and deposited in the rotating mould.
- After the materials have all been placed within the mould, the speed of rotation is increased to intensify the centrifugal forces in order to compact the solid materials, and remove all air bubbles or pockets from the laminate. The temperature of the mould and the materials within it is increased by spraying hot water onto the outside of the mould. Heating increases the action of the catalysts, and the exothermic reaction continues to add further heat. At completion of curing, the mould is cooled with water and the pipe is extracted.
- After curing the pipe ends are trimmed and bevelled and a coupling is mounted onto one end of each pipe.

4.1.5 Resin transfer moulding

The description of this and the following process can only be considered as a very basic summary. Major in-house developments by many producers have produced variations and refinements of these processes to fit specific qualities of raw materials and the requirements of the particular applications.

Resin transfer moulding, or RTM (often referred to as resin injection), employs a matched male/female mould, which is filled with glass fibre

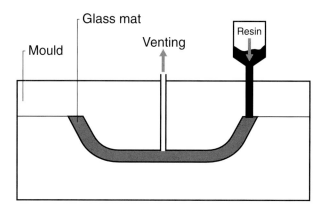

Figure 7. Resin transfer moulding (RTM).

reinforcement tailored closely to the dimensions of the mould, using chopped strand mat or woven rovings. Alternatively one can use a glass fibre preform, obtained by spraying up together chopped fibres and a binding material onto a form which matches closely the dimensions of the mould.

Once the tool is closed, the resin/hardener/additive mix is injected into the mould, chasing out the air via judicially placed regulated vents. Depending upon the resin formulation and the construction material of the mould, the mould may be heated or the material may be left to cure at ambient temperature. The process allows close control of dimensions, weight, thickness of laminate and glass-to-resin ratios. The moulds can be gel-coated on one or both sides prior to insertion of the glass fibre reinforcement, with the possibility of producing high-gloss, smooth surfaces if the mould itself is of high quality. This process requires relatively small capital investments and is ideally suited to small to medium quantity production runs.

4.1.6 Compression moulding

Compression moulding, or press moulding, again requires a matched male/female mould. The process may use fibre reinforcement placed into the mould as described above, the resin mix being poured onto the reinforcing material and dispersed into the reinforcement via the closing action of the mould under low pressure. Alternatively, and depending upon the size and configuration of the mould, the manufacturer may opt for a dough moulding compound.

Cure may be carried out at ambient temperature, especially for short production runs using a composite mould. For larger production runs, heated metallic moulds are used which offer the combined advantage of closer control of the curing cycle, as well as decreasing the time of the production cycle.

The advantages of this process are the same as indicated above, with associated higher capital costs but improved production efficiency, of special interest for longer production runs.

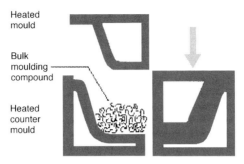

Figure 8. Compression moulding.

4.1.7 Pultrusion

Pultrusion is a highly automated manufacturing process for the production of structural profiles. The process involves the drawing of continuous glass fibre rovings, impregnated with a wide range of resin systems, through high-precision forming dies, a subsequent heated die to obtain final dimensions, and curing and a final cutting device. Standard shapes include I-beams, wide-flange beams, round, square and rectangular hollow sections, round, square and flat bars, channels and angles.

Different resin systems are available, from standard isophthalic polyester for general usage, to epoxy or vinyl ester for improved structural properties or chemical resistance, up to phenolic systems for high-temperature and fire-resistant applications.

The nature of the production process and basic raw materials provide the profiles with high axial strength, whereas the possibility of including woven rovings, chopped strand mats and surface veils allow the manufacturer to vary mechanical properties, improve cross-layer or tangential properties and obtain excellent surface aspects in line with the requirement of the particular form or application.

Compared to steel, the structural profiles offer considerable advantages in that they are corrosion-resistant in even the most aggressive environments, have a low-weight, high-strength ratio, offer a high grade of dimensional stability and are nonconductive.

Typical dimensions of standard products include angles and channels up to 200 mm wide, cylindrical hollow section up to 100 mm in diameter, square hollow section up to 100×100 mm, I-beams up to 500 mm wide $\times 200$ mm deep, and flat sheet or shallow forms up to 1500 mm wide. Special tooling can be made for many forms on a customer or project basis where the inherent properties of pultruded profiles offer the engineer a novel way to solve many structural problems in a cost efficient and aesthetic manner.

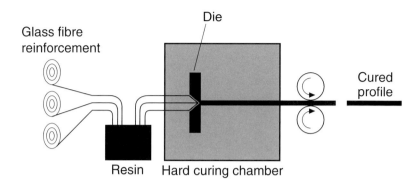

Figure 9. Pultrusion process.

Two lesser-known derived production processes exist, pull winding which combines filament winding with pultrusion and has been used with some success to manufacture continuous pipe or cylindrical hollow section with improved hoop properties, and pull forming, whereby the material is finally formed, in an uncured form, after having first passed the pre-forming die. This final forming is obtained by pressing or squeezing the uncured material into a mould, in which it is finally cured.

4.2 Thermoplastics

4.2.1 Injection, compression and transfer moulding

These moulding operations account for approximately 25% of all plastic manufacturing, and consist of a continuous process of filling a cavity with pressurised heated materials. The moulding materials are generally in solid form (powder, granules, flakes or as a more or less malleable paste-moulding compound).

Compression and transfer moulding, two of the main techniques used for thermosets as discussed in more detail above, are relatively unimportant with thermoplastic materials. The main drawback being the requirement to cool the mould before each demoulding operation, this operation being largely incompatible with the high-frequency moulding operations generally required for thermoplastic moulding.

Injection moulding, however, although used for some applications with thermosets, is by and large reserved for such thermoplastics as polystyrene, copolymers, polyolefins, cellulosics, acrylics, vinyls, linear polyesters, polyphenylenes and polysulphones, for example. Under the effects of high pressure, a material which has been fluidised by heating is introduced at high speed into a closed and relatively cool mould. The mould is maintained in a closed position by a hydraulic or mechanical system, the opening of which at the end of the moulding cycle also serves to eject the moulded item.

The exact quantity of raw material required is generally prepared by an Archimedes-type screw within a heated cylinder, with an axial displacement of the screw to fill the mould. The pressure at the level of injection into the mould is generally in the range of 500 to 2000 bar, depending on the type of material being used, the shape and number of mould and dimensional tolerances, for example. The machines are generally rated as a function of the closing force and their volume. The highest performance machines in the market today allow a moulding capacity of around 150 kg, associated with a closing force of around 10000 tonnes.

Typical products obtained by this process are pipefittings, nozzles, small covers and caps.

68 Thermoplastics

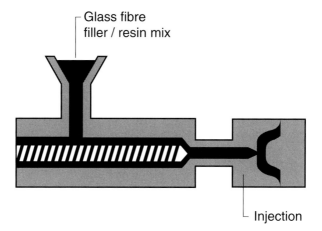

Figure 10. Injection moulding.

4.2.2 Extrusion

Extrusion is by far the production method which accounts for the largest volume of plastic products (approximately 40%) and is used for all thermoplastic pipe and profile production. This process is characterised by a continuous production of a product which has a relatively small cross-section compared to length, although pipe is extruded in some cases up to a diameter of 2 m.

The raw material, in bulk, in either powder or granulate form, flows continuously into the heated barrel of the extruder, in which turn one or more screws. The heated material fluidises, and the screw pushes it through the head of the machine which is shaped to the required form and dimension. Cooling and pulling tools complete the production line, at the end of which an automatic cutting device cuts the finished product into the required lengths.

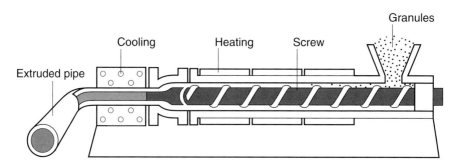

Figure 11. Extrusion.

Practically all thermoplastic products can be extruded, and include PVC, PVC-C, PP, all forms of polyethylene, ABS and PVDF. For some specific applications two or more different materials can be co-extruded, forming a material which has different physical properties throughout the different areas of its structure.

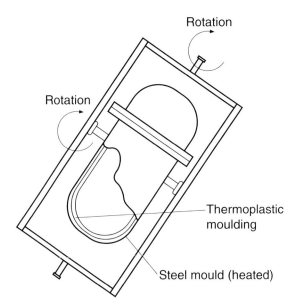

Figure 12. Rotational moulding.

4.2.3 Rotational moulding

This type of moulding is used to manufacture small and medium production runs of hollow forms such as small tanks, septic tanks and containers.

A fixed volume of polymer, either as a fine powder or in a semi-liquid form is introduced into a closed mould which rotates around two axes located at 90° to each other, at speeds between 10 and 50 rotations per minute. The raw material is distributed uniformly over the total surface of the mould, and gels as the mould is heated to a temperature of 200 to 400°C. A subsequent intense cooling of the mould causes solidification, and the finished object can then be removed from the mould.

Initially this process was used to manufacture small objects, but recent developments have permitted the moulding of polyethylene tanks with a volume of 25 m^3. Developments are under way with ABS, polyamides and polycarbonates, as well as with glass fibre reinforcements and foaming agents.

Production times tend to be relatively long, but this can be compensated for by correspondingly cheap moulds. Single- or double-skin mouldings are possible, permitting consideration of integral insulation or increased rigidity of thin wall constructions.

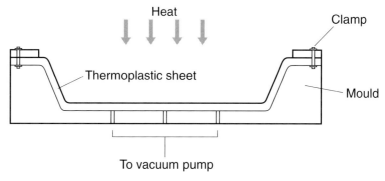

Figure 13. Thermoforming.

4.2.4 Thermoforming

Thermoforming allows for the production of objects of a specific form from a heated thermoplastic sheet. The most significant thermoforming to have been developed is vacuum forming from a wide range of large thermoplastic sheets. The main process consists of blocking by suction on the bottom of a relatively cold mould, a thermoplastic sheet softened by exposure to infrared rays, but keeping a relative elasticity, rather like that of a rubber material. The sheet has to be maintained fixed around the edge of the mould by a mechanical device. The finished form can be removed as soon as it has cooled down. For some materials or applications the vacuum is replaced by blowing via a bell, thereby pushing the softened sheet into the mould. Various combinations of negative and positive pressures are also used, as is the use of a die to push the sheet into the mould.

Moulds can be made from a variety of materials, including wood or plaster, and as such short runs or prototypes can be manufactured at very low costs. This technique lends itself to the manufacture of relatively large forms with a low wall thickness, such as shallow tanks or covers, or can be used in conjunction with other forming or welding technologies discussed below.

4.2.5 High-frequency welding

High-frequency welding is used for instantaneous sealed assembly of thin sheet or film form of thermoplastics such as PVC, cellulose acetate and ABS. It cannot be used for polyolefins and fluorocarbons.

A high-frequency welding machine consists of a high-frequency generator with electrodes which will disperse the energy into the jointing area of the material to be assembled. Welding times can vary between 0.1 and 30 seconds with a force of 500 W up to 40 kW, at a frequency up to 50 MHz. Interesting applications are to be found in the construction of flexible tank covers for odour control in wastewater treatment plants.

4.2.6 Boiler work (welded tanks and other vessels)

Boiler work is perhaps not the most accurate term to describe the construction of tanks and other vessels such as gas scrubbers, reactors, silos and other process vessels, but it has been borrowed from the traditions of the similar industry of metallic boiler work. In actual fact, the similarities of metallic and thermoplastic vessel construction are numerous, and as a general rule if a vessel can be built in steel it can also be built in thermoplastics or a combination of thermoplastic with a lightweight composite reinforcement. This activity comprises the cutting, forming and welding together of components to form a finished construction.

Thermoplastics can be cut with most commonly used hand or powered saws, such as band saws, circular and jig saws, and a typical toolbox will contain roughly the same type of tools as a carpenter, supplemented by the tools required to heat and weld. Welded constructions can be cylindrical, conical, square or rectangular, but the inherent properties of most plastic materials lend themselves more easily and economically towards a circular section.

Sheets can be formed at ambient or high temperature, depending upon the material being used, the diameter of the vessel being built and the thickness of the sheet being formed. When heat is required the procedure generally entails placing the sheet inside a temperature-controlled oven until the material's softening point has been achieved, and then transferring the softened sheet to a forming tool, generally built in wood. The sheet maybe clamped to the form, and in all cases is left to cool to ambient temperature. The formed sheet can then be trimmed and prepared for welding. A radius can be formed onto a sheet, or a full dished end for a tank or vessel can be obtained using similar techniques.

Ways of welding the different components together are various, again depending upon materials, wall thickness, weld and vessel configurations for example, but can be summarised as hot gas welding, hot gas extrusion welding and fusion welding of sheet and tube, all as discussed in more detail in Chapter 7.

4.2.7 Machining

Machining of thermoplastics, from sheet, rod or tube, is generally limited to short or medium size production runs for which excessive tooling costs exclude a moulding technique. It can also be used to rework a moulded product in order to obtain fine dimensional tolerances which cannot be obtained with standard mouldings.

When machining thermoplastics one must pay attention to the fact that thermoplastics are poor thermal conductors, and when too much heat is generated they can become brittle or soften. The advance of the machine tool needs to be limited, and the machining area needs constant cooling. The slowing of the advance can in most cases be compensated to some extent by a higher rotational speed. Thermoplastics containing fibre reinforcement require special attention, as the abrasive effect of the fibres will blunt normal tooling. Tungsten carbide or diamond-coated cutting tools are often required.

4.2.8 Dip moulding or coating

This technique involves heating of a support or object, generally in metal although glass products can also be treated in this manner. The item to be coated is heated and then dipped into a bath of either a solid or fluidised coating material, followed generally by a curing programme inside an oven. Typical coating materials are Rilsan, PVC, polypropylene or polyethylene, PVDF and PTFE. Thickness of coatings can vary between 0.5 up to several millimetres depending upon the coating material being used and the form of the object being coated.

Typical applications are coating of steel flanges, supports and handling devices which will be in contact with an aggressive environment.

CHAPTER 5

Design, Production and Testing Standards

The extent, range and scope of content of standards for the design, manufacture, installation and testing of thermoplastic and thermosetting or composite pipe systems and vessels is as large and varied as the industries themselves, and to make matters more complicated several of the large national and international users of composite materials have issued their own standards. In the field of composite materials the advantage of possessing varied and flexible production processes has to some extent limited the development of large-scale standardised production in favour of particular client or industry standards related to very specific applications.

It is important to identify and specify the required standards to purchase or manufacture a specific product, but it is equally important to resist the temptation to over-specify the number of applicable standards, which will often result in a conflict between two or more of the specified standards. In order to minimise this risk the author recommends that the specification uses only one set of national or international standards, i.e. all ASTM, BS or ISO, for example.

North America, due to the scale of its own domestic market combined with a large export market, has developed the broadest and most in-depth series of standards, whereas Europe is still largely dependent upon national standards. European standards do exist, and more are under development. In the meantime, the position in Europe is somewhat confused. European standards (EN and ETS) are automatically taken up as national standards by all country members of the European Union, with an associated and obligatory suppression of any divergent standards. It will require several more years for this process to be completed, but due to the cross-border business developed by major manufacturers, a certain homogeneity of design is being readily adopted. International standards (ISO and CEN) are adopted by each member country on a voluntary and non-obligatory basis, and may exist in parallel with national European and other major national standards for many years to come.

International standards (ISO) by their very nature are the most recent applicable documents and in spite of the non-obligatory nature of their application as or within a national standard, many of them are being adopted as either a national or European standard. We have for this reason attempted to identify as many examples as possible of applicable standards concerning materials and equipment covered by this handbook. Where possible, standards having been created or adopted by two separate organisations, have been noted a such (for example ISO BS 10931).

As the issue of standards is both complex and ongoing, misunderstanding and confusion will continue for many years to come. Should readers meet with confusion due to an ongoing process after publication of this book, the author intends to maintain an up to date database on this issue and would be pleased to examine any further requirements of the reader.

We indicate below, for each market sector and for each geographical zone, some of the main and most frequent standards referred to and used in the water-related market. Although none of the listings can be considered in any way exhaustive, an examination of the most significant standards will refer the reader back to more detailed material, component or specific behaviour testing and requirements.

The scope and content of most of the standards listed are evident from the title. We have, however, commented on some of the standards for the sake of clarity, or because of the relative importance or significance of a particular standard.

Further details on applicable standards can be found in later chapters where these standards are directly related to a specific application such as pipeline and sewer renovation and irrigation.

5.1 Composite Pipe Systems

5.1.1 North American standards

Design, production, installation and testing are covered by standards issued by the following two main organisations.

ASTM. The American Society for Testing and Materials has developed and published the most extensive range of standards, testing requirements and procedures for all forms of plastic materials. Most documents related to standards, testing or approval of reinforced plastics refer back to a wide range of ASTM testing procedures, which are often used in Europe, the Middle East, Africa, the Pacific area as well as in North and South America. The following listing includes those which are most often specified and used.

ASME. The American Society of Mechanical Engineers, founded in 1880, now has 36 Technical Divisions, including Aerospace, Materials, Materials Handling and Production Engineering. It sets many industrial and manufacturing standards. This organisation is of significant importance for both pressurised and non-pressurised composite vessels via their qualification programmes under the headings ASME Code, Section X and ASME RTP-1.

AWWA. The American Water Works Association has developed standards specific to the usage of various materials in water treatment and distribution, including composite materials as included in the following listing, of which the most significant on a worldwide basis is the AWWA standard for glass fibre pressure pipe, AWWA C950.

API. The American Petroleum Institute was formed in 1919 in order to standardise engineering specifications, and has today developed some 500 equipment and operating standards used around the world. This organisation is of significant interest for the standards relating to high- and low-pressure composite pipe systems under the headings API 15HR and API 15LR.

For some very specific applications, such as buried fire mains, these are frequently covered by other organisations, such as Factory Mutual. Most specific or end user specifications rely heavily on ASTM standards for testing of raw materials and finished products.

5.1.1.1 Specifying and testing of composite pipe systems

ASTM D 570. Standard test method for Water Absorption of Plastics.

ASTM D 2310. Standard Classification for Machine Made Reinforced Thermosetting Resin Pipe.

Having manufactured and tested the pipe to various ASTM standards, and having obtained specific mechanical properties, ASTM D 2310 allows the manufacturer to classify the pipe, via a codified system, in a way which allows the end user to make a considered comparison of the product, either in comparison with his own design requirements, or against a second supplier.

A typical coded classification could read RTRP-12EF, where RTRP means reinforced thermosetting resin pipe. The first digit, 1, confirms the pipe is made by filament winding (a 2 would indicate centrifugal casting). The second digit, 2, indicates a polyester resin (1 is epoxy, 3 phenolic and 4 furane). The first letter concerns whether the pipe has a liner or not, and whether it is reinforced with a veil or similar. In the case of the above example, the E defines a reinforced polyester liner. The second letter, in this case F, concerns the hydrostatic design basis, or HDB, obtained when testing the pipe to ASTM D 2992, procedure A or B. In the case of our example the F designates a minimum cyclic value of 8000 psi.

ASTM D 2996. Standard Specification for Filament Wound Reinforced Thermosetting Resin Pipe, 1–16 inches in diameter (25–400 mm).

This standard allows a more refined specification of a pipe to be established, using a coded system as defined above. The classification to ASTM D 2996 is an extension to ASTM D 2310, and cannot stand on its own.

Using the same pipe as defined above, the fuller specification would now read RTRP-12EF1-3112, where the additional digits, in the order as written, add the following information: 1 confirms that the ASTM D 2992 test was run on pipe with free unrestrained ends; 3 confirms a minimum value of 40 000 psi on a short-term burst test to ASTM D 1599; 1 confirms a minimum value of 8000 psi for longitudinal tensile strength to standard ASTM D 2105; 1 confirms a minimum tensile modulus of 2×10^6 psi when tested to ASTM D 2105; and 2 confirms a minimum stiffness factor of 1000 in^3 lb/ln^2 when tested to ASTM D 2412.

To summarise, the above standards allow both the client and the manufacturer to establish performance requirements in a standardised and codified form, which then allows a considered and valid comparison of similar or identical products.

ASTM D 2517. Standard Specification for Reinforced Thermosetting Resin Pipe.

ASTM D 2997. Standard Specification for Centrifugally Cast Reinforced Thermosetting Resin Pipe.

ASTM D 3262. Standard Specification for Reinforced Plastic Mortar Sewer Pipe.

ASTM D 3517. Standard Specification for Fiberglass Pressure Pipe.

ASTM D 3754. Standard Specification for Fiberglass Sewer and Industrial Pressure Pipe, 8–144 inches in diameter (200–3600 mm).

ASTM D 4024. Standard Specification for Reinforced Thermosetting Resin Flanges. This is based on a codified system for filament wound and moulded flanges, similar to the code system described above.

ASTM D 4161. Standard Specification for Fiberglass Pipe Joints using Flexible Elastomeric Seals.

ASTM D 5686. Standard Specification for 'Fibreglass' (Glass Fibre Reinforced Thermosetting Resin) Pipes and Fittings, Adhesive Bonded Joint Type Epoxy Resin, for Condensate Return Lines.

AWWA C 950. Standard for Fiberglass Pressure Pipe for Water Service, 1–144 inches in diameter (25–3600 mm). This is the most widely used standard for buried pipe systems handling both potable water and sewage. The scope of the standard includes the manufacture and testing of pipe and jointing systems for use in both above and below ground water and effluent systems.

As well as conventional RTRP the standard includes reinforced plastic mortar pipe (RPMP), which covers glass fibre reinforced materials which incorporate fillers such as siliceous sand. Resin systems include both epoxy and polyesters, and basic diameters, pressure classes and stiffness classes may form the basis for extrapolation to material characteristics outside of the actual specification.

Appendices also provide some guidance on design and installation of pipe manufactured to this standard.

API 15 LR. Specification for Low Pressure Fiberglass Line Pipe, 2–16 inches in diameter, up to 1000 psi (cyclic)/25–300 mm, 70 bar.

API 15 HR. Specification for High Pressure Fiberglass Line Pipe, 1–8 inches in diameter, above 1000 psi (cyclic)/25–300 mm, 70 bar.

API 15 AR. Specification for Fiberglass Tubing.

These three API standards have been developed specifically for oil field applications, but are often used for handling water rather than crude oil, and as such are of importance to this handbook. They all rely heavily on basic testing to ASTM D 2992, with subsequent and variable interpretations and extrapolations, and we include further comments under this standard below.

MIL-P-28584A. Specification for Epoxy Resin Pipe and Fittings from 2 to 12 inches diameter for use in continuous service at 125 psi and 250°F (50–300 mm/8.5 bar/120°C. Often used for condensate return lines and insulated pipe systems used for district heating.

5.1.1.2 Recommended practices

ASTM D 3567. Standard Practise for Determining Dimensions of Reinforced Thermosetting Resin Pipe and Fittings.

ASTM D 2563. Standard Practise for Classifying Visual Defects in Glass Reinforced Plastic Laminate Parts. This defines three different levels of acceptance for both the liner and structure of laminates, both with and without the possibility of carrying out remedial work after an initial inspection.

ASTM D 3839. Standard Practise for Underground Installation of Flexible Reinforced Thermosetting Resin Pipe and Reinforced Plastic Mortar Pipe.

AWWA C 950. Appendix C — Installation (for AWWA C 950 — see above).

API RP 15 L4. Recommended Practise for Care and Use of Reinforced Thermosetting Resin Line Pipe.

API RP 15 A4. Recommended Practise for Carc and Use of Reinforced Thermosetting Resin Casing and Tubing.

5.1.1.3 Test methods

(a) Tensile properties:

ASTM D 638. Standard Test Methods for Tensile Properties of Plastics.

ASTM D 1599. Short Term Hydraulic Failure Pressure of Plastic Pipe, Tubing and Fittings.

ASTM D 2290. Apparent Tensile Strength of Ring or Tubular Plastics and Reinforced Plastic Pipe and Tube by Split Disk Method.

ASTM D 2105. Longitudinal Tensile Properties of Reinforced Thermosetting Plastic Pipe and Tube.

  (b) Compressive properties:

ASTM D 695. Standard Test Methods for Compressive Properties of Rigid Plastics.

  (c) Flexural properties:

ASTM D 790. Standard Test Methods for Flexural Properties of Unreinforced and Reinforced Plastics and Electrical Insulating Materials.

ASTM D 12925. Measuring Beam Deflection of Reinforced Thermosetting Plastic Pipe under Full Bore Flow.

(d) Long-term pressure containment:

ASTM D 1598. Time to Failure of Plastic Pipe under Constant Internal Pressure.

ASTM D 2143. Cyclic Pressure Strength of Reinforced Thermosetting Plastic Pipe.

ASTM D 2992. Obtaining Hydrostatic Design Basis for Reinforced Thermosetting Resin Pipe and Fittings, Procedure A, Cyclic/Procedure B, Static.

This is probably the most important standard in use for the design of composite pipe systems, for both the direct definition of a particular wall thickness and also as a basis from which similar or associated structures and configurations may be validated.

This recommended practice sets out to determine the long-term hydraulic design basis (HDB) for internal pressure rating of glass fibre piping and is by far the most commonly known standard within the industry.

It is worth knowing, however, that the ratio of outside diameter to wall thickness should be 10:1 or more for it to be applicable. This standard presents two ways to establish the pressure rating, Procedure A (cyclic) and Procedure B (static). The static of Procedure B is the most commonly used in Europe mainly because of the equipment and laboratories available to perform it.

To the knowledge of the author, only two or three manufacturers in Europe have cyclic equipment available in-house and there is only one fully equipped independent testing house, whereas in the USA many manufacturers have equipment to perform their own cyclic tests, to Procedure A in-house.

Procedure A consists of exposing a minimum of 18 identical pipe samples or fittings, or both, to cyclic internal pressures at a cycle rate of 25 cycles/minute and at several different pressures. Temperature of the liquid inside the pipe and the air outside should be controlled and constant.

The long-term hydrostatic strength of pipe and fittings is obtained by an extrapolation to 150×10^6 cycles (11.4 years) of a log–log plot of the linear regression line for hoop stress verses cycles to failure. Failure is defined as pipe weeping or any other catastrophic failure or rupture. Failures should happen within the following ranges:

Cycles			Failure points
1000	to	10 000	at least 3
10 000	to	100 000	at least 3
100,000	to	1 000 000	at least 3
1 000 000	to	10 000 000	at least 3
After 15 000 000			at least 1
Total			at least 18

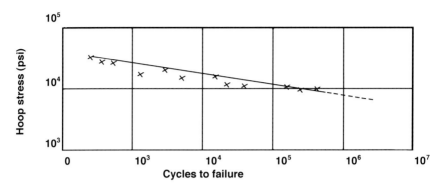

Figure 14. Regression curve for an epoxy pipe.

Procedure B consists of exposing a minimum of 18 identical pipe samples or fittings, or both, to constant internal hydrostatic pressures at different pressure levels in a controlled environment and measuring the time to failure for each pressure level. The temperature of the liquid inside the pipe and the air outside should be controlled and constant.

The long-term hydrostatic strength of pipe and fittings is obtained by an extrapolation to 100 000 h (11.4 years) of a log–log linear regression line for hoop stress versus failure. Failures should happen during the following time frames:

Hours			Failure points
10	to	1000	at least 4
1000	to	6000	at least 3
After 6000			at least 3
After 10 000			at least 1
Total			at least 18

Should failure not occur within the specific cyclic/pressure time frame then the test is stopped, the projected level of pipe performance revised and the test recommenced.

Under the possible conditions from initiating the requirement to test and receiving the final report, a period of 15 months will have passed. Should a rerun be necessary then a period of 20 to 26 months could be considered as normal. Once the test run is complete, the test data are controlled, and following procedures as set out within the standard, a lower confidence limit, taking into consideration the spread of results, is computed.

Due to the log–log nature of the regression line, it is common to project the results to 15, 20, 25 and even 50 years. The extrapolated hydrostatic design basis (HDB) is then calculated using a factor of service/safety determined by the manufacturer, client or approval body.

Various factors are used, the most common and those adopted by various approval bodies, including ASTM, are:

- Procedure A: 0.80/1.00
- Procedure B: 0.50/0.56

These factors are actually under review by an API committee, and it is possible that less conservative factors will be applied in the near future.

One can readily understand that the ASTM 2992 tests, which form the basic design data of most worldwide specifications, cannot be carried out at short notice and without due care and consideration. This places some manufacturers at a disadvantage until such time as sufficient testing data has been accumulated.

The extent, duration and costs of the ASTM 2992 test precludes it from being used to test the total range of products available from each manufacturer and by consequence, further extrapolation or associated testing is required. The basis of this associated testing is to substantiate the relative performance of pipe, fittings and joints over a total similar product range. Most product specifications or standards will define groups of diameters from which basic testing will be considered as valid. For example:

Diameter 25 mm to 80 mm	Test diameter 80 mm
Diameter 100 mm to 150 mm	Test diameter 150 mm
Diameter 200 mm to 400 mm	Test diameter 400 mm
Diameter 450 mm to 800 mm	Test diameter 800 mm

However, if the basic manufacturing process and raw materials are identical over a total range, for example diameter 50 mm to 1000 mm, then the following scenario could be agreed and applied:

- Complete testing to ASTM D 2992 of pipe of diameter 100 mm.
- Partial testing to ASTM D 2992 (first 1000 hours) of diameters 150 mm and 400 mm.
- Short-term testing of all the above to ASTM D 1599.
- Selective short-term testing of various pipes, fittings and joints of a series of diameters between 50 mm and 1000 mm.

The basic idea being to prove that as short-term test results of all material are within an agreed level of performance, one can assume that as production technology and raw materials are constant throughout the total range, then the long-term behaviour of all components will be in line with the extrapolation of the basic long-term testing programme.

The above is only one scenario and agreement should be sought between the end user and manufacturer before proceeding with any testing. However, because of the cost and duration of the basic test, certain end users will accept a one-off full log term test on a small-diameter pipe system, usually 100 mm, with all the supporting evidence of performance based upon a selection of reduced tests and instantaneous, short-term test results.

API 15 LR, as published in 1986, was the first independent standard to impose a wide range of performance testing on a complete range of pipe, fittings and assemblies, and as such has been totally or partially incorporated into many other national or company standards and specifications.

API 15 LR imposes a rigid framework within which both basic initial ASTM 2992 long-term testing and considerable short-term and instantaneous testing of all physical and some thermal properties are combined, covering the full range of pipe, fittings and assemblies, in each and every diameter and configuration.

From these test programmes a complete pipe system can be homologated by the supplier, which certifies the performance ratings (temperature and pressure) of the total system. It should be noted that API delivers a licence to the manufacturer, enabling the manufacturer to label his pipe systems with the API 15 LR label. Before obtaining this licence, the manufacturer is required to obtain a quality certificate, API QI. Subsequent controls by API as to the use of API labels are controlled via the API QI quality programme. A similar, but more constraining programme, API 15 HR has been set up for higher pressure pipe systems (above 1000 psi, 70 bar).

(e) Pipe stiffness:

ASTM D 2412. External Loading Characteristics of Plastic Pipe by Parallel Plate Loading.

(f) External pressure:

ASTM D 2924. External Pressure Resistance of Reinforced Thermosetting Resin Pipe.

(g) Chemical resistance: note these standards are equally applicable to composite tanks and vessels, as described below.

ASTM C 581. Standard Practise for Determining Chemical Resistance of Thermosetting Resins used in Glass Reinforced Structures intended for Liquid Service.

ASTM D 3615. Chemical Resistance of Thermoset Molding compounds used in the Manufacture of Moulded Fittings. (This standard has been discontinued by ASTM and has not been replaced. Some manufacturers have retained this document as part of their production and quality control systems.)

ASTM D 3681. Chemical Resistance of Reinforced Thermosetting Resin Pipe in a Deflected Condition.

Specifically designed for larger diameter buried pipe handling water or effluent. Requires a minimum of 18 pipe samples which are deflected to various levels to induce different strain levels, whilst partially in contact with a solution of sulphuric acid at 5%. The time to failure is recorded and a strain regression curve is generated in much the same way as the hydrostatic design basis is determined in ASTM D 2992.

(h) Acoustic emission testing: note that these standards are equally applicable to composite tanks and vessels, as described below.

ASTM E 569. Recommended Practise for Acoustic Emission Monitoring of Structures Intended for Liquid Service.

ASTM E 1316. Definition of Terms Relating to Acoustic Emissions.

ASTM E 650. Recommended Practise for Mounting Acoustic Emission Contact Sensors.

ASTM E 750. Recommended Practise for Measuring Operating Characteristics of Acoustic Emission Instrumentation.

5.1.1.4 Fire or flame resistance

Fire resistance and/or flame spread may be a significant consideration in the design and manufacture of plastic equipment. Two well-known standards allow testing and classification of plastic materials.

ASTM D 635. Standard Test Method for Rate of Burning and/or Extent of Time of Burning of Plastics in a Horizontal Problem.

ASTM E 84. Standard Test Method for Surface Burning Characteristics of Building Materials.

5.1.2 Individual national standards — Europe

Practically all European countries have published their own standards, all of which should be surpassed by equivalent European standards as and when they may become available. We reserve our attention to the major national standards as issued by France, Germany and the UK.

5.1.2.1 France; standards issued by AFNOR (L'Association Francaise de Normalisation) under the heading T as being at the experimental stage, and which could be transformed into a European standard during the qualification procedure

T57 200. Pipes and Fittings of composite glass thermosetting materials. General review. Description. Classification. Characteristics.

T57 201. Pipes and Fittings in FRP. Test to determine the hoop rigidity.

T57 202. Reinforced plastic pipes. Sealing ring type joints for installation under pressure or not. Suitability for use.

T57 203. Glass fibre reinforced plastic pipes. Dimensions.

T57 205. Glass fibre reinforced plastic pipes. Test method for short-term resistance, under pressure, to rupture.

T57-206. Glass fibre reinforced plastic pipes. Glass epoxy resin pipes for the transport of hot water under pressure. Characteristics and test methods.

T57-207: Glass fibre reinforced plastic pipes. Collection of basic data for the dimensional calculation of pipes and fittings under constant internal pressure. Test method.

T57-208. Fibre reinforced plastic pipes. Design and dimensioning of cemented socket assemblies.

T57-209. Fibre reinforced plastic pipes. Underground installation of flexible pipelines with or without pressure.

T57-213. Fibre reinforced plastic pipes. Resistance determination under cyclic internal test pressure. Test method.

5.1.2.2 Germany; standards issued by the Deutsches Institut für Normung under the heading DIN or DIS

DIN 16 867. Glass fibre reinforced polyester resin (UP-GF) pipes. Fittings and joints for chemical pipelines. Technical delivery conditions.

DIN 16 868: Glass fibre reinforced unsaturated polyester resin (UP-GF) pipes. Part 1: Wound, filled, dimensions. Part 2: Wound, filled. General quality.

DIN 16 869. Centrifugally cast filled fibre reinforced unsaturated polyester resin (UP-GF) pipes. Part 1: Dimensions. Part 2: General quality requirements, testing.

DIN 16 870-1. Wound glass fibre reinforced epoxy pipes; dimensions.

DIN 16 871. Centrifugally cast glass fibre reinforced epoxy pipes; dimensions.

DIN 16 964: Wound glass fibre reinforced polyester resin (UP-GF) pipes, general quality requirements. Testing.

DIN 16 965. Parts 1, 2, 4 and 5: Wound glass fibre reinforced polyester resin pipes, types A, B, D, and E; dimensions.

DIN 16 966.

Part 1 — glass fibre reinforced polyester resin pipes. Fittings and joints. General Quality requirements, testing.
    Part 2 — Elbows. Dimensions.
    Part 4 — Tees and nozzles. Dimensions.
    Part 5 — Reducers. Dimensions.

Part 6 — Flanges and Seals. Dimensions.

Part 7 — Pipes, fittings and joints, bushings, flanges, flanged and butted joints. General quality Requirements, testing.

Part 8 — Laminated joints. Dimensions.

DIN 19565-1. Centrifugally cast and filled polyester resin glass fibre reinforced (UP-GF) pipes and fittings for buried drains and sewers, dimensions and technical delivery conditions.

DIN 53 769.

Part 1 — Testing of glass fibre reinforced plastic pipelines, determination of the adhesive shear strength of type B pipeline components.

Part 2 — Testing of glass fibre reinforced plastic pipes; long-term hydrostatic pressure test.

Part 3 — Testing of glass fibre reinforced plastic pipes; short-term flattening test and flattening endurance.

Part 6 — Testing of glass fibre reinforced plastic pipes; testing of pipes and fittings under pulsating conditions.

DIN 54 815. Pipes of filled polyester resin moulding materials. Part 1: Dimensions, materials, designation. Part 2: Requirements, testing.

5.1.2.3 UK; standards issued by BSI (the British Standards Institute) under the BS heading.

BS 3974. Specification for Pipe Supports. Part 1: Pipe hangers, slider and roller type supports. Part 2: Pipe clamps, cages, cantilevers and attachments to beams. Part 3: Large-bore, high-temperature, marine and other applications.

BS 5350. Method of Test for Adhesives. Part C5: Determination of bond strength in longitudinal shear.

BS 5480. Specification for Glass Fibre Reinforced Plastics (FRP) Pipes and Fittings for Use for Water Supply or Sewage. Part 1: Dimensions, materials and classifications. Part 2: Design and performance requirements.

BS 6464. Specification for Reinforced Plastics Pipes, Fittings and Joints for Process Plants.

BS 7159. Design and Construction of Glass Reinforced Plastics (FRP) Piping Systems for Individual Plants or Sites.

BS 8010: Code of Practice for Pipeline. Section 2.5. Glass reinforced thermosetting plastics.

5.1.3 European and cross-border standards (EN)

EN NF DIN 705. Plastic piping systems. Glass reinforced thermosetting plastics (GRP) pipes and fittings. Methods for regression analyses and their use.

EN NF DIN 761. Plastic piping systems. Glass reinforced thermosetting plastics (GRP) pipes. Determination of the creep factor under dry conditions.

EN NF DIN 1115. Plastic piping systems for underground drainage and sewerage under pressure. Glass reinforced thermosetting plastics (GRP) based on polyester resin.

    Part 1: General.
    Part 2: Pipes with flexible, reduced articulation or rigid joints.
    Part 3: Fittings.
    Part 4: Ancillary equipment.
    Part 5: Fitness for purpose of the system.
    Part 6: Recommended practice for installation.

EN NF DIN 1119. Plastic piping systems. Joints for glass reinforced thermosetting plastics (GRP) pipes and fittings. Test methods for leaktightness and resistance to damage of flexible and reduced articulation joints.

EN NF DIN 1120. Plastic piping systems. Glass reinforced thermosetting plastics (GRP) pipes and fittings. Determination of the resistance to chemical attack from the inside of a section in a deflected condition.

EN NF DIN 1225. Plastic piping systems. Glass reinforced thermosetting plastics (GRP) pipes and fittings. Determination of the creep factor under wet conditions and calculation of the long-term specific ring stiffness.

EN NF DIN 1226. Plastic piping systems. Glass reinforced thermosetting plastics (GRP) pipes. Test method to prove the resistance to initial ring deflection.

EN NF DIN 1227. Plastic piping systems. Glass reinforced thermosetting plastics (GRP) pipes. Determination of the long-term ultimate relative ring deflection under wet conditions.

EN NF DIN 1228. Plastic piping systems. Glass reinforced thermosetting plastics (GRP) pipes. Determination of initial specific ring stiffness.

EN NF DIN 1229. Plastic piping systems. Glass reinforced thermosetting plastics (GRP) pipes. Test methods to prove the leaktightness of the wall under short-term internal pressure.

EN NF DIN 1393. Plastic piping systems. Glass reinforced thermosetting plastics (GRP) pipes. Determination of initial longitudinal tensile properties.

EN NF DIN 1394. Plastic piping systems. Glass reinforced thermosetting plastics (GRP) pipes. Determination of the apparent initial circumferential tensile strength.

EN BS NF DIN 1447. Plastic piping systems. Glass reinforced thermosetting plastics (GRP) pipes. Determination of long-term resistance to internal pressure.

EN NF DIN 1448. Plastic piping systems. Glass reinforced thermosetting plastics (GRP) components. Test methods to prove the design of rigid locked socket and spigot joints with elastomeric seals.

EN NF DIN 1449. Plastic piping systems. Glass reinforced thermosetting plastics (GRP) components. Test methods to prove the design of cemented socket and spigot joints.

EN DIN 1450. Plastic piping systems. Glass reinforced thermosetting plastics (GRP) components. Test methods to prove the design of bolted flanged joints.

EN DIN 1636. Plastic piping systems for non-pressure drainage and sewerage. Glass reinforced thermosetting plastics (GRP).
Part 1: General.
Part 2: Pipes with flexible reduced articulation or rigid joints.
Part 4: Ancillary equipment.

EN NF DIN 1638. Plastic piping systems. Glass reinforced thermosetting plastics (GRP) pipes. Test methods for the effects of cyclic internal pressure.

EN DIN 1796. Plastic piping systems for water supply with or without pressure. Glass reinforced thermosetting plastics (GRP) based on polyester resin (UP).
Part 1: General.
Part 2: Pipes with flexible, reduced articulation or rigid joints.
Part 4: Ancillary equipment.
Part 5: Fitness for purpose of the system.
Part 6: Recommended practice for installation.

EN NF 1862. Plastic piping systems. Glass reinforced thermosetting plastics (GRP) pipes. Determination of the relative flexural creep factor following exposure to a chemical environment.

5.1.4 International standards (ISO)

ISO DIS 7370. Glass fibre reinforced thermosetting plastics (GRP) pipes and fittings. Nominal diameters, specified diameters and standard lengths.

ISO DIS 7509. Glass reinforced thermosetting plastics (GRP) pipes. Determination of time to failure under sustained internal pressure.

ISO 7510. Plastics piping systems. Glass reinforced thermosetting plastics (GRP) pipes and fittings. Test methods to prove the leaktightness of the wall under short-term internal pressure.

ISO 7511. Glass fibre reinforced thermosetting plastics (GRP) pipes and fittings. Test methods to prove the leaktightness of the wall under short-term internal pressure.

ISO 7684. Plastics piping systems. Glass reinforced thermosetting plastics (GRP) pipes. Determination of the creep factor under dry conditions.

ISO 7685. Plastics piping systems. Glass reinforced thermosetting plastics (GRP) pipes. Determination of initial specific ring stiffness.

ISO DIS 8483. Glass reinforced thermosetting plastics (GRP) pipes and fittings. Test method to prove the design of bolted flanged joints.

ISO DIS 8513. Glass reinforced thermosetting plastics (GRP) pipes. Determination of initial longitudinal properties.

ISO 8521. Plastics piping systems. Glass reinforced thermosetting plastics (GRP) pipes. Determination of the apparent initial circumferential tensile strength.

ISO DIS 8533. Glass reinforced thermosetting plastics (GRP) pipes and fittings. Test method to prove the design of cemented wrapped joints.

ISO 8572. Pipes and fittings made of glass reinforced thermosetting plastics (GRP). Definitions of terms relating to pressure, including relationships between them, and terms for installation and jointing.

ISO 8795. Plastic pipe systems for the conveyance of water intended for human consumption. Migration assessment. Determination of migration values for plastic pipes. (Note: valid for thermosetting and thermoplastic materials.)

ISO 10465. Underground installation of flexible glass reinforced thermosetting resin (GRP) pipes.
  Part 1: Installation.
  Part 2: Comparison of static calculation methods.
  Part 3: Installation parameters and application limits.

ISO 10466. Plastics piping systems. Glass reinforced thermosetting plastics (GRP) pipes. Test method to prove the resistance to initial ring deflection.

ISO DIS 10467. Plastic piping systems for pressure and non-pressure drainage and sewerage. Glass reinforced thermosetting plastics (GRP) based on unsaturated polyester (UP) resins.

ISO DIS 10468. Plastics piping systems. Glass reinforced thermosetting plastics (GRP) pipes. Determination of the long-term specific ring creep stiffness under wet conditions and calculation of the wet creep factor.

ISO DIS 10471. Plastics piping systems. Glass reinforced thermosetting plastics (GRP) pipes. Determination of the long-term ultimate bending strain and the long-term ultimate relative ring deflection under wet conditions.

ISO DIS 10639. Plastic piping systems for water supply with or without pressure. Glass reinforced thermosetting plastics (GRP) based on unsaturated polyester (UP) resins.

ISO 10928. Plastics pipe systems. Glass reinforced thermosetting plastics (GRP) pipes and fittings. Methods for regression analysis and their use.

ISO 10952. Plastics pipe systems. Glass fibre reinforced thermosetting plastics (GRP) pipes and fittings. Determination of the resistance to chemical attack from the inside of a section in a deflected condition.

ISO DIS 14828. Plastics pipe systems. Glass reinforced thermosetting plastics (GRP) pipes. Determination of the long-term specific ring relaxation stiffness under wet conditions and calculation of the wet relaxation factor.

5.1.5 Japanese standards

The Japanese Industrial Standards (JIS) Committee is working in close co-operation with international organisations such as ISO, and in many cases a JIS standard maybe identical or very similar to an ISO standard.

Many of the standards listed below exist in an official translation in the English language, especially those concerning thermoplastic pipe systems.

JIS K 7013. Fibre reinforced plastic pipes.

JIS K 7014. Fittings and joints for fibre reinforced plastic pipes.

JIS K 7020. Glass reinforced thermosetting plastics (GRP) pipes and fittings. Methods for regression analysis and their use.

JIS K 7030. Pipes and fittings made of glass fiber reinforced plastics (GRP). Definitions of terms relating pressure, including relationship between them, and terms for installation and jointing.

JIS K 7031. Plastics piping systems. Glass reinforced thermosetting plastics (GRP) pipe and fittings. Test methods to prove the leaktightness of the wall under short-term internal pressure.

JIS K 7032. Plastics piping systems. Glass reinforced thermosetting plastics (GRP) pipes. Determination of initial specific ring stiffness.

JIS K 7033. Plastics piping systems. Glass reinforced thermosetting plastics (GRP) pipes. Determination of initial tensile properties.

JIS K 7034. Plastics piping systems. Pipes made of glass reinforced thermosetting plastics (GRP). Determination of the resistance to chemical attack for the inside of a section in a deflected condition.

JIS K 7035. Plastics piping systems. Glass reinforced thermosetting plastics (GRP) pipes. Determination of the creep factor under wet conditions and calculation of the long-term specific ring stiffness.

JIS K 7036. Plastics piping systems. Glass reinforced thermosetting plastics (GRP) pipes and fittings. Test methods to prove the design of bolted flanged joints.

JIS K 7037. Plastics piping systems. Glass reinforced thermosetting plastics (GRP) pipes. Determination of the apparent initial circumferential tensile strength.

JIS K 7038. Plastics piping systems. Glass reinforced thermosetting plastics (GRP) pipes. Test method to prove the resistance to initial ring deflection.

JIS K 7039. Plastics piping systems. Glass reinforced thermosetting plastics (GRP) pipes. Determination of the long-term ultimate bending strain and calculation of the long-term ultimate relative ring deflection, both under wet conditions.

JIS K 7033. Plastics piping systems. Glass reinforced thermosetting plastics (GRP) pipes and fittings. Test methods to prove the design of cemented or wrapped joints.

5.2 Composite Tanks and Vessels

Whereas the volume, scope and content of standards related to pipe systems are huge, the corresponding standards for both composite and thermoplastic vessels are very limited in number, but by their scope and detailed content do cover practically all the requirements of the market. As very large vessels become more common perhaps some effort will be required to examine this particular market sector in more detail. We go on to examine the actual standards developed in North America and Europe.

5.2.1 North America

ASME Boiler and Pressure Vessel Code, Section X: Fibre Reinforced Plastic Pressure Vessels.

This code specifies requirements of raw materials, production technologies, two different design procedures and criteria for both prototype and on line pressure testing, all contained within a quality system specific to the scope of the code itself.

The actual scope of Section X is as follows:

- establishes types and minimum properties of raw materials which can be used;
- specifies procedures for testing the mechanical properties of laminates; and
- defines two distinct methods whereby the design of a pressure vessel may be qualified, as follows: Class 1 design — this procedure validates the design of a vessel via a comprehensive production and destructive testing programme (see below); Class 2 design — this procedure establishes a programme of mandatory design procedures and subsequent non-destructive testing.

Authors Note: the terms destructive and non-destructive should be understood as follows.

'Destructive' is read as meaning that the vessel has undergone test procedures to the extent that even if destruction is not achieved, the damage caused to the vessel is such that it cannot be used as a pressure vessel.

'Non destructive' is understood as meaning no significant reduction in the characteristics of the vessel have been achieved during a successful test programme defined to prove the fitness for purpose of the vessel under test.

Class 1 design is most useful to define the characteristics of a prototype prior to producing the same vessel in long production runs of identical or similar conception. Class 2 design is essentially for one off or very limited construction runs, where it is not feasible to scrap a single vessel with a destructive test programme.

Section X also:

- Suggests some basic and non-obligatory design methods for Class 1 vessels.
- Defines mandatory design methods for Class 2 vessels.
- Defines the manufacturing procedures which may be used.
- Limits the design of end closures, nozzles and supports and the methods by which they may be attached to the vessel.
- Defines test methods and procedures to be used when testing prototype vessels.
- Establishes a method to define and control production methods used to manufacture Class 1 prototype and production vessels, including definition of which deviations may be accepted or which may require re-qualification of the vessel.
- Establishes methods to ensure that no essential deviation from approved fabrication procedures has occurred.
- Establishes rules for final testing, inspection and associated reports.
- Defines the requirements for stamping and marking the vessel.

Permitted resin systems are specified as polyester, vinyl ester, epoxy, furane and phenolic, and minimum levels for tensile strength and interlaminar shear strength are specified for each resin type.

Maximum pressure ratings are 150 psi for bag moulded, centrifugally cast and contact moulded vessels, 1500 psi for filament wound vessels, and 3000 psi for filament wound vessels with polar boss openings (approximately 10, 100 and 200 bar respectively) with service temperature limitations of $-20°C$ up to $122°C$.

The basic test procedures for proving the performance level of Class 1 prototype vessels, which probably accounts for 99% of all stamped or non-stamped vessels made in accordance with this standard, are:

- Vessel thickness: 12 measurements to an accuracy of $\pm 2\%$ of true thickness.
- Percentage glass/resin to ASTM D 2584.
- Vessel weight measured to within $\pm 1\%$.
- Visual examination to equivalent of ASTM D 2563.

- Tensile strength (determined from the results of the hydrostatic pressure test described below).
- Barcol hardness.
- Volumetric expansion.
- Cyclic pressure test: 100 000 cycles from zero to design pressure at minimum 66°C or higher if the vessel has a higher rated service pressure.
- Hydrostatic pressure test at 66°C or at the design temperature if higher than this. The test pressure to be applied at a uniform rate in order that six times the design pressure is reached in not less than one minute.

Should the vessel be equipped with an inner liner, the cyclic pressure test shall be carried out as follows: 10 000 cycles at the minimum service temperature of the vessel; 90 000 cycles at 66°C or the maximum design temperature if greater.

Perhaps the greatest criticism which can be made of this standard, as far as a Class 1 vessel is concerned, is the fact that a manufacturer can obtain a certification without actually having produced and tested a vessel to the requirements of the standard. The manufacturer is, however, required to manufacture and test at least one prototype vessel before actually selling vessels, which are stamped to certify conformity to the standard.

ASME RTP-1. A relatively new development for the qualification of a fabricator to design, manufacture, control and test FRP vessels, such as tanks and gas scrubbers. Based on the much older procedure required to obtain the RP stamp for ASME Section X — high pressure vessels, this procedure requires the audit of the company, essentially around a quality system applicable to this type of production, but also the manufacture, control and testing of a prototype typical vessel containing all standard components and assembly types normally required in this type of construction. The audit is carried out by an ASME inspector, assisted by an authorised inspector as per the ASME Section X procedure. The qualified fabricator is listed by ASME as a qualified supplier, and is subject to follow up audits on a triannual basis.

ASTM D 3299. Standard specification for filament wound glass fibre reinforced thermosetting resin chemical-resistant tanks.

ASTM D 4097. Standard specification for contact moulded glass fibre reinforced thermosetting resin chemical-resistant tanks.

5.2.2 Europe

The following three national standards have been developed by the industry over the last 25 years or so, and although the approach may differ, the actual results from applying each of the standards is relatively similar. BS 4994 is perhaps more conservative and calculations to this standard will probably show heavier wall thickness for a similar laminate construction. It is generally agreed that BS 4994 will form the basic structure of the new European standard for composite vessels. BS 4994 is detailed as part of the examination of the composite vessel market in Chapter 7.

5.2.2.1 France

NFT 57 900. Code of Construction for Tanks and Apparatus in FRP (a national standard developed from a code developed essentially by the manufacturers in the 1970s).

5.2.2.2 Germany

AD Merkblatt N1. Fibreglass Reinforced Plastic Vessels.

5.2.2.3 UK

BS 4994. Specification for the Design and Construction of Vessels and Tank in Reinforced Plastics.

5.2.3 Japan

JIS K 7012. Glass fiber reinforced thermosetting resin chemical-resistant tanks.

5.3 Thermoplastic Pipe Systems

Due to the quantity and scope of standards applicable to thermoplastic pipe systems we have limited our listing to some of the most significant ASTM standards applicable in North America, a selection of the most significant national standards in the UK, Germany and France, together with the ISO standards which have been or are being adopted in place of national standards throughout the European community. The present situation is somewhat complicated by the process of suppression of national standards, and the modification and replacement by European and/or ISO or international standards.

For further information, see the reference to the web sites of all the major standard organisations.

5.3.1 North America

ASTM D 1599. Standard test method for short-term hydraulic failure pressure of plastic pipe, tubing and fittings.

ASTM D 1785. Standard specification for polyvinyl chloride (PVC) plastic pipe, schedules 40, 80 and 120.

ASTM D 2241. Standard specification for polyvinyl chloride (PVC) plastic pipe, SDR series.

ASTM D 2282. Standard specification for acrylonitrile–butadiene–styrene (ABS) plastic pipe (SDR-PR) based on outside diameter.

ASTM D 2321. Standard practice for underground installation of thermoplastic pipe for sewers and other gravity flow applications.

ASTM D 2447. Standard specification for polyethylene (PE) plastic pipe, schedules 40 and 80, based on outside diameter.

ASTM D 2466. Standard specification for polyvinyl chloride (PVC) plastic pipe fittings, schedule 40.

ASTM D 2467. Standard specification for polyvinyl chloride (PVC) plastic pipe fittings, schedule 80.

ASTM D 2837. Standard test method for obtaining hydrostatic design basis for thermoplastic pipe materials.

ASTM D 3000. Standard specification for polybutylene (PB) plastic pipe (SDR-PR) based on outside diameter.

ASTM F 441. Standard specification for C-PVC plastic pipe schedules 40 and 80.

ASTM F 442. Standard specification for C-PVC plastic pipe schedule SDR-PR.

ASTM F 645. Standard guide for the selection, design and installation of thermoplastic water pressure piping systems.

ASTM F 771. Standard specification for polyethylene (PE) thermoplastic high-pressure irrigation pipeline systems.

ASTM F 948. Standard test method for time to failure of plastic piping systems and components under constant internal pressure with flow.

ASTM F 1668. Standard guide for construction procedures for buried plastic pipe.

ASTM F 1675. Standard practice for life-cycle costs analysis of plastic pipe used for culverts, storm sewers and other buried conduits.

5.3.2 Europe

Represented by the three largest users of thermoplastic pipe systems, the UK, Germany and France.

5.3.2.1 UK national standards

BS 3505. Unplasticised polyvinyl chloride (PVC-U) pressure pipes for cold potable water.

BS 3506. Specification for unplasticised PVC pipe for industrial use.

BS 4346. Part 1: Joints and fittings for use with unplasticised PVC pressure pipes — solvent weld. Part 2: Joints and fittings for use with unplasticised PVC pressure pipes — mechanical joints.

BS 5391. Specification for ABS pressure pipe. Pipe for industrial use.

BS 5392. Specification for ABS fittings for use with ABS pressure pipe. Fittings for use with pipe for industrial uses.

BS 5556. General requirements for dimensions and pressure ratings for pipes of thermoplastic materials (metric series).

BS 6437. Polyethylene pipes (type 50) in metric diameters for general purposes.

BS 6572. Specification for blue polyethylene pipes up to nominal size 63 for below ground use for potable water.

BS 6730. Specification for blue polyethylene pipes up to nominal size 63 for above ground use for potable water.

We also include a listing of WIS standards, that is Water Industry Specifications issued in the UK by the Water Research Centre, and available via their publications department in Swindon (see bibliography). Apart from being a standard for the industry these documents are also of interest to the inexperienced engineer wishing to obtain the maximum of advice on pipe selection, installation and use.

WIS 4-24-01. Specification for mechanical fittings and joints including flanges for polyethylene pipes for the conveyance of cold potable water for the size range 90 to 1000 mm inclusive made of metal or plastics or a combination of both.

WIS 4-31-05. Solid wall concentric external rib-reinforced PVC-U sewer pipe.

WIS 4-31-06. Specification for blue unplasticised PVC pressure pipes, integral joints and post-formed bends for cold potable water (underground use).

WIS 4-31-07. Specification for unplasticised PVC pressure fittings and assemblies for cold potable water (underground use).

WIS 4-31-08. Molecular oriented polyvinyl chloride (MOPVC) pressure pipes for underground use.

WIS 4-32-03. Specification for blue polyethylene (PE) pressure pipes for cold potable water (nominal sizes 90 to 1000 mm for underground or protected use).

WIS 4-32-05. PE pipes for sewer linings (non pressure applications).

WIS 4-32-08. Specification for site fusion of PE 80 and PE 100 pipe and fittings.

WIS 4-32-09. Specification for black polyethylene pressure pipes for potable water above ground or sewage (nominal sizes 90 to 1000 mm).

WIS 4-32-10. Non-circular PE sewer linings.

WIS 4-32-11. Thermoplastic end load-resistant mechanical fittings for polyethylene pipes of nominal sizes ‹63 mm.

WIS 4-32-13. Specification for blue higher performance polyethylene (HPPE/PE 100) pressure pipes (nominal sizes 90 to 1000 mm) for underground or protected use for the conveyance of water intended for human consumption.

WIS 4-32-14. Specification for PE 80 and PE 100 electro fusion fittings for nominal sizes up to and including 630 mm.

WIS 4-32-15. Specification for PE 80 and PE 100 spigot fittings and drawn bends for nominal sizes up to and including 1000 mm.

WIS 4-32-16. Butt fusion jointing machines.

WIS 4-33-01. Polypropylene encapsulated steps for use in manholes and access chambers.

WIS 4-37-02. Design against surge and fatigue conditions for thermoplastic pipes.

5.3.2.2 German national standards

DIN 4266-1. PVC-U, PE-HD and PP drainage pipes for landfill requirements, testing and control.

DIN V 4279-9. Internal pressure test of pressure pipelines for water - Pressure pipes of low density polyethylene PE-LD, pressure pipes of high-density polyethylene PE-HD (PE80 and PE100), pressure pipes of cross-linked polyethylene PE-X and pressure pipes of unplasticised PVC-U.

DIN 8061. Unplasticised polyvinyl chloride pipes, chemical resistance of pipes and fittings of PVC-U.

DIN 8062. Pipes of rigid PVC-U — dimensions.

DIN 8072. Pipes of low-density PE — dimensions.

DIN 8073. Pipes of low-density PE — general quality requirements, testing.

DIN 8074. Pipes of high-density polyethylene (HdPe) PE 63/PE 80/PE 100 — dimensions.

DIN 8075. Pipes of high density polyethylene (HdPe) PE 63/PE 80/PE 100 — testing.

DIN 8077. Specification for polypropylene pressure pipe, PP-H 100/PP-B 80/PP-R 80 — dimensions.

DIN 8078. Specification for polypropylene pressure pipe, PP-H 100/PP-B 80/PP-R 80: quality requirements and test methods.

DIN 8079. Chlorinated PVC-C pipes — PVC-C 250 dimensions.

DIN 8080. Chlorinated PVC-C pipes. General quality requirements and testing.

DIN 16887. Determination of the long-term hydrostatic pressure resistance of thermoplastic pipes.

DIN 16891. Pipes of ABS or acrylonitrile–styrene–acrylester (ASA); dimensions.

DIN 16892. Cross-linked polyethylene (PE-X) pipes; general quality requirements, testing.

DIN 16893. Cross-linked polyethylene (PE-X) pipes; dimensions.

DIN 16961. Thermoplastic pipes and fittings with profiled outer and smooth inner surfaces; dimensions.

DIN 16962. Specification for polypropylene pressure fittings. Types, dimensions and general quality requirements and test methods.

DIN 19533. Pipelines of high-density PE (high-density polyethylene) and low-density PE (low-density polyethylene) for drinking water supply; pipes, pipe connections and fittings for pipelines.

DIN 19534. Pipes and fittings of PVC-U with socket for elastic sealing ring joints for sewerage. Part 2: Technical specifications for delivery. Part 3: Quality control and installation. Part 102: Assessment of conformity (proposal for a European standard).

DIN 19537. High-density polyethylene (HdPe) pipes and fittings for drains and sewers.

DIN 19566. Thermoplastic pipes and fittings with profiled wall and smooth pipe inside for sewers and drains. Part 1: Dimensions. Part 2: General requirements; testing.

DIN 16963. High-density polyethylene (HdPe) fittings — dimensions.

DIN 19533. Pipe systems of high- and low-density polyethylene for drinking water.

DIN 19582. PVC pressure pipe for water supply.

5.3.2.3 French national standards

NF T 54 017. Plastics. Non-plasticised PVC pipes and fittings for domestic water drainage installations. Specifications.

NF T 54 023. Plastics. Non-plasticised PVC pipes. Determination of conventional water absorption.

NF T 54 028. Plastics. Non-plasticised PVC pipeline fittings. Glued connections. Dimensional characteristics.

NF T 54 038. Plastics. Non-plasticised PVC pipes and fittings. Assembly by sealing rings for pressure pipelines. Dimensional.

NF T 54 039. Plastics. Non-plasticised PVC pipes and fittings. Assembly by sealing rings for pressure pipelines. Fitness for use.

NF T 54 063. Plastics. Polyethylene pipes for drinking water supply networks. Specifications and test methods.

NF T 54 071. Plastics. Low-density polyethylene PE 32 pipes for drinking water services. Specifications and test methods.

NF T 54-085. Reticulated polyethylene pipes for pipelines under pressure. Specifications.

NF T 54-086. Unplasticised polyvinyl chloride (PVC-U) pipe for underground irrigation. Specifications.

5.3.3 Japanese standards

JIS K 6741. PVC-U pipes.

JIS K 6742. PVC-U pipes for water works.

JIS K 6743. PVC-U pipe fittings for water works.

JIS K 6761. Polyethylene pipes for general purposes.

JIS K 6762. Double-wall polyethylene pipes for water supply.

JIS K 6769. Cross-linked polyethylene (XPE) pipes.

JIS K 6770. Cross-linked polyethylene (XPE) pipe fittings.

JIS K 6776. Chlorinated polyvinyl chloride (CPV-C) pipes for hot and cold water supply.

JIS K 6777. Chlorinated polyvinyl chloride (CPV-C) pipe fittings for hot and cold water supply.

JIS K 6778. Polybutene (PB) pipes.

JIS K 6779. Polybutene (PB) pipe fittings.

JIS K 6780. Profile wall polyethylene pipes.

JIS K 6787. Cross-linked polyethylene (XPE) pipes for water works.

JIS K 6788. Cross-linked polyethylene (XPE) pipe fittings for water works.

JIS K 6792. Polybutene (PB) pipes for water works.

JIS K 6795. Cross-linked polyethylene (XPE) pipes — effect of time and temperature on the expected strength.

JIS K 6796. Pipes and fittings made of cross-linked polyethylene (XPE) — estimation of the degree of cross-linking by determination of the gel content.

5.3.4 Cross-border standards

5.3.4.1 International standards (ISO) with indication of their eventual adoption by other major European countries and organisations. (EN/BS/NF/DIN)

ISO 161-1. Thermoplastic pipes for the conveyance of fluids — nominal outside diameters and nominal pressures. Part 1: metric series.

ISO 161-2. Thermoplastic pipes for the conveyance of fluids — nominal outside diameters and nominal pressures. Part 2: inch-based series.

ISO 265-1. Pipes and fittings of plastic materials — fittings for domestic and industrial waste pipes — basic dimensions — PVC-U.

ISO 1167. Thermoplastic pipes for the conveyance of fluids — resistance to internal pressure — test methods.

ISO 2536. Unplasticised polyvinyl chloride (PVC) pressure pipes and fittings, metric series. Dimensions of flanges.

ISO EN DIN 3126. Plastics pipes — measurement of dimensions.

ISO DIS 3212. Polypropylene (PP) pipes. Resistance to internal pressure; specification.

ISO 3213. Polypropylene (PP) pipes. Effect of time and temperature on expected strength.

ISO 3458. Assembled joints between fittings and polyethylene (PE) pressure pipes; test of leakproofness under internal pressure.

ISO 3459. Polyethylene (PE) pipes; joints assembled with mechanical fittings; internal under-pressure test.

ISO 3501. Assembled joints between fittings and polyethylene (PE) pressure pipes; test of resistance to pull out.

ISO 3503. Assembled joints between fittings and polyethylene (PE) pressure pipes; test of leakproofness under internal pressure when subjected to bending.

ISO 3603. Fittings for unplasticised polyvinyl chloride (PVC) pressure pipes with elastic sealing ring type joints; pressure test for leakproofness.

ISO 3604. Fittings for unplasticised polyvinyl chloride (PVC) pressure pipes with elastic sealing ring type joints; pressure test for leakproofness under conditions of external hydraulic pressure.

ISO 4059. Polyethylene (PE) pipes—pressure drop in mechanical pipe-jointing systems — method of test and requirements.

ISO 4065. Thermoplastic pipes — universal wall thickness table.
ISO 4191. PVC-U pipes for water supply — recommended practice for laying.

ISO 4422. Pipes and fittings made of unplasticised polyvinyl chloride (PVC-U) for water supply.
  Part 1: General.
  Part 2: Pipes.
  Part 3: Fittings and joints.
  Part 4: Valves and ancillary fittings.
  Part 5: Fitness for purpose of the system.

ISO 4427. Polyethylene pipes for water supply. Specifications.

ISO 4433. Thermoplastic pipes. Resistance to liquid chemicals Classification:
  Part 1: Immersion test method.
  Part 2: Polyolefin pipes.
  Part 3: PVC-U, high-impact poly (vinyl chloride) (PVC-HI) and chlorinated poly (vinyl chloride) (PVC-C) pipes.
  Part 4: Poly(vinylidene fluoride) (PVDF) pipes.

ISO 4435. PVC-U pipes and fittings for buried drainage and sewage systems.

ISO 6259. Thermoplastic pipes. Determination of tensile properties.
  Part 1: General test method.
  Part 2: Pipes made of unplasticised polyvinyl chloride (PVC-U), chlorinated polyvinyl chloride (PVC-C), and high-impact polyvinyl chloride (PVC-HI).
  Part 3: Polyolefin pipes.

ISO TR 7073. Recommended techniques for the installation of unplasticised polyvinyl chloride (PVC-U) buried drains and sewers.

ISO TR 7074. Performance requirements for plastic pipes and fittings for use in underground drainage and sewage.

ISO 7279. Polypropylene (PP) fittings for pipes under pressure; sockets for fusion using heated tools; metric series.

ISO 7349. Thermoplastic vessels; connection references.

ISO 7508. PVC-U valves for pipes under pressure — basic dimensions — metric series.

ISO 8242. Polypropylene valves for pipes under pressure — basic dimensions — metric series.

ISO 8361. Thermoplastic pipes and fittings. Water absorption.
  Part 1: General test method.
  Part 2: Test conditions for unplasticised polyvinyl chloride) (PVC-U) pipes and fittings.
  Part 3: Test conditions for ABS pipes and fittings.

ISO 8584. Thermoplastic pipes for industrial applications under pressure. Determination of the chemical resistance factor and of the basic stress.

Part 1: Polyolefin pipes.
Part 2: Pipes made of halogenated polymers.

ISO 8772. High-density polyethylene (PE-HD) pipes and fittings for buried drainage and sewerage systems; specifications.

ISO 8773. Polypropylene pipes and fittings for buried drainage and sewage systems; specification.

ISO DIS 8795. Plastic piping systems for the transport of water intended for human consumption. Migration assessment. Determination of migration values for plastic pipes and fittings and their joints.

ISO DIS 9080. Thermoplastic pipes for the transport of fluids — methods of extrapolation of hydrostatic stress rupture data to determine the long-term hydrostatic strength of thermoplastic pipe materials.

ISO DIS 9356. Fused polyolefin pipe assemblies. Resistance to internal static pressure. Test method.

ISO 9393. Thermoplastic valves — pressure test methods and requirements.
    Part 1: general.
    Part 2: Test conditions and basic requirements for Pe, PP, PVC-U and PVDF valves.

ISO EN DIN 9967. Thermoplastic pipes; determination of creep ratio.

ISO EN DIN 9969. Thermoplastic pipes; determination of ring stiffness.

ISO 10146. Cross-linked polyethylene (PE-X) pipes — effect of time and temperature on the expected strength.

ISO 10358. Plastic pipes and fittings — combined chemical resistance tables.

ISO 10501. Thermoplastic pipes for the transport of liquids under pressure — calculation of head losses.

ISO 10508. Thermoplastic pipes and fittings for hot and cold water systems.

ISO BS 10931. Plastic piping systems for industrial applications. Poly (vinylidene fluoride):
    Part 1: General.
    Part 2: Pipes.
    Part 3: Fittings.
    Part 4: Valves.
    Part 5: Fitness for purpose of the system.

ISO 11647. Fusion compatibility of polyethylene (PE) pipes and fittings.

ISO BS 11922-1. Thermoplastic pipes for the conveyance of fluids — dimensions and tolerances. Part 1: metric series.

ISO BS 11922-2. Thermoplastic pipes for the conveyance of fluids — dimensions and tolerances. Part 2: inch-based series.

ISO EN NF DIN 12162. Thermoplastic materials for pipes and fittings for pressure applications — classification and design — overall service (design) coefficient.

ISO 12176. Plastic pipe and fittings — equipment for fusion jointing polyethylene systems.
  Part 1: Butt fusion.
  Part 2: Electrofusion.
  Part 3: Fusion operator badge.

ISO 12230. Polybutene (PB) pipes — effect of time and temperature on the expected strength.

ISO EN BS DIN 13478. Thermoplastic pipes for the conveyance of fluids — determination of resistance to rapid crack propagation (RCP) — full scale test.

ISO EN BS DIN 13479. Thermoplastic pipes for the conveyance of fluids — determination of resistance to crack propagation — test method for slow crack growth on notched pipes (notch test).

ISO EN NF DIN 13760. Plastic pipes for the conveyance of fluids under pressure — Miner's rule — calculation method for cumulative damage.

ISO EN DIN 13845. Plastic piping systems — elastomeric sealing ring type socket joints for use with unplasticised polyvinyl chloride (PVC-U) pipes. Test method for leaktightness under internal pressure and with angular deflection.

ISO EN DIN 15493. Plastics piping systems for industrial applications — ABS, PVC-U, and PVC-C. Specifications for components and piping systems. Part 1: metric series.

ISO DIS 15874. Plastic piping systems for hot and cold water installations — polypropylene (PP).
  Part 1: General.
  Part 2: Pipes.
  Part 3: Fittings.
  Part 5: Fitness for purpose.
  Part 7: Assessment of conformity.

ISO DIS 15875. Plastic piping systems for hot and cold water installations — cross-linked polyethylene (PE-X):
  Part 1: General.
  Part 2: Pipes.

Part 3: Fittings.
Part 5: Fitness for purpose.
Part 7: Assessment of conformity.

ISO DIS 15876. Plastic piping systems for hot and cold water installations — polybutylene (PB).
Part 1: General.
Part 2: Pipes.
Part 3: Fittings.
Part 5: Fitness for purpose.
Part 7: Assessment of conformity.

ISO DIS 15877. Plastic piping systems for hot and cold water installations — chlorinated polyvinyl chloride (PVC-C).
Part 1: General.
Part 2: Pipes.
Part 3: Fittings.
Part 5: Fitness for purpose.
Part 7: Assessment of conformity.

5.3.4.2 European standards, (EN) with indication of their eventual adoption by other major European countries and organisations (BS/NF/DIN)

EN NF DIN 638. Plastic piping and ducting systems — thermoplastic pipes — determination of tensile properties.

EN DIN 762. Plastic piping and ducting systems; joints with elastomeric seals; test methods for retention of, and damage to, sealing rings.

EN DIN 763. Plastic piping and ducting systems; Injection moulded thermoplastic fittings — test method for visually assessing effects of heating.

EN DIN 852/1. Plastic piping systems for the transport of water intended for human consumption — migration assessment. Part 1: Determination of migration values of plastic pipes.

EN NF DIN 921. Plastic piping systems — thermoplastic pipes — determination of resistance to internal pressure at constant temperature.

EN DIN 1042. Plastic piping systems; fusion joints between polyolefin pipes and/or fittings; determination of resistance to internal pressure at constant temperature.

EN DIN 1046. Plastic piping and ducting systems; plastics systems outside building structures; recommended practice for installation above and below ground.

EN NF DIN 1053. Plastic piping systems; thermoplastic piping systems for non-pressure applications. Test method for watertightness.

EN NF DIN 1277. Plastic piping systems. Thermoplastic piping systems for buried non-pressure applications. Test methods for leaktightness of elastomeric sealing ring type joints.

EN DIN BS 1401. Plastics piping systems for non-pressure underground drainage and sewerage - unplasticised polyvinyl chloride (PVC-U). Part 1: Specification for pipes, fittings and the system. Part 3: Guidance for installation.

EN DIN 1437. Plastic piping systems; piping systems for underground drainage; test method for resistance to combined temperature cycling and external loading.

EN BS NF DIN 1466. Plastic piping and ducting systems. Thermoplastic pipes. Determination of ring flexibility.

EN DIN 1452. Plastic piping systems for water supply. Unplasticised polyvinyl chloride (PVC-U). Part 2: Pipes.

EN BS DIN 1852. Plastic piping systems for non-pressure underground drainage and sewerage — polypropylene (PP). Part 1: Specification for pipes, fittings and the system.

EN NF DIN 1979. Plastic piping and ducting systems — thermoplastic spirally formed structured wall pipes — determination of the tensile strength of a system.

EN BS NF DIN 12107. Plastics piping systems — injection moulded thermoplastic fittings, valves and ancillary equipment — determination of the long-term hydrostatic strength of thermoplastics materials for injection moulding of piping components.

EN DIN 12108. Plastics piping systems — recommended practice and techniques for the installation inside building structures of pressure piping systems for hot and cold water intended for human consumption.

EN DIN 12201. Plastic piping systems for water supply —polyethylene (PE). Part 2: Pipes.

EN DIN 12202. Plastics piping systems for hot and cold water — polypropylene (PP).
   Part 1: General.
   Part 2: Pipes.
   Part 3: Fittings.
   Part 5: Fitness for purpose.
   Part 7: Assessment of conformity.

EN NF DIN 12293. Plastics piping systems — thermoplastic pipes and fittings for hot and cold water — test method for the resistance of mounted assemblies to temperature cycling.

EN DIN 12294. Plastics piping systems — thermoplastic pipes and fittings for hot and cold water — test method for leaktightness under vacuum.

EN NF DIN 12295. Plastics piping systems — thermoplastic pipes and fittings for hot and cold water — test method for resistance of joints to pressure cycling.

EN DIN 12318. Plastic piping systems for hot and cold water — polyethylene (PE-X). Part 2: Pipes.

EN DIN 12319. Plastic piping systems for hot and cold water — polybutylene (PB). Part 2: Pipes.

EN NF DIN 12731. Plastic piping systems for hot and cold water — chlorinated polyvinyl chloride (PVC-C):
    Part 1: General.
    Part 2: Pipes.
    Part 3: Fittings.
    Part 5: Fitness for purpose.
    Part 7: Assessment of conformity.

5.4 Thermoplastic Tanks and Vessels

At the present time the author has little knowledge of North American standards which could be applied to this activity, other than a basic standard ASTM D 1998 — standard specification for polyethylene upright storage tanks. However, a European norm is currently under discussion, and when published will fill a void in European standardisation.

DVS 2205. In Europe at the present time the only significant documentation in this field is the German standard for the design of thermoplastic containers, reference DVS 2205. This document is split into five sections, and making references to PP/HdPe/PVC and PVDF as construction materials.

Part 1 — determination of permissible stress values, and which considers the creep strength of the selected material against the design life of the vessel, a joint factor depending upon the type of welding used and the material in question, a corrosion index which takes into account the corrosion resistance of the material being used and the fluid being stored, and finally a coefficient based upon the design temperature.

Part 2 — determination of wall thickness taking into account the static head pressure, assembly of the shell to a rigid base, safety factors depending upon the working conditions and the environment in which the finished vessel will be installed. Nozzle design and local reinforcement is considered, together with three possible bottom configurations, flat, conical and dished, making reference to AD Merkblatt B2 for the calculation of conical bottoms.

Part 3 — design of welded joints in containers made from thermoplastics.

Part 4 — considers the flanged joint design.

Part 5 — considers the design of rectangular tanks or containers.

The following documents should be referred to as complementary to DVS 2205.

DVS 2203 — Requirements and testing methods of thermoplastic containers.

DVS 2206 — Testing of components in thermoplastic materials.

DVS 2209 — Welding of thermoplastic materials — processes and features of extrusion welding.

Note DVS=Deutscher Verband für Schweissertechnik eV Düsseldorf or the German Association for Welding Technology.

5.5 Composite Profiles and Associated Constructions

Pultrusion of composite profiles and their subsequent assembly and use, although highly developed over the last 20 years or so, is another area which would seem to be somewhat lacking in design and manufacturing standards.

By their very nature, composite profiles are destined in the main to be used as or in structural load-bearing constructions, and standards must consider both the properties of the pultruded section and the means and performance of the assembly of the different items into a complete structure. Work is ongoing in this domain, and it would seem that official standards will be available shortly to the end user, both in North America and in Europe.

In the USA, a joint committee of the American Society of Civil Engineers and Pultrusion Industry Council section of the Composite Fabricators Association is working on a standard for the design of pultruded glass fibre reinforced composite structures. In the meantime many actual testing standards may be used to define the properties of pultruded sections, and many of these tests and standards will be incorporated into the above-mentioned standard.

Typical ASTM testing standards involved are:

ASTM D 2734. Test method for obtaining void content of reinforced plastics.

ASTM D 2990. Test method for tensile, compressive, and flexural creep and creep rupture of plastics.

ASTM D 3647. Standard practice for classifying reinforced plastic pultruded shapes according to composition.

ASTM D 3914. Standard test method for in-plane shear strength of pultruded glass reinforced plastic rod.

ASTM D 3917. Standard specification for dimensional tolerances of thermosetting glass reinforced plastic pultruded shapes.

ASTM D 3918. Standard terminology relating to reinforced plastic pultruded products.

ASTM D 4385. Standard practise for classifying visual defects in thermosetting reinforced plastics pultruded products.

ASTM D 4475. Standard test method for apparent horizontal shear strength of pultruded reinforced plastic rods by the short beam method.

ASTM D 4476. Standard test method for flexural properties of fibre reinforced pultruded plastic rods.

Japan has two relevant standards in this field.

JIS K 7011. Glass fibre reinforced plastics for structural use.

JIS K 7015. Pultruded fibre reinforced plastics.

The situation in Europe is as follows. Whilst numerous material testing standards exist in most major countries, much in line with those listed above from ASTM, a new European standard specifically for the manufacture and assembly of pultruded profiles, under the temporary classification Pr EN 13 706 -P1. P2. P3 is now in circulation for a six-month period for final comments. The target date for the end of this enquiry period was fixed at 9 March 2000. Under these conditions it can be expected that publication may be achieved in early 2001. One significant development contained within this project is an attempt to classify pultruded sections into different categories depending upon their respective levels of overall mechanical properties.

CHAPTER 6

Using Non-metallic Materials with Potable Water

6.1 Laws and Regulations

At the time of writing this handbook, many bodies, some of them referred to below, were in the process of updating and harmonising standards and publication of information. Any reader considering undertaking any activities concerning materials approved for contact with potable water is strongly recommended to enter into contact with the relevant authorities in order to check the up-to-date situation.

Many materials may have an adverse affect on the quality of drinking water, and unsuitable materials used to construct pipe systems or storage tanks for example, can cause problems with taste and appearance, as well as allowing the leaching of dangerous substances from the material into the water. In some cases, they may also promote microbiological growth. The legislation and regulations in place in Europe and North America apply to both metallic and non-metallic materials, and most of them have either been upgraded in recent years or are currently going through this process. In the field of metallic materials the widespread and very recent use of lead piping systems is now causing extremely costly replacement and renovation programmes in many countries, and within the non-metallic area, the previous widespread use of asbestos is also causing some concern. The European Directive dated 3 November 1998 requires a general and progressive decrease in the amount of lead in potable water, with a maximum of 10 µg/l in 2013. Governments and regulatory bodies may perhaps be criticised today due to the new over-reactive regulations, perhaps based on recent discoveries concerning the use of materials previously considered as safe. These new regulations may in some cases be considered as being theoretical in nature and contrary or at least exaggerated when compared to an actual situation. However, when considering the long-term requirements of materials used to distribute or stock potable water and history, this relatively conservative approach may at the end of the day be correct. What is perhaps more disturbing is the difference in requirements from one country to another, even within the European Union, although recent and future directives in this field should progressively lead to some form of conformity, and a subsequent decrease in the cost for suppliers of testing their materials and products. At the present time there is very limited acceptance of testing by any european country concerning testing carried out within another member state of the European Union.

As a result of differences between standards in the different member countries, and because of the apparent absence of any harmonisation between the countries, the European Commission set up a working group in 1998, including members from France, Germany, The Netherlands and the UK, to develop a prototype European Approval System, based on the harmonisation of the national standards in the countries represented by the above-mentioned members. The working group, under the designation RG-CPDW (Regulatory Group — Construction Products Drinking Water) started work in 1999, with the initiation of a five year programme.

It is perhaps equally significant that the leading controlling organisations in Europe and North American (the WRc in the UK and NSF in North America) have recently formed a joint venture, WRc-NSF Ltd, combining the know how and experience of the two companies concerning the effects of different materials on the quality of drinking water.

The UK, The Netherlands and North America are the only three major areas which publish and regularly update listings of materials and products which meet the requirements of their legislation concerning non-metallic materials in contact with potable water. The NSF programme in North America is the only scheme which incorporates quality control, and which is subject to audit during both the original examination of the material and during the annual follow-up checks and controls. Some European countries, for example The Netherlands and Denmark, carry out random checks to verify that standards and material specifications are maintained, whereas other countries apply a limit to the validity of an approval, generally three or five years, after which new testing and approval is required.

Both the European Union and individual member states have issued so-called 'positive lists' which define chemicals or chemical compositions which may be used in the manufacture of materials destined to be in contact with potable water. Compliance with one or both positive listings is mandatory, and may be in some cases sufficient for applications in some countries.

6.2 Comparison of Standards in Europe and North America

We summarise actual conditions in the four major European countries forming the committee for harmonisation referred to above. Further details of the testing methods, requirements for testing and many other details concerning all the European counties can be obtained from the WRc (see listing below) via their guide *European Approval Systems — Effects of Materials on Water Quality*. We also list further contacts for information on approvals or testing requirements for most of the European countries and North America in Section 6.3.

6.2.1 France

Recent modifications in the legislation in France require that all material to be used for public storage or handling of potable water should be tested and approved by one of the three laboratories listed below.

The overall procedure requires completion of a four stage test programme.

- A submission concerning all the components and ingredients used in the manufacture of the product or material to be tested, with confirmation that all ingredients are in the list of materials approved for contact with potable water, i.e. that all materials are on the 'positive list'. As a similar list exists at a European level, it is preferred that the ingredients to be used figure in both positive listings.
- A rapid screening which confirms that the ingredients are within the listing of approved materials.
- Cytotoxicity tests which allow appreciation of any eventual cytotoxicity problem introduced into the water by the contact with the material under test.
- An in-depth, detailed screening analysis to determine any migration of organic or mineral compounds from the material into the water.

6.2.2 Germany

Existing German standards are under review with an expectation that the new standards will follow the lines traced by British and French legislation. Actual testing programmes in place are listed below:

6.2.2.1 KTW recommendations

These recommendations form the basis of a voluntary acceptance scheme used within Germany for materials in contact with drinking water. Compliance is obtained through a declaration of composition of the product (assessed against a positive list of ingredients), together with a series of individual tests.

This assessment/testing of products is undertaken on the basis of the proposed use, with an in-service exposure of the product playing an important role. Using the appropriate surface area/volume exposure ratio the product is assessed for its effect on water quality after each of three sequential 72 hour leaching periods for the following aspects:

- Water quality assessments — clarity, colour, odour, flavour and tendency to foam.
- Chemical determinants — chlorine demand, migration of organic carbon, and other requirements which may vary with the product type, its proposed use and its components.

6.2.2.2 DVGW W270 testing

This test method, currently under review, was originally developed for products designed for use in water treatment and distribution, such as storage tanks, pipes and pipe linings.

The test is broken down into two stages:

- Large panels of the material being tested are placed into tanks through which is flowing de-chlorinated tap water at supply temperature.
- An assessment of microbiological growth is taken after three and six months exposure, and includes: colony counts; microscopic examination and measure of biomass (volume).

6.2.3 The Netherlands

Materials for use in contact with potable water in The Netherlands must be covered by a certificate concerning the toxicological aspects of the material (Attest Toxicologische Aspecten or ATA), issued by KIWA, The Netherlands Waterworks Testing and Research Institute.

The procedure is based upon conformity to the positive listing, or an individual assessment of a material not on the positive list. Migration tests are carried out in accordance with the conditions set out in the European Directive.

6.2.4 UK

The UK regulation divides product testing into two groups as follows:

- Testing of non-metallic materials for use with potable water in accordance with the Water Byelaws Scheme tests of effect on water quality (BS6920: 1996); and
- the more comprehensive testing of toxicological requirements to ensure legal compliance with regulation 25 of the Water Supply (Water Quality) Regulations 1989.

To summarise, all products and materials which will be in contact with potable water are required to complete the first stage of testing. Material which will be used by a public water supply company in the preparation or conveyance of water will then in most probability be required to complete the second stage of testing.

The initial test programme includes the taste and appearance of water in contact with the material for 14 days and the assessment of the ability of the material to support the growth of microorganisms after a period of 7 to 8 weeks' immersion. The test programme continues with the extraction of substances which may be of concern to public health, and is completed with an extraction test to demonstrate that no toxic metals will leach from the product into the water.

The Water Byelaws Scheme is operated by the WRc Evaluation and Testing Centre Ltd, on behalf of the water suppliers of the UK, and products and materials which satisfy the above requirements are listed in the *Water Fittings and Materials Directory* published every six months by the WRc.

The second stage of the approval system is operated by the Drinking Water Inspectorate (DWI) in London, and is a requirement for practically all products or materials in contact with potable water within the installations of a public water supply company. The application is examined by a committee, and if the application is successful an approval letter is issued which sets out any conditions attached to the approval. This stage of the approval system requires the following type of information:

- product description and outline of use;
- outline description of the manufacturing process;
- detailed specification of the product, including full chemical composition and technical specification of each ingredient;

- description of the product use;
- evidence of the product's effect on water quality;
- toxicological datal
- full details of curing systems of thermosetting materials; and
- instructions for use.

6.2.5 European legislation and directives

As a general rule, and whilst the industry awaits a pan-European standard, the testing carried out in one member state of the European Union will in general not be accepted by another member state, especially within those states having developed the most comprehensive testing schemes.

As mentioned above, a 'positive listing' of chemicals which can be used in materials in contact with potable water exists and represents an initial level of approval to the requirements of the European market.

A recent directive, dated 3 November 1998 lays down the basis for obtaining a uniform minimum quality of potable water throughout the industry, even if some of the requirements need not be met until the second decade of the 21st century. No such directive concerning the materials in contact with potable water has yet been identified.

6.2.6 North America

Two significantly different procedures are used in the USA, based upon standards and testing procedures developed by the FDA (Food and Drug Administration) and NSF International (National Sanitation Foundation). As far as non metallic materials in contact with potable water are concerned it would appear that the NSF listing is used in a predominant fashion throughout North America in that more than 80% of the individual states require, where possible, products to comply with the relevant NSF standards.

6.2.6.1 FDA

The FDA requires that, under the provision of the Federal, Food, Drug and Cosmetic Act, all materials in contact with potable water shall comply with current FDA regulations. The company requesting FDA approval submits information concerning the chemical identification and composition of the components, together with its physical, chemical and biological properties. The manufacturer is also required to describe the intended use of the product, how it shall be installed, used and serviced.

Data is also included to show how and to what extent the product may directly or indirectly influence the quality or characteristics of the water with which it is in contact.

Most manufacturers supplying products and materials for handling potable water in conformity with the requirements of the FDA supply statements such as resins and curing agents used in the manufacture of 'product X' are defined as acceptable with the US Food, Drug and Cosmetic Act as listed under 21 CFR Part 177 Subpart C Section 177.2280 and 21 CFR Part 175 Subpart C Section 175.300.

6.2.6.2 NSF Standard 61

In 1984 the US Environmental Protective Agency (US EPA) announced the privatisation of its drinking water additives products registration programme. In 1985 the EPA awarded NSF a co-operative agreement for development of health effects standards and a products certification programme. NSF brought together a consortium of organisations in which federal and state regulatory agencies and the water utilities participated in the development of two consensus, health effects standards. Today, the NSF Drinking Water Additives Program tests and certifies drinking water treatment chemicals and drinking water system components against those standards to ensure that these products do not contribute contaminant effects to drinking water that could cause adverse health effects.

NSF evaluates drinking water system components to the requirements of ANSI/NSF Standard 61: Drinking Water System Components — Health Effects. Products covered under this standard include, amongst others, pipes and fittings, cements and coatings, filter media, potable water materials and device components. This standard sets criteria to determine which contaminants migrate or extract into drinking water, and whether the contaminant levels are below the maximum allowable levels.

NSF Standard 61 differs from all other programmes in that it incorporates a plant visit/audit together with a material quality control system both at the time of the original audit, as well as during the annual and compulsory follow-up audit. The audit and subsequent material quality control is carried out upon the basis of random selection of materials by the NSF staff.

The total procedure runs as follows:

- application by the client;
- submission of formulation, toxicology and product use information;
- formulation review by NSF toxicology group;
- plant audit and sample collection by NSF staff;
- chemical and microbiological testing by NSF;
- final toxicology evaluation; and
- listings agreement.

The last four are included within the annual audit and follow-up programme.

NSF Standard 61 is broken down into seven different sections, of which the following are of significance for the material described in this handbook.

- Section 4 'Pipes and related products', e.g. fittings, pipes, tubing and well screens.
- Section 5 'Protective barrier material', e.g. coatings and linings.
- Section 6 'Joining and sealing materials', e.g. adhesives and gaskets.
- Section 8 'Mechanical devices', e.g. pumps and valves.

It is useful at this point to draw to the attention of the reader the existence of NSF Standard 14. Whereas NSF 61 is concerned with a large range of products and materials and their influence on the quality of water in contact with these products, NSF 14 limits the range of application to plastic pipe systems. NSF 14 includes the consideration of drain, waste and vent (DWV), sewer and other non-potable water products. NSF 14 also includes conformance to the requirements of a recognised and specified performance standard such as ASTM D 2241. As all potable water materials and products complying with NSF 14 must also comply with NSF 61 health effect requirements, manufacturers of pipe systems have the option of listing these products in both Standard 14 and 16 listings; however, most choose to list their products only under NSF 14.

To further differentiate and explain the two different standards, the reader should note that DWV pipe certified by NSF bears the NSF-dwv listing mark. The Mark designates that the product conforms to the requirements of NSF 14 and ASTM standards such as D2665 — Standard Specification for Poly Vinyl Chloride (PVC) Plastic Drain, Waste and Vent Pipe and Fittings.

NSF Standard 14 provides for minimum quality control test requirements for evaluating pipe during production. The ASTM standard provides the minimum performance requirements for DWV pipe. These include minimum hydrostatic burst, impact resistance, pipe stiffness, deflection loads and crush resistance. It is important to note that the certification requirements of DWV and potable water (PW) pipe are not the same, even though a product can be evaluated and certified against both set of requirements. The critical differences are:

- Health effects. DWV pipe is not reviewed for health effects. As a result, the materials used may not be acceptable for use in drinking water applications.
- Hydrostatic design stress. DWV pipe materials are not required by Standard 14 or ASTM D 2665 to be pressure rated.

If a pipe product certified for a DWV application is used in a potable water application, there is no certainty that the materials used are suitable for drinking water contact, or that the product will be able to perform in a pressure application.

If a product has been evaluated and certified for both PW and DWV applications, it will meet all the requirements of both applications. All NSF listed pipe bears the NSF-pw or NSF-dwv mark, and a product certified for both applications will bear both NSF marks.

6.3 Testing Organisations or Official Bodies Where Relevant Information Can Be Obtained

6.3.1 Austria — ÖVGW certification

Österreichische Vereinigung für das Gas und Wasserfach. Postfach 26, Schubertring 14, A-1010 Wien. Tel: +43 1 513 15 88-0/Fax: +43 1 513 15 88-25.

6.3.2 Belgium

Belaqua, Chaussée de Waterloo 255 bte. 6, BE 1060 Bruxelles. Tel: +32 2 5374302/Fax: +32 2 539 2142.

6.3.3 Denmark

Several different organisations hold responsibility for different products and technologies. The two most important are:

ETA-Danmark A/S, PO Box 54, DK 2971 Hoersholm. Tel: +45 45 76 20 20/Fax: +45 45 76 33 20.

Danish Standards Association, Products Certification Department, Baunegardsvej 73, DK 2900 Hellerup.

6.3.4 Finland

Ministry of the Environment, Housing and Building Dept, PO Box 399, FIN 0012 Helsinki. Tel: +358 0160 5687/Fax: +358 0160 5541.

6.3.5 France

Université Henri Poincaré — Nancy 1, Faculté de Médecine, Laboratoire d'Hygiène et de Recherche en Santé Publique, 118 bis Rue Gabriel Péri,

54515 Vandoeuvre Cedex. Tel: +33 3 83 50 36 36/Fax: +33 3 83 57 90 75.
Institut Pasteur de Lille, 1 rue du Professeur Calmette, 59019 Lille Cedex. Tel: +33 3 20 87 77 30 /Fax: +33 3 20 87 73 83

Centre de Recherche et de Contrôle des Eaux de la Ville de Paris, 144 & 156 Rue Paul Vaillant Couturier, 75014 Paris. Tel: +33 1 40 84 77 88/Fax: +33 1 40 84 77 66.

6.3.6 Germany

Technologiezentrum Wasser (TZW), DVGW, Karlsruher Strasse 84, D-76139 Karlsruhe. Tel: +49 721 9678 0/Fax: +49 721 9678 101.

6.3.7 Italy

Instuto Superiore di Sanita, Viale Regina Elena 299, 00161 Roma. Tel: +39 064464990/Fax: +39 064469938.

6.3.8 The Netherlands

KIWA, PO Box 70, 2280 AB Rijswijk. Tel: +31 704144400/Fax: +31 704144420.

6.3.9 Norway

Norwegian Food Control Authority, PO Box 8187, N 0034 Oslo. Tel: +47 22579900/Fax: +47 22579901.

6.3.10 Spain

Subdireccion General de Sanidad Ambiental, Direccion General de Salud Publica, Ministerio de Sanidad y Consumo, Paseo del Prado 18-20, E-28071 Madrid. Tel: +34 915962084/Fax: +34 915964409.

6.3.11 Sweden

Svenskt Byggodkännade AB, PO Box 553, S 371213 Karlskrona. Tel: +46 45520600/Fax: +46 45520688.

6.3.12 UK

ITS Cranleigh (UK) Ltd, Manfield Park, Cranleigh, Surrey, GU6 8PY. Tel: +44 1483 268800/Fax: +44 1483 267579.

Law Laboratories Ltd, Shady Lane, Great Barr, Birmingham, B44 9ET. Tel: +44 990 903060/Fax: +44 121 3667003.

The Water Quality Centre, Spencer House, Manor Farm Road, Reading, RG2 0JN. Tel: +44 118 9236214/Fax: +44 118 9236373.

WRc plc, Henley Road, Medmenham, Marlow, Bucks, SL7 2HD. Tel: +44 1491 571531/Fax: +44 1491 579094.

WRc Evaluation and Testing Centre, Fern Close, Pen-Y-Fan Industrial Estate, Oakdale, Gwent, NP1 4EH. Tel: +44 1495 248 454/Fax: +44 1496 249234.

6.3.13 North America

NSF International, 3475 Plymouth Road, Ann Arbor, MI 48105, USA. Tel: +1 734 769 8010/Fax: +1 734 769 0109.

CHAPTER 7

Products and Applications — Composite and Thermoplastic Pipe Systems

Pipe systems are by far the largest application of composite and thermoplastic materials for handling water, and vary in dimension from 10 mm to 3000 mm, with pressure ratings from zero for gravity flow lines up to 200 bar or more for high-pressure water injection lines in oilfield enhanced recovery schemes. Temperatures of the fluids being handled can vary from practically 0°C in many applications up to 120°C for district heating schemes, whereas some composite based materials will perform at temperatures as low as −20°C. Thermoplastics such as polyethylene and PVDF allow for working temperatures as low as −40°C.

Most composite and thermoplastic materials can handle a wide range of qualities of water, including fresh water, seawater, demineralised or deionised water, industrial effluent with a wide range of chemicals, domestic wastewater and sewerage. The choice of the type of material most adapted to the application in question will depend upon temperature and pressure ratings, support design and cost for above ground installations, pipe stiffness and its capacity to withstand backfill loading and eventually traffic loading for buried systems, the cost of supply and installation, including consideration of the various assembly techniques available.

All composite and thermoplastic pipe systems are highly resistant to both the fluid being handled inside the pipe, as well as the external environment in which the system is installed. Protective coating of buried lines, and painting of above ground systems are not required, nor is any cathodic protection. The light weight reduces the needs for heavy lifting devices, and the most recent developments in jointing technology makes installation faster than with conventional materials.

7.1 Composite Materials — Filament Wound Epoxy Pipes and Fittings

7.1.1 General information

Epoxy-based filament wound pipes and fittings offer the highest physical properties of all commercially available composite pipe systems. Their strength and resilience, combined with their elevated heat distortion temperature make them the ideal product for a wide range of applications. Their level of chemical resistance is similar to that of a good polyester, and is well below that of vinyl ester or thermoplastic systems when one is considering the handling of chemicals.

Epoxy resins, whether cured with aromatic, cycloaliphatic or anhydride based hardeners, lend themselves to high-speed winding and curing, even when very heavy wall thickness is involved. Epoxy resins are not subject to shrinkage during cure, nor do they exhibit high exothermic reactions, both of which would cause delamination during the curing cycle.

Due, however, to the requirement of heating both the resin and the mandrels to achieve full impregnation and cure, the level of investment to produce high-grade epoxy pipe systems is very high, especially when compared with a similar plant for polyester-based products. There are only a handful of companies in the world capable of supplying a full range of epoxy-based products, three having their origins in the USA, the others having their origins in Europe and the Middle East. All are included in the directory of companies in Chapter 14.

Epoxy-based pipes and fittings are probably both the most expensive and at the same time the cheapest form of composite pipe systems available on the market today. The extremely high level of performance allows the production of very thin wall pipe in small to medium diameter systems, with an extremely efficient production process. The non-shrink properties of filled epoxy resins also allows very economic compression moulding of small-diameter fittings. Typical examples of these being the Red Thread 11 range of pipe and Green Thread range of fittings commercialised by Smith Fiberglass Products Company.

Large-diameter pipe and fittings, for working pressures of 50 bar or more, represent the other extreme of the price range, where epoxy competes with high grades of duplex or super duplex steel, especially in applications in the petrochemical industry and in onshore and offshore oil production.

7.1.2 Design basis

Design or service conditions of all types of composite pipe systems are all based upon a combination of design temperature, that is the temperature of the fluid being handled, and design pressure, which is generally the maximum pressure to which the total system may be exposed. The design and construction of buried pipe, especially for diameters above 400 mm, is also very dependent upon soil or live loads, and the relative stiffness of the pipe being used. National and international standards which cover the above-mentioned considerations are listed and described in some detail in Chapter 5.

Epoxy-based systems tend to be defined on the basis of long-term pressure testing, for example to ASTM D 2992, and the resultant product defined in relation to an extended service life of perhaps 25 or 50 years, with a relatively small residual safety factor. Polyester- and vinyl ester-based systems tend to be designed on the basis of a short-term or instantaneous pressure testing, such as ASTM D 1599, and with relatively high safety factors of between 6 and in exceptional cases as high as 16.

7.1.3 Product range

Epoxy-based pipe systems are available in a range of diameters from 1 inch (25 mm) up to 48 inch (1200 mm). Pressure ratings are limited as a function of service temperature and diameter, but pressures of up to 25 bar are normal throughout the total range, with pressures up to 200 bar for specific application in small diameters.

Pipes are generally supplied in lengths of 6, 9 or 12 m, and usually incorporate either integral or bonded spigot/socket ends for assembly, as described below.

A wide range of standard fittings are available, and non-standard fittings may be supplied by the manufacturer, or manufactured by experienced and qualified specialist manufacturers and installation companies. Small-diameter, low-pressure fittings are often made by some form of compression or resin transfer moulding, whereas medium to large-diameter and high-pressure fittings are usually made by filament winding. Larger, low-pressure fittings are often made by cutting and assembling pipe sections, using laminating techniques similar to those described below for butt and strap jointing.

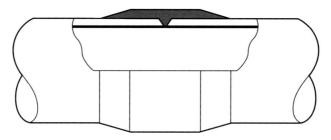

Figure 15. Butt and strap joint.

7.1.4 Jointing systems

A wide range of jointing systems are available as follows:

Butt and strap (see Figure 15) — a standard in field joints for polyester or vinyl ester resin systems as detailed below. It is only used with epoxy systems to prefabricate fittings or spools or to secure a tie-in in the field. Due to the nature of the resins and hardeners used, together with the requirement to heat cure the finished joint, this type of assembly should be left for specialised and qualified labour.

Socket adhesive bonding (see Figure 16) — an adhesive bonded joint which uses either a tapered spigot into a straight bell, or more commonly a straight spigot into a belled socket. The use of a straight socket removes the requirement to have specialised pipe shaving equipment on site.

Conical/conical adhesive bonding (Figure 17) — an adhesive bonded joint that uses matched conical sockets and a shaved conical spigot. This type of bond produces the assembly with the highest mechanical properties, but requires specialised pipe shaving equipment on site during installation. These tools are sold or rented out by the pipe manufacturer.

Threaded and bonded joints (Figure 18) — a joint similar to the previous except for the fact that factory-made spigots and sockets incorporate a matched thread which facilitates assembly of the two pipes together, and at the same time maintains the integrity of the joint during curing. This type of joint is currently available in 2 inch to 12 inch diameter pipe (50–300 mm).

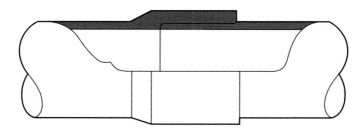

Figure 16. Socket adhesive joint.

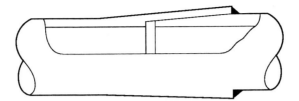

Figure 17. Conical spigot and socket joint.

Figure 18. Threaded and bonded joint (TAB joint).

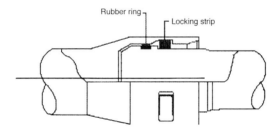

Figure 19. Mechanical lock joint with an O-ring seal.

O-ring or lip-ring seals with a locking device (Figure 19) — a joint used especially for medium to large-diameter pipe, especially for buried conditions. A single lip ring, or a single or double O-ring forms the seal between a male spigot and a female socket, and the axial loading is restrained by a metallic or thermoplastic locking key, also inserted between the spigot and the socket. Should the buried pipe be capable of absorbing end loading due to the compacted soil around the pipe and eventually thrust blocks at changes in direction, the total system may be installed without the use of a locking key.

Various proprietary jointing systems such as Pronto-Lock and Pronto-Lock II marketed by Ameron International, or the TMJ from Smith Fiberglass

Products Company combine threaded connections with an O-ring seal for fast positive make-up and sealing on site in all weather conditions.

Full face flanges or Van Stone (collar plus loose flange)-type flanges to all international bolt hole standards are also available from the manufacturers. These flanges are generally bonded onto the pipes and fittings.

Adhesives for bonded joints must always be supplied by the manufacturer of the pipe system, as each supplier has developed adhesives and packaging to suit each specific product and range of dimensions. Epoxy adhesives generally require heating in order to both accelerate the cure time and to obtain the highest mechanical properties possible. Ambient temperature cure of many epoxy adhesives is, however, possible, especially if service temperatures are also limited to the ambient temperature, but if used particular care must be taken to avoid movement of the pipe during this time (which may be as long as 24 hours). The TAB (threaded and bonded) assembly developed by Smith Fiberglass Products Company allows ambient temperature cure whilst maintaining the pipes firmly locked together.

7.1.5 Installation

Although many aspects of composite pipe installation are common to or similar to those required to install conventional pipe systems, all composite pipe systems, and especially epoxy-based systems, require specific attention, experience and training. The use of mechanical jointing systems using O-ring or lip ring seals removes the requirement for specialised crews to bond or laminate joints. Tie-ins are, however, a normal requirement on site, allowing final dimensional adjustment of pipe systems to fit up to pumps and vessels. This requirement, together with the need to maintain an overall awareness on site of the relative fragility of composite systems compared to steel or concrete, leads the author to strongly recommend the inclusion of a least one qualified composite pipe fitter in every crew installing this type of material. Should the installation use only bonded or butt and strap joints then, of course, the proportion of qualified labour should be increased to meet the requirements of each particular project.

It is always difficult to estimate the time required to install a new type of pipe system, and some assistance may be obtained from the main suppliers. As an example we include some basic information taken from the *Engineering and Design Guide* published by Smith Fibreglass Products Company under the heading Suggested Labour Units for Installation. The original document is published with reference to imperial dimensions and has been transposed into metric units by ourselves with some rounding up of the figures obtained (see Table 4).

This document then proceeds with an example, which we have transcribed as follows:

Table 4. Pipe and joint make up — adhesive bonded glass fibre reinforced pipe and fittings. *Courtesy of Smith Fibreglass Product Company.*

Nominal pipe size (mm)	Unloading and stockpiling[1] (min/m)	Placing in hangers[2] (min/m)	Cutting (min/operation)		Tapering[3] (min/operation)		Unit joint make up[4] (min)
			Hand saw	Mech. saw	Hand tool	Power tool	
25	0.1	2.3	1	*	1	1	1
40	0.13	2.3	1	*	1	1	1
50	0.17	2.3	1.5	*	1.5	1	1.5
80	0.23	2.3	2	1	2.5	1.5	2
100	0.23	2.7	2.5	1.8	3.5	1.5	3
150	0.66	3.3	4.5	2.5	5	2.5	4
200	1.35	4	*	3	*	12	5
250	1.35	4.6	*	3.5	*	12	6
300	1.65	5.6	*	4	*	12	8
350	1.65	4.3	*	4.5	*	16	10
400	1.65	4.3	*	5	*	16	12

[1] Unloading and stockpiling figures are based on one-worker time, total time for two workers should be doubled.
[2] Placing in hangers figures are based on one worker operation for 25/100 mm, two workers for 150/250 mm, three workers for 300/400 mm, total times should be figured by multiplying figures given by workers needed per operation.
[3] Tapering figures are given per worker; using power tools, two workers will be needed. The 200/400 mm figures are based on use of the 200/400 mm grinding tapering tool.
[4] Each joint make up figure includes cleaning, applying adhesive and properly engaging the joint. Allow 4 minutes for mixing each adhesive kit.).

Bill of material — 150 mm diameter

124 m of pipe in 15 sections
6 off 90° elbows
2 off 45° elbows
2 off flanges
2 off 150×80 mm saddles
8 off adhesive kits

Labour required to install piping

1. Unloading and stockpiling — 124 m @ 0.66 min/m × 2 workers = 164 minutes
2. Cuts to make (power saw) — estimate 18 cuts @ 2.5 min/cut = 45 minutes
3. Tapering operation (power tools) — estimate 18 tapers @ 2.5 min/taper × 2 workers = 90 minutes
4. Adhesive mixing — 8 adhesive kits @ 4.0 min/kit × 1 worker = 32 minutes
5. Joint make up: 27 off for pipe and 18 off for fittings and flanges — 45 off in total @ 4.0 min/joint × 3 workers = 540 minutes
6. Saddles — 2 saddles × 6.0 min each to mark and cut holes, 10.0 min each to prepare pipe surfaces, 3.0 min each for joint make up and 4.0 min each for banding = 46 minutes

Total time = 917 minutes = 15.30 hours computed labour
Actual labour based on 6 hours of productive labour for each 8 hours worked:
a. Man days = 15.30 hours/6 hours = 2.55 man days or
b. Man hours = 15.30 hours/6 hours × 8 = 20.40 man hours

The information detailed above is of an indicative nature only, and does not take into account any specific characteristics of a job site or labour qualifications and efficiency, but does allow readers with experience in handling conventional materials to make their own comparisons.

7.1.6 Supporting documentation

Of all the manufacturers of composite pipe systems, those companies making epoxy-based systems provide the highest level of technical documentation, including product specifications, design and engineering manuals and installation guides. Some of these documents, including highly developed calculation tools can now be obtained in CD format, or can be downloaded directly from the manufacturer's web site. For example, Smith Fiberglass Products Company's *Success by Design*, which includes piping design, engineering software, CAD drawings and an electronic catalogue.

7.1.7 Typical suppliers, products, designation, sizes and pressure/temperature ratings

Ameron International, Fiberglass Pipe Group, has plants in Burkbennet and Mineral Wells, Texas, USA, The Netherlands, Singapore and Malaysia, together with a 40% owned affiliate in Saudi Arabia, and produces a wide range of epoxy-based pipe systems, the most important of which are:

- Bondstrand 2000 — an aromatic amine cured system, with pressure ratings from 10 to 30 bar, for continuous operating temperatures up to 120°C in diameters 25 to 400 mm. Recommended for handling potable, wastewater and sewage systems, cooling and deionised water and general drainage systems. Bondstrand 2000 meets the requirements of US Federal regulations 21CFR175.105 and 21CFR177.2280 for conveying foodstuffs when joined with Bondstrand RP6B epoxy adhesive.
- Bondstrand 2400/3400 — an amine-cured system with pressure ratings from 10 to 50 bar, for continuous operating temperatures up to 93°C in diameters 50 to 1000 mm. Recommended for drainage and buried fire mains, salt and brackish water, oilfield water injection lines and general wastewater and sewage lines.
- Bondstrand 2000MP — an aromatic amine-cured system with a pressure rating of 8.6 bar, for continuous operating temperatures up to 120°C in diameters 50 to 250 mm. Recommended for hot and chilled water, district heating schemes and steam condensate, and certified to the US military specification MIL-28584B. Ameron also produce a 2000M-FP version, in diameters 25 to 250 mm, complete with an adhesive bonded intumescent coating for fire protection and impact resistance, specifically designed for offshore and above ground fire protection and any industrial piping at risk to fire exposure.

- Bondstrand 3000A — an aromatic amine-cured system with pressure ratings from 10 to 30 bar, for continuous operating temperatures up to 99°C in diameters 50 to 1000 mm. Recommended for cooling and potable water, boiler feed water, replacement of steel piping and all applications involving water mains and water treatment.
- Bondstrand 3200A — an aromatic amine-cured system with pressure ratings up to 14 bar, for continuous operating temperatures up to 99°C in diameters 50 to 400 mm. Recommended for buried fire mains, municipal waste, brine and brackish water, and is listed by both UL and FM for underground fire protection systems.
- Bondstrand 4000 — an aromatic amine-cured system with pressure ratings up to 10 bar, for continuous operating temperatures up to 120°C in diameters 25 to 400 mm. Recommended for handling industrial waste, solvents and slurries.

Ameron also produce a range of specialised pipe systems using polysiloxane phenolic resins, cycloaliphatic amine-cured epoxies as well as a range of vinyl ester and polyester resins, the latter being reviewed in following sections.

Fibercast Company has one plant in Sand Springs, Oklahoma, USA, and produces a limited range of filament wound epoxy-based pipe systems, essentially 350 to 1800 mm diameter pipe to complement their more exhaustive range of centrifugally cast pipes which are reviewed below. Fibercast are now part of the Denali Corporation, alongside the Plasticon group of companies in Europe.

The Fiberdur-Vanck group has only been formed recently, bringing together three significant companies under the umbrella of the B.G.A (Beteilgungsgesellschaft Aachener Region mbH), itself a subsidiary of Eschweiler Bergwerks-Vereins AG, has plants in Aldenhoven, Germany (the former Fiberdur pipe plant), in Staffelstein, Germany (the former Vanck tank and vessel unit) and in Udine, Italy (the former Vetroresina pipe and tank plant), together with smaller units in France and Hungary and produces in the Aldenhoven plant a range of epoxy-based pipe systems:

- Amine-cured epoxy, diameter 350–1000 mm for a pressure rating of 6 bar.
- Amine-cured epoxy, diameter 150–1000 mm for a pressure rating of 10 bar.
- Amine-cured epoxy, diameter 25–300 mm for a pressure rating of 16 bar.

All the above system are available with two different inner liner thicknesses, depending upon the corrosion resistance required. Based upon the above product range, the group also supplies two specialised product lines:

- Thermothan pipe system, a pre-insulated pipe system, diameter 25 to 300 mm, rated 16 bar, comprising an epoxy pipe, polyurethane foam insulation and a polyethylene outer casing, destined for district heating schemes.
- Navicon pipe system, an electrically conductive pipe system, diameter 25 to 800 mm, for pressure ratings of 4 to 16 bar, at temperatures from −50 to 80°C. This product line is destined essentially for offshore and onboard applications where the build-up of static electricity within the system could be dangerous.

Future Pipe Industries (FPI), a group comprising Emirates Pipe Industries, Gulf Eternit Industries Co Ltd, and the former Wavin Repox unit has production plants located in The Netherlands and the UAE, and sales offices throughout Europe and the Middle East.

The group manufactures a wide range of composite pipe systems, including the amine-based epoxy system GRE Wavistrong range. This product line includes diameter 25 to 1200 mm, with pressure ratings of 8, 10, 12, 16, 20, 25 and 32 bar depending upon diameters.

FPI offers different resin formulation to handle standard products or potable water, as well as an electrically conductive version for maritime applications. They also offer an extensive range of jointing systems, for both restrained and non-restrained applications.

Smith Fiberglass Products Company, has plants in Little Rock and Wichita, USA and Harbin, China. Products include:

- Red Thread II — diameter 50 to 600 mm, rated 16/30 bar at 99°C, an unlined epoxy using bell×spigot adhesive or TAB (threaded and bonded) joints, offers a choice between filament wound or compression moulded fittings. The pipe system is designed to handle water, effluent and mildly aggressive chemicals. Red Thread II is one of the oldest pipe specifications available, and in spite of extremely competitive pricing is qualified to numerous standards, including API 15 LR, Factory Mutual for buried fire mains and UL for handling petroleum products (equipment of service stations). Red Thread II is also certified NSF for use in contact with potable water.
- Green Thread — diameter 25 to 600 mm, rated 10/20 bar at 107°C, a lined (0.9 mm) epoxy using a bell×spigot adhesive joint. An improved level of corrosion resistance with a higher service temperature, often used in industrial wastewater treatment plants.
- Blue Streak — diameter 50 to 350 mm, rated 14/30 bar at 66°C, an unlined epoxy using bell×spigot adhesive or TAB joints. Designed to handle water, effluent and mildly aggressive chemicals, essentially at ambient temperature, at extremely competitive price levels.

High-pressure pipe, tubing and casing for oilfield (essentially water injection) and water-well applications are available from three companies in North America: Ameron and Smith Fiberglass already mentioned above, and Fiber Glass Systems, located in Antonio, Texas.

Ameron, via their acquisition of Centron Corp, Mineral Wells, Texas, USA, commercialise:

- line pipe in 2, 3 and 4 inch diameter for 500 to 800 psi pressure ratings;
- surface pipe in $1\frac{1}{2}$ to 8 inch diameter for 300 to 1250 psi ratings;
- tubing in $1\frac{1}{2}$ to $4\frac{1}{2}$ inch diameter for 1200 to 2500 psi ratings, and
- casing in 41/2 to 95/8 inch diameter for pressure ratings of 650 to 2000 psi.

Smith Fiberglass offer both anhydride and amine-cured epoxy-based high-pressure pipe systems, notably:

- Blue Streak — anhydride-cured high-pressure systems in $1\frac{1}{2}$ to 6 inch diameter, for pressures up to 3500 psi at temperatures up to 80°C.
- Red Thread — aromatic amine-cured epoxy high-pressure systems in $1\frac{1}{2}$ to 6 inch diameter, for pressures up to 3500 psi at temperatures up to 99°C.

Fiber Glass Systems offer a wide range of tubing, pipe and casing, for pressures up to 4000 psi and in a range of diameters up to 8 inch (or more for casings). Table 5 summarises the range and properties of the principal product lines available.

7.1.8 Recent developments

The Fiberglass Pipe Group of Ameron International has recently and dramatically extended the frontiers of the use of epoxy pipe system with the introduction of two new product lines. Bondstrand steel strip laminate (SSL) pipe for pipeline applications using the Coil-Lock mechanical joint designed for applications such as high-pressure water injection lines (up to 330 bar) for onshore or offshore oilfield applications; and Bondstrand series 3400 also with the Coil-Lock mechanical joint bridges the gap between conventional low-pressure Bondstrand pipe such as the 2000 series and the SSL product line.

The Bondstrand SSL pipe is made of multiple steel straps encased between inner and outer glass fibre reinforced epoxy jackets, the number of layers varying between 3 and 10 depending upon the pressure rating, designated by the SSl rating, example SSL5/5 layers.

It is difficult to qualify new hybrid composite products against existing standards for composite pipe systems, but Ameron states that the laminate used to manufacture Bondstrand SSL complies with the requirements of the standard API HR, that safety factors exceed API 5L requirements by 20%, and

136 Composite Materials — Filament Wound Epoxy Pipes and Fittings

Table 5. Epoxy pipe systems for handling water, product range and properties

Manufacturer/product name: Ameron, Fiberglass Pipe Group; trademark: Bondstrand

Series	Maximum temp (°C)	Diameter (mm)	Pressure (bar)	Jointing systems[1]	Comments/main applications
2000	121	25–400	11–31	QL	General purpose adhesive bonded system
2400	93	50–1000	10–75	TT/KL	For oilfield and fire protection systems
2000MP	121	50–250	10–15	QL	For hot water and steam condensate
2000M-FP	121	25–400	10–15	QL	With an external intumescent coating
3000A	99	50–400	10–30	TT/PL	For general industrial and water applications
3200A	99	50–400	13.6	TT/PL	For underground fire protection systems
3300A	99	200–300	20	TT/PL	Industrial, brackish and seawater applications
3400	99	50–1000	40–140	Coil-Lock®	Handling water at high pressure
4000	121	25–400	10–30	QL	For very corrosive and erosive fluids
SSL	93	200–1000	25–330	Coil-Lock®	SSL: steel strip laminate/maximum pressure

Manufacturer/product name: Fibercast Co

Centricast III EP	107	40–350	20–10	Adhesive joint	A basic corrosion resistant pipe system
Centricast Plus RB2530	121	25–350	20–10	Adhesive joint	Improved impact, and bending and strength
F-Chem	121	25–350	10	Adhesive joint	Filment wound with a thicker internal liner
FM	121	100–350	10	Adhesive joint	FM approved pipe for buried fire mains

Manufacturer/product name: Fiberdur-Vanck

6 bar	110	350–1000	6	Adhesive joint	All three standard product ranges are
10 bar	110	150–1000	10	Adhesive joint	available with two different liner
16 bar	110	25–300	16	Adhesive joint	thicknesses.
Thermotan	110	25–300	16	Adhesive joint	Pre-insulated pipe system
Navicon	100	25–800	16	Adhesive joint	Electrically conductive sysytem

Manufacturer/product name: Fibre Glass Systems

Line Pipe	93	40–300	20–275	BxS/API 8rd	Max rating 100 mm/270 bar;200 mm/135 bar;300 mm/70 bar
Downhole tubing	93	$1\frac{1}{2}$–$9\frac{1}{2}$	70–275	API 8rd/10rd	
Downhole casing	93	$5\frac{1}{2}$–$9\frac{5}{8}$	70–270	API 8rd/10rd	

Manufacturer/product name: Future Pipe Industries, Wavistrong

8	EST: 90/ESN: 110	350–1200	8	RSLJ/BxS	EST : standard tensile resistant epoxy
12.5	EST: 90/ESN: 110	200–1200	12.5	RSLJ/BxS	ESN: standard non tensile resistant epoxy
16	EST: 90/ESN: 110	200–800	16	RSLJ/BxS	Also available
20	EST: 90/ESN: 110	150–600	20	RSLJ/BxS	EWT: as EST but for potable water
25	EST: 90/ESN: 110	100–600	25	RSLJ/BxS	EWN: as ESN but for potable water
32	EST: 90/ESN: 110	25–300	32	RSLJ/BxS	Electrically conductive (CST or CSN)

Manufacturer/product name: Smith Fiberglass Products Company

Red Thread II	99	50–400	15–30	TAB/BxS	Basic general purpose system
Red Thread II PP	99	200–400	31	TMJ	PP, Performance Plus: improved rating
Green Thread	107	25–400	15–30	BxS	Improved chemical resistance
Blue Streak EP	80	50–350	12–30	TAB/BxS/SFT	Basic general purpose system
TBS	80	40–150	70–240	SFT/API 8-round	High-pressure anhydride system
TRT	99	40–150	70–240	SFT/API 8-round	High-pressure amine system
SDT	80	40–175	70–200	API 8-round	Down hole tubing

1.QL: Quick-Lock; PL: Pronto-Lock; KL: Key Lock are all mechanical joints whereas TT: taper/taper is an adhesive bonded joint.; RSLJ: rubber seal lock joint of diameter 80 through 1200 mm; BxS of diameter 25 through 400 mm.; TAB: threaded and bonded; BxS bell×spigot adhesive bond; TMJ: threaded mechanical joint; SFT: Smith fast thread, API 8-round: API 8 round threads per inch.

that there is a minimum safety factor of four on the joint. Based on experience with the same resin system in conventional epoxy pipe systems, the corrosion resistance to fluids such as oil and gas, carbon dioxide, hydrogen sulphide and hot saline water, as well as aggressive soil conditions is well established. The product line is pressure rated up to 330 bar for 200 mm diameter and 70 bar for 1000 mm diameter at 21°C. Pressure de-rating factors are supplied for a full range of operating temperatures up to 93°C (0.718), with, for example, 0.901 and 0.804 for 50 and 75°C respectively.

Pipe in diameter 200 to 1000 mm is supplied in lengths of approximately 11.8 m, with a range of pup joints available for make-up in the field. The range of fittings include standard filament wound elbows, at 1.5, 3 and 6D and reducers. ANSI and ISO flanges, tapping couplings, closure pieces and tees are available in coated or alloy steel.

The jointing technology, Coil-Lock utilises the proven capability of the Bondstrand Key-Lock joint in a helical coil configuration in order to multiply the shear loading capacity due to the high operating pressures.

The hydrostatic seal is obtained by means of a standard double elastomeric O-ring. A round nylon or aluminium key is used within a thread form as a helical coil. As an option, polysulphide or polythioether adhesive sealant can be injected between the double O-ring for an additional seal in critical applications.

The Bondstrand 3400 series also with the Coil-Lock mechanical joint bridges the gap between conventional low-pressure Bondstrand pipe such as the 2000 series and the high-pressure SSL product line. The 3400 series uses a conventional glass fibre reinforced filament wound laminate, associated with the Coil-Lock joint developed for very high pressures. The 3400 series is designed for pressure ratings of 40 to 140 bar at ambient temperature, in a size range of 100 to 400 mm. Pressure ratings are downgraded as service temperatures increase.

7.2 Composite Materials — Large Diameter Filament Wound and Centrifugally Cast Polyester and Vinyl Ester Pipes and Fittings

7.2.1 General information

Polyester- and vinyl-ester based filament wound and centrifugally cast pipe systems can be broken down into two main groups: on the one hand, process pipe systems for chemical, water treatment and power plants, usually in small to medium size diameters and having a relatively high proportion of fittings, and on the other hand, pipe destined principally to transport large volumes of water or effluent from one point to another, usually in the medium to large diameter range and with a low proportion of fittings. It is perhaps natural that the two different areas of application has generated two different types of fabrication and companies.

The different types of vinyl ester resins offer possibilities of exceedingly high corrosion resistance, high-temperature resistance and non-inflammable properties, depending upon the actual resin being used. Polyester resins also offer a wide range of similar properties, but are much more limited than vinyl esters as far as corrosion and temperature resistance are concerned. Polyester resins are, however, highly satisfactory for all applications with water, up to the temperature limitations of each specific resin. Price levels of polyester compared to vinyl ester, perhaps in the range of 1 to 2 or 3 for general purpose applications, allow applications with composite materials which would be too costly with vinyl esters.

Vinyl ester and polyester resins are much more sensitive to curing agents than epoxy, and very small differences in the level of, for example catalysts will have a significant effect on the characteristics of the finished laminate. Both families of resins use similar peroxide-based catalysts and cobalt accelerators and promoters.

When we look at the market for larger diameter pipe, especially that pipe designed specifically for the water and effluent market, the number of companies qualified for this market is relatively low, and for probably 80% of the market, all companies are using two different patented and licensed technologies, that is those developed by Hobas (centrifugal cast pipe) and Owens Corning Engineered Pipe Systems (continuous filament winding).

Typical diameters are from 200 mm up to 3700 mm for filament wound pipe, and up to 2400 mm for centrifugal cast pipe, with a wide range of pressure ratings from 4 to 25 bar, and in a variety of stiffness classes.

7.2.2 Design basis

The larger diameter range of pipe systems designed specifically for handling water and effluent are designed on a basis which combines ASTM and AWWA standards, or the equivalent national standards for many European countries.

Typical and most significant standards are AWWA C 950, ASTM D 3262, ASTM D 3517 and ASTM D 3754, with similar standards in other countries, such as:

- Belgium, NBN T 41-101 and 102;
- Germany, DIN 16 869 and 19 565;
- Italy, UNI 9032 and 9033;
- Sweden, SS 3622 and 3623; and
- Japan, JIS A5350 and at the present time an ISO committee, TC 138, is close to issuing a draft standard for water and sewer pipe, for consultation purposes.

Typical product tests include short- and long-term pressure testing to ASTM D 1599 and ASTM D 2992, and because this range of product tends to be large in diameter, with a relatively thin and flexible wall, samples are subject to long-term sustained strain in a corrosive environment to ASTM D 3681, designed to simulate underground septic sewer conditions such as could be found in tropical zones.

7.2.3 Product range

The overall range of diameters and pressure ratings has been defined above. This larger diameter water and effluent pipe is supplied in three different configurations.

7.2.3.1 Continuous filament wound pipe — Owens Corning Engineered Pipe Systems, trademark: Flowtite

Due to the continuous nature of the production process, pipe sections can be made as long as practical for handling and delivery. In most cases the pipe is delivered in 12 m sections, although in some locations it can be delivered in 18 m lengths.

Separate Flowtite couplings are supplied with each pipe. The structural part of this joint is manufactured in the same material as the pipe, and is equipped with elastomeric joints. The joint has been developed and tested to the requirements of ASTM D 4161, and the resultant angular deflections allowable on each joint are as follows:

- diameter 100–500 mm, 3.0°;
- diameter 600–900 mm, 2.0°;
- diameter 1000–1800 mm, 1.0°; and
- diameter 2000–2400 mm, 0.5°.

The Flowtite couplings are not designed to withstand axial loading. Pressurised systems need to be restrained either by compacted backfill or, if necessary, by thrust blocks at each change in direction.

7.2.3.2 Reciprocal filament wound pipe

Supplied by Fibaflow Reinforced Plastics, UK, trademark Fibaflow; Iniziative Industriali, Italy, trademark Sarplast; Protesa, Spain; Sguassero, Italy; and Smith Fiberglass Products Company, USA, trademark Big Thread.

In most cases the pipe is delivered in 12 m sections, which corresponds to the length of the mandrels used in production. A male and female spigot/socket joint configuration is generally an integral part of the pipe, usually obtained by machining pipe ends which have extra thickness built into them during production. Sealing is obtained via single or double O-rings, and in the case of double O-rings some manufacturers propose a system whereby the space between the two O-rings is pressurised immediately after assembly, thereby ensuring that the joint is correctly seated and sealed. This process allows a buried pipe to be backfilled prior to hydrostatic pressure testing of the line, with little or no risk that a joint will leak at that time.

Most manufacturers can supply pipes which incorporate locking devices in order to restrain the pipe, and in this case thrust blocks are not required to maintain the system in place.

7.2.3.3 Centrifugal cast pipe — Hobas and licensed companies operating the Hobas license

The pipe is delivered in 6 m sections, which corresponds to the length of the mandrels used in production.

Standard Hobas pipe uses FWC couplings for both pipe-to-pipe and pipe-to-fitting assembly. These composite couplings have an elastomeric lining, generally in EPDM.

The FWC joint complies with the requirements of ISO 8639, which means that it remains sealed under severe deflection, with internal or external pressurisation, and combinations of all three conditions. Because the centrifugal casting produces a pipe with a smooth external finish, it may be cut at any point along its 6 m length, and the FWC coupling attached.

A wide range of standard fittings are available, and non standard fittings may be supplied by the manufacturer, or manufactured by experienced and qualified specialist manufacturers and installation companies. Larger, low-pressure fittings are often made by cutting and assembling pipe sections, using laminating techniques similar to those described for butt and strap jointing.

7.2.4 Installation

Many aspects of medium to large diameter composite pipe installation are common to or similar to those required to install conventional pipe systems. Handling and lifting operations are similar, although the lighter weight of composite materials compared to concrete or lined steel affords some advantages. Jointing and sealing systems are similar for both material systems, and labour specifically qualified to handle composite materials is hardly ever required.

7.2.5 Supporting documentation

Supporting documentation concerning the engineering of the pipe system, together with relevant instructions concerning all aspects of installation are available from all suppliers, with perhaps the most comprehensive literature being published by the two groups present on an international basis.

7.3 Composite Materials — Small to Medium Diameter Filament Wound Polyester and Vinyl Ester Pipes and Fittings

7.3.1 General information

Polyester- and vinyl ester-based filament wound and process pipe systems for chemical and industrial applications, water treatment and power plants are by far the most widely used composite pipe systems, probably manufactured in close on 100 locations worldwide, ranging from small companies employing less than 20 staff, up to very large companies forming divisions of major international corporations.

Investment levels required to produce small and medium diameter polyester and vinyl ester pipe systems are relatively low, which has led to the creation of literally hundreds of companies on a worldwide basis who can manufacture this type of product, most of them using labour-intensive contact moulding methods. Quality and quality control are as important with these products as they are with epoxy-based products, and we recommend that the reader procures products from companies able to demonstrate their knowledge and experience with the types of products they produce, and especially for pipe systems handling water or mild effluent, to limit their choice to filament wound products. Typical diameter ranges are 25 to 600 mm, with pressure ratings of 6, 10 and 16 bar, for temperatures up to 60 to 95°C depending upon the type of resin being used.

7.3.2 Design basis

Design conditions of process composite pipe systems tend to be based on short-term instantaneous pressure testing, with a safety factor of between 4 and 16 depending upon national standards and service conditions. Hydrostatic testing is carried out at ambient temperature, and factors of reduction are applied against higher service or design temperatures, alongside the safety factors indicated above.

It is important that pipe, fittings and jointing systems are all included in the test programme in order that the complete pipe system maintains overall minimum specified safety factors.

Larger companies also carry out long term testing on their polyester- and vinyl ester resin-based products, in line with those defined for epoxy-based systems, and the availability of testing to ASTM D 2992 or similar should be considered favourably by a potential buyer as a sign of seriousness on the part of the producer.

The UK has developed two standards already referred to in Chapter 5:

- BS 6464. Specification for Reinforced Plastics Pipes, Fittings and Joints for Process Plants.
- BS 7159. Design and Construction of Glass Reinforced Plastics (FRP) Piping Systems for Individual Plants or Sites.

These two standards are probably the most developed and constraining standards developed to date for composite process pipe systems, and have met with little success outside of the UK. In the author's opinion they should be reserved for fitting intensive process pipe systems handling highly aggressive and corrosive fluids, and their application for handling water and mild effluent should be discouraged.

Probably the most specific set of standards for this type of product is to be found within the German DIN documents, referred to in Chapter 5, and which we can develop here in more detail as we examine the product range.

DIN 16 965, for glass fibre reinforced pipes is split into five parts as follows:

- part 1, pipes designated as type A, with an FRP inner liner minimum 1.0 mm thick;
- part 2, pipes designated as type B, with a thermoplastic inner liner (dual laminate);
- part 4, pipes designated as type D, with an FRP inner liner minimum 2.5 mm thick; and
- part 5, pipes designated as type E, with 2.5 mm FRP inner liner and an anti-corrosion structural layer fabricated from alternate layers of chopped strand mat and woven roving.

DIN 16 965 specifies that the structural laminate should provide a safety factor of 6 against the nominal pressure rating when the pipe is subject to a short-term internal pressure test. Pipe systems are pressure rated at ambient temperature, and DIN 16 867 should be used to define the relationship between maximum service pressure and operating temperatures above ambient. It is also important that the supplier can demonstrate the effective safety factor on pipe, fittings and joints, whether supplied in adhesive bonded or butt and strap configurations. The design and configuration of all fittings and assemblies is defined by DIN 16 966.

7.3.3 Product range

Using the types A, B and D pipe from above as being the most representative of all pipe grades for use in handling water or mild effluent, the overall range of diameters and pressure ratings can defined as:

- nominal pressure range 6 bar — diameter 250 to 1000 mm;
- nominal pressure range 10 bar — diameter 250 to 600 mm; and
- nominal pressure range 16 bar — diameter 20 to 400 mm.

The process pipes are generally supplied in lengths of 6, 9 or 12 m, and may be supplied with plain ends for assembly by butt and strap, or may incorporate either integral or bonded spigot/socket ends for adhesive bonding.

Typical and qualified suppliers are:

Ershigs in the USA
Keram Chemie in Germany
Plasticon-Kialite in The Netherlands
Plastilon in Finland
Speciality Plastics in the USA.

7.4 Composite Materials — Small to Medium Diameter Centrifugally Cast Epoxy and Vinyl Ester Pipes and Fittings

To the author's knowledge, only one significant company and plant is today manufacturing centrifugal cast epoxy and vinyl ester pipe and fittings. The company concerned, one of the pioneers of the industry, having produced the first machine-made FRP in the world in 1948, is Fibercast Co, located in Sand Springs, Oklahoma, USA.

The company manufactures both epoxy and vinyl ester systems in diameters of 1 to 14 inch (25 to 350 mm) in various grades which reflect different services temperatures, corrosion resistance and durability. All centrifugally cast pipes have, due to the manufacturing technology, a thick, pure resin inner liner which confers a high level of corrosion resistance to the finished product.

The product range includes:

Centricast III EP, epoxy-based composite for service temperatures up to 105°C;

Centricast Plus RB-2530, epoxy composite-based for service temperatures up to 120°C;

Centricast III VE, vinyl ester-based composite for service temperatures up to 95°C;

Centricast Plus CL-2030, vinyl ester-based composite for service temperatures up to 95°C;

Novacast VE-150, a vinyl ester-based composite for service temperatures up to 105°C;

FM, an FM approved pipe which meets the requirements of Factory Mutual's Research Corp, Class 1614 Approval Standard for Underground Fire Protection; and

Dualcast, a double containment pipe system using both centrifugal casting and filament winding, in either epoxy or vinyl ester resins depending upon the nature and temperature of the fluids being handled.

7.5 Composite Materials — Filament Wound Epoxy and Vinyl Ester Pipe Fittings

All producers of filament wound pipe also manufacture epoxy resin-based fittings by a filament winding process, with or without the incorporation of uni- or multi-directional woven roving. Some manufacturers reserve filament wound products for their high-pressure fittings, and use contact or compression moulding for applications at a lower pressure. Other companies, such as the Ameron group, produce practically all fittings by a filament winding process, and in order to serve their international market, via their pipe plants in the USA, Europe, the Middle East and the Far East, Ameron has created a plant in Singapore, specifically for the manufacture of the majority of its fittings requirements.

Although the same material, production and testing standards are used worldwide, especially for epoxy-based systems, dimensional standards and jointing configurations have not followed, and in most cases this has created a barrier to the creation of companies specialising entirely in the manufacture of composite fittings. A few companies have now taken this task on board, assisted by a certain consensus within the pipe industry to proceed towards common dimensions and jointing configurations, especially in Europe. These companies are listed in the directory of suppliers, but their actual scope of supply is in general limited to flanges.

Within the vinyl ester pipe and fitting market, little or negligible progress has been made in standardising dimension, wall thickness and jointing configuration. By consequence there has been no development of companies dedicated to the supply of fittings for this market, leaving each supplier to make their own. Mechanisation via filament winding of fittings is required not only to improve their performance but also, via automated and sustained production runs, decrease costs and improve the overall economy of the pipe system. Much work remains to be carried out in this field, with perhaps the most significant developments and highest quality fittings coming from the Plastilon production unit in Imatra, Finland.

7.6 Composite Materials — Resin Transfer and Compression Moulded Epoxy, Vinyl Ester and Polyester Fittings

Due to the lack of standard dimensions and the different jointing configurations developed by the major pipe manufacturers, it has been difficult for companies wishing to enter the fittings market. With an ever increasing standardisation of diameter, wall thickness and joint configuration, especially in Europe, one can expect a more active business development in the future. Some of the pioneer companies in this field are listed in the directory of companies in Chapter 10.

7.7 Thermoplastic Pipes

The family of thermoplastic pipes is the largest of all the plastic pipe families, with numerous reputable and qualified manufacturers in all industrialised or semi-industrialised countries. Most thermoplastic pipes are extruded and the nominal diameter is based upon a calibrated external diameter. Two major international dimensional standards exist, imperial dimensions to ANSI/ASTM standards in North America and metric standards to national or EN standards throughout Europe. Most other areas have adopted one of the above standards, thereby permitting practically a worldwide standardised industry, much as the metallic pipe industry has developed over the last century or so. The UK maintains in many cases dual dimensional standards, based on their previous imperial standards and the more recent metric dimensions.

Thermoplastic pipe systems generally offer a wide range of service temperatures, excellent abrasion and corrosion resistance, together with a range of easy-to-use jointing systems. They are generally non-toxic and taint free, and many are approved for handling potable water. Pressure ratings are usually defined for a nominal service life of 50 years at ambient temperature. Most thermoplastic pipes are, however, highly sensitive to service temperature, and nominal pressure ratings have to be de-rated for both fluid and ambient temperatures above 20°C.

The smooth internal surface of thermoplastic pipes means lower pressure losses and savings in energy costs, and/or possibly the use of smaller bore pipe compared to conventional materials. Due to the smooth, non-stick surface, and the general lack of corrosion, pressures losses are kept to their initial low level throughout the lifetime of the pipe system.

Expansion of thermoplastic pipe, due to changes in either ambient or fluid temperature, or both, can be considerable and needs to be analysed in depth, especially for pipe installed above ground. This analysis must be carried out in conjunction with the design of supporting systems. The rigidity of the pipe will be influenced by the fluid and ambient temperatures, and together with the specific gravity of the fluid being handled, may have considerable influence on the support system and support spans being used.

7.7.1 Extruded pipe

7.7.1.1 PVC

The oldest (first industrial application for water mains around 1955) and perhaps the most commonly found material handling water, although the various grades of polyethylene have to some extent superseded the older material. PVC pipe is generally available in diameters of 10 up to 630 mm (1/4 to 24 inch), and with pressure ratings varying from gravity flow up to 16 bar. Pipe and fittings made from standard PVC are generally supplied in a grey colour, although other colours as well as fully transparent material are also available.

PVC pipe systems may be assembled by either mechanical joints with an elastomeric O-ring or lip ring type seal or by cold solvent adhesive bonding.

PVC pipe systems are pressure rated at 20°C and are subject to a considerable de-rating as temperatures increase, approximately as follows:

40°C — de-rating factor 0.70;
50°C — de-rating factor 0.45;
60°C — de-rating factor 0.15.

Although PVC is a relatively rigid material, distances for supports on above ground systems are also extremely variable as a function of service temperature, with a de-rating factor of approximately 0.9 for a temperature increase from 20 to 40°C, and continuous supports are recommended for service temperatures above 40°C.

Performance of PVC pipe, as far as temperature and pressure are concerned, can be improved considerably by over-winding them with a composite laminate (see dual laminate pipe systems below).

7.7.1.2 PVC-C or chlorinated PVC

This is a modified form of PVC which improves on the temperature and chemical resistance of normal PVC. This material can be used to handle many fluids up to a temperature of 100°C, and is especially useful for condensate return lines and many applications handling industrial effluent. A complete range of pipe and fittings is available over a diameter range of $\frac{1}{4}$ inch to 12 inch, with a basic pressure rating dependent upon diameter, but offering 10 or 16 bar over the total range.

PVC-C is also influenced by temperature, with approximate de-rating factors as follows:

30°C — de-rating factor 0.92;
50°C — de-rating factor 0.65;
70°C — de-rating factor 0.40;
90°C — de-rating factor 0.20.

Support centres are less influenced by temperature than with normal PVC, with a de-rating of approximately 50% between 20 and 80°C. PVC-C pipe systems are assembled with a specific solvent adhesive, although the material itself (available in sheet, bar and welding rod) can be welded using the welding techniques discussed below.

7.7.1.3 Bi-oriented PVC or otherwise strengthened PVC pipe

Most of the principal manufacturers of PVC pipe systems have developed an enhanced performance product, which allows a combination of thinner wall thickness and lower weight, together with an improved ductility and performance. An associated higher level of impact resistance than that of standard PVC allows this new-generation material to compete with PeHd, which has taken a large share of the water distribution market, due essentially to its resilience and ductility.

The new product line developed by most companies is designated as being bi-axial or bi-oriented and is commercialised under trade names such as Apollo Bi-axial from Wavin, BiOroc 25 from DYKA Plastic Pipe Systems and BO from Alphacan.

The technology varies slightly from one manufacturer to another, but all are based on the principle of retreating an already extruded pipe with axial and radial stretching at a predetermined and controlled temperature inside a close tolerance mould. This treatment rearranges the molecular structure of the PVC in such a way as to redistribute and increase the load-carrying capacities of the pipe, whilst maintaining a 2:1 ratio between hoop and axial strength in order to accommodate the normal distribution of stress in a pipe under pressure.

This approach results in a thinner wall thickness, less raw material, higher strength and impact resistance, and a slightly increased inner diameter which will have a beneficial effect on flow rates, pressure drops and energy consumption.

Typical figures concerning the above are, in comparison with standard PVC: admissible stress multiplied by approximately 2; and impact resistance multiplied by up to 8. Compared to a ductile iron pipe with a similar performance level, the weight is divided by approximately 6.

Typical product ranges include diameters of 100 to 300 mm for a nominal pressure rating of 10 to 25 bar at 20°C.

Another approach to improving the performance of standard PVC pipes has been taken by the company Uniplas Ltd, who have developed a system of both pipe and fittings, under the name Aquaforce PVC-A (PVC-Alloy). The manufacturing process involves the combination of an impact-resistance modifier (CPE, chlorinated polyethylene) with the basic PVC. The CPE ensures that the pipe has improved toughness and ductility, whilst reducing the wall thickness by approximately 30% when compared to standard PVC.

Uniplas Ltd supply a full range of six diameters, from 100 to 400 mm, with four different pressure ratings, and a choice of 92 bends made from the same material.

All of these pipes are generally supplied in blue or with a blue strip, and all meet the requirements of many of the national standards for products in contact with potable water. Many of the above systems also benefit from new improved socket designs and elastomeric seals.

7.7.1.4 Polyethylene

Polyethylene pipe systems are available in various grades or types, as detailed below. It should also be noted that within each specific classification of polyethylene pipes, specific grades exist for potable water, for industrial and municipal effluent, for effluent pipe inside and outside of buildings, for irrigation, and also perforated or slotted pipe used as soak away pipes in municipal waste land-fill sites.

HdPe or high-density polyethylene, with sub-designations such as HdPe 80, HdPe 100 or HPPe 100.

MdPe or medium density polyethylene, and LdPe or low-density polyethylene, both of which are also classified with the same numbering system as used on HdPe.

Polyethylene pipe generally benefits from various coded indications as to performance as follows:

- SDR (standard dimension ratio). The relationship between the pipe diameter (outside) and the wall thickness, and therefore the pressure rating is defined by the standard dimension ratio. The ratio between the outer diameter of the pipe and the wall thickness remains constant for all pipe diameters for a given pressure rating.
- MRS (minimum required strength). The MRS results from the ISO classification for polyethylene pipe, based upon a 10 000 hour test involving a number of samples, and extrapolated to 50 years (extrapolation method ISO/DP 9080.2).

LdPe, MdPe and HdPe are generally required to have a minimum tensile strength of 6.3 MPa at 20°C at 50 years which is equal to an MRS of 8.0, and is defined by the figure 80 immediately after the definition of the type of polyethylene, e.g. HdPe 80. Lower grades do exist, defined as 32, 50 and 63.

HPPe, under the same system, is required to have a minimum tensile strength of 8.0 MPa at 20°C and at 50 years, equal to an MRS of 10.0, and defined by the figure 100, e.g. HPPe 100.

The classification system developed at ISO/CEN defines minimum values only, and other factors need to be taken into account when defining the maximum allowable service pressure of a particular pipe system, and this same system recommends a minimum design factor of 1.25 to take into account various unknown factors.

Maximum sustained working pressures for water at 20°C with a design factor of 1.25 and based upon a minimum service life of 50 years are as follows.

For LdPe, MdPe and HdPe 80:

SDR 40 — pressure 2.5 bar
SDR 33 — pressure 3.2 bar
SDR 26 — pressure 4.0 bar
SDR 17 — pressure 6.0 bar
SDR 11 — pressure 10.0 bar
SDR 7 — pressure 16.0 bar

For HPPe 100
SDR 40 — pressure 4.0 bar
SDR 26 — pressure 6.0 bar
SDR 17 — pressure 10.0 bar
SDR 11 — pressure 16.0 bar

For working temperatures over and above 20°C, the pressure rating and/or the service life will require de-rating, as suggested below, although exact figures should be obtained from the manufacturer.

Temperature (°C)	Pressure rating factor	Minimum service life (years)
20	1.0	50
30	0.88	25
40	0.80	25
50	0.56	25
60	0.56	10
70	0.49	5
80	0.34	5

Other factors to be considered when working specifically with polyethylene pipe systems are as follows.

Notch sensitivity. This is the consideration of when a pipe is notched or scored during handling or installation, the area in question may over time become brittle and fail. Although no exact figures are presently available it is suggested that pipe with notches up to a maximum depth corresponding to 10% of the wall thickness may be installed.

Fracture and crack propagation. Crack propagation will not occur if the pipe is full of water, but can occur if the pipeline contains 10% or more of air, and would seem to be limited to medium to large diameter pipe, for example

250 mm diameter and above for PE80 and 500 mm and above for PE100 pipe.

Thermal expansion. Polyethylene has very high coefficients of expansion, which may be of concern both during installation and in service. The inherent flexibility of the material may be used in many buried installations, whereby the pipe may be slightly snaked in the trench, and changes in length, especially before backfilling is complete, will be absorbed due to the snaking. When installing a new pipeline, the line should be allowed to stabilise at ambient temperature before making final tie-ins, and partial backfilling of the pipe, outside of the area of joints, will minimise the effect of ambient temperature variations. Approximate rates of expansion are as follows:
Pe80 — 0.15 mm/m/°C; and Pe100 — 0.13 mm/m/°C.

Pipe bending radii. Due to the extreme flexibility of polyethylene pipe it may be bent to achieve a change in direction, rather than using a bend. Bending radii are functions of ambient temperature, pipe diameter and the SDR rating. As a guide, and under optimum ambient conditions, it is possible to achieve on the thickest wall SDR, a minimum bending radius of 15 times the outer diameter of the pipe. With SDR11 and 17 the radius needs to be increased to 25 times the diameter, or 35 during cold periods. These two values should be further increased by 50% for thin wall pipe with SDR ratings of 26 and 33.

Pipes are available from 16 to 1600 mm diameter, in pressure ratings of up to 20 bar depending upon the diameter.

Pipe is generally available in coils containing up to 150 m for diameters up to 63 mm, and up to 100 m for diameters up to 180 mm.

All dimensions are available in straight lengths which vary between 5 and 12 m depending upon the manufacturer, quantity of pipe or transport considerations.

Some companies are able to extrude lengths of 30 m should, for example, transportation by rail be available, and a few can extrude lengths of several hundred metres when the pipe can be floated to the installation site either by river or sea. This possibility is of great interest for outfall lines or seawater intake lines for cooling systems in industrial plants and power stations.

Together with the basic grades of polyethylene pipe detailed above, other grades having different specific properties are also available.

7.7.1.5 Cross-linked HdPe or PE-X

Cross-linked polyethylene pipe is designed for higher service pressures and temperatures than those which can be achieved with standard polyethylene. Alongside the standard high-density polyethylene the manufacturer incorporates a cross-linking agent, which under the influence of higher temperatures and pressures used in the manufacture of HdPe, creates chemical bonds between the long molecular chains of the polyethylene, to create a three-dimensional network.

PE-X can withstand temperatures as low as −100°C and as high as 110°C, shows no signs of brittleness at temperatures as low as −140°C. Cross-linked polyethylene is used in underfloor heating installations, in a limited diameter range of 6 to 25 mm.

7.7.1.6 PB

Polybutene is also used extensively in underfloor heating schemes.

7.7.1.7 PP

Polypropylene pipe is available in a wide range of diameters, from 10 to 1000 mm, with pressure ratings from 2.5 to 16 bar. The characteristics of polypropylene as compared to polyethylene show improved chemical resistance and higher service temperatures, but impact and especially low-temperature properties are considerably lower. Polypropylene is rarely specified for handling water, but can be found essentially in chemical process and effluent applications.

PPs is a fire-retardant grade of polypropylene, available only in a restricted range of diameters and wall thicknesses, often specified for ventilation systems handling corrosive gas.

7.7.1.8 PVDF

Polyvinylidene fluoride is the most widely used of the new-generation of fluoropolymers in pipe form. PVDF is an expensive material, reserved for applications requiring corrosion resistance greater than that of PVC or PP, and for applications with ultra-pure water (using a high-purity formulation designated PVDF-HP and a specific welding procedure). PVDF pipe is available in 16 to 225 mm diameter, in two pressure ratings, 10 and 16 bar.

7.7.1.9 Dual laminate pipe systems

These combine the corrosion resistance of thermoplastic materials with the strength and impact-resistance of composite materials. A bonded structural laminate will also allow higher service temperatures than with the thermoplastic material only, although the possible effect on the level of corrosion resistance at higher temperatures must be considered. The most common dual laminate pipe systems are based on PVC-U or PVC-C, polypropylene and PVDF. Dual laminates using high-density polyethylene, essentially for abrasion-resistant applications such as slurry transport, and PFA, FEP and E-CTFE, for exceptional chemical resistance are also available.

Typical and qualified suppliers are:

Garlway-Plasticon in the UK
Kialite -Plasticon in The Netherlands
Keram Chemie in Germany
MB Plastics in the UK
Ollearis in Spain
Plastilon in Finland.

7.7.1.10 Thermoplastic lined steel pipe systems

These find many applications within the petrochemical, pharmaceutical, chemical and fine chemicals industries, and the lower grades, using polypropylene, polyvinylidene chloride or polyvinylidene fluoride may find limited applications handling deionised water where high levels of purity must be maintained, within water treatment plants where physical or impact damage to pipe systems is a possibility, and in all other applications where the combined physical properties of steel and the corrosion resistance of thermoplastics are required.

More information can be obtained via reference to the dominant applicable standards and specifications as defined by:

ASTM F423 — Standard specification for polytetrafluoroethylene (PTFE) plastic lined ferrous metal pipe, fittings and flanges.

ASTM F491 — Standard specification for polyvinylidene fluoride (PVDF) plastic lined ferrous metal pipe, fittings and flanges.

ASTM F492 — Standard specification for polypropylene (PP) plastic lined ferrous metal pipe, fittings and flanges.

ASTM F781 — Standard specification for perfluoro-alkoxyalkane copolymer (PFA) plastic lined ferrous metal pipe, fittings and flanges.

ASTM F1545 — Standard specification for plastic lined ferrous metal pipe, fittings and flanges. This standard covers many of the requirements for PTFE, PVDF, PP, PFA, FEP and PVDC lined pipe.

Typical and qualified suppliers are:

3P in France

Resistoflex in the USA, who have recently acquired the previous Dow Plastic Lined Piping Products from the Dow Chemical organisation.

7.7.2 Spiral wound pipe

Extruded large-diameter pipe is largely the domain of high-density polyethylene, with the maximum external diameter of 1600 mm for extruded pipe being more than doubled by this process.

In order to increase the annular rigidity of polyethylene pipes, essentially for low-pressure or non-pressurised drainage systems, various techniques and

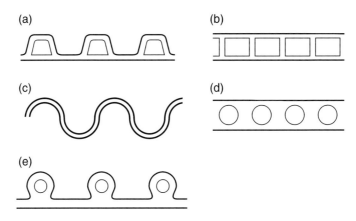

Figure 20. Spiral wound pipe wall configurations.

configurations have been developed to manufacture annular corrugated pipe. The basic process involves the extrusion of a solid or hollow profile onto a rotating mandrel in a spiral configuration. The welding of successive spiral profiles is obtained automatically as the two surfaces make contact. The pitch, profile shape and dimensions, and individual wall thickness can all be varied to suit the design criteria of each specific product. As well as extending the diameter range of polyethylene pipes, and diameters of up to 3500 mm are available, highly rigid small diameter pipes, as little as 200 mm diameter are also available. A wide range of profile configurations means that pipe is available with corrugated or smooth surfaces on the inside, outside, or both (Fig. 20 above).

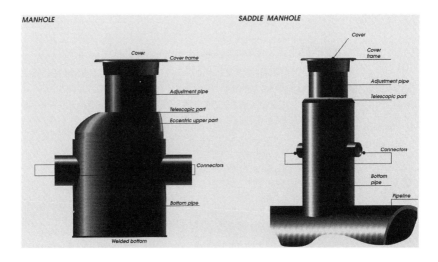

Figure 21. Typical polyethylene manhole construction. Courtesy of KWH Pipe Ltd.

Assembly of this type of pipe is via a spigot/socket connection, with sealing achieved either by elastomeric joints or welding. Applications include culverts, storm sewers, gravity flow drains, seawater inlets and effluent outfalls, and relining of existing sewers. Fully structural manholes can be built from the same material and welded into the overall system (Figure 21 above).

Typical and qualified suppliers are:

- Bauku in Germany, with licensed production in Japan, Italy, Australia, Malaysia and Taiwan
- Hancor in the USA
- KWH Pipe in Finland, but with plants also in Denmark, Portugal, Canada, Thailand, India, Malaysia and China
- Polypipe Civils in the UK, with subsidiaries in France and Poland.

7.7.3 Assembly of thermoplastic pipe systems

7.7.3.1 Solvent cement bonding

PVC-U, PVC-C and ABS pipe systems can all be assembled using spigot socket joint and a solvent cement. Pipes are supplied with a thermoformed socket at one end and a plain end at the other, whereas fittings are supplied generally with sockets at both ends. The solvent cement is usually supplied by the manufacturer of the pipe system along with an associated cleaning fluid. It is important to use specified cleaning fluids rather than standard solvents, as not only do the specified cleaning agents degrease the surfaces to be bonded, but they also penetrate into the material causing it to swell, and improve the bond strength of the assembled joint.

Significant points concerning solvent cement bonding are:

- check that the type of solvent cement corresponds to the type or grade of pipe being installed;
- check that the lifetime of the cement has not been over run;
- the two components must be assembled in one continuous and rapid movement whilst the cement is still soft and wet; and
- assembled components should be left undisturbed for 24 hours prior to moving, filling or testing.

All the major suppliers give detailed instructions concerning the handling, storage, cutting and preparation of pipe ends, assembly, installation and testing of their products, which we do not intend to repeat here. We have however drawn up a list of the most common problems met on-site during installation.

- Bond failure caused by lack of cleaning of the surfaces to be assembled, with possible contamination by grease, dirt or humidity.

FOR FITTINGS 75mm-125mm

1. Cut the pipe to the required length.
   Draw the nut, collar and O-ring on to the end of the pipe to a distance of about twice the pipe diameter (put the white split-ring aside for later use)
   NOTE: Use silicaon to lubricate the pipe, O-ring and inside of the fitting

2. Insert the pipe into the body of the fitting until it meets the interior step of the fitting boby. Then draw the O-ring and the collar close to the body of the fitting.

3. Tighten the nut with the wrench until O-ring and collar enter the fitting and reach end-postion
   Unscreww nut from fitting.

4. Now open the white split-ring and mount on pipe with larger side against collar. Ensure that the collar and the split-ring meet the body of the fitting. Screw the nut tightly towards the body of the fitting. For final tightening use a PLASSON wrench.
   NOTE: Although the nut should be closed tightly, there is no need for it actually to meet the body.

Figure 22. Assembly of a compression joint. Courtesy of Plasson Ltd.

- Bond failure caused by uneven distribution of the cement, or forgetting to apply the cement to both surfaces to be joined.
- Pipe was not cut square, not cleaned up after cutting, or has not been pushed in completely, all of which mean that an insufficient bonding surface has been created.
- The cement was allowed to dry before the two components were assembled.
- An attempt was made to adjust the fit-up immediately after completion of the joint, thereby breaking the seal which had already been made and dried.
- The can of cement had been left open for too long, causing evaporation of the solvent component, or cleaning agent from a contaminated brush has been mixed into the cement.
- Whereas twisting of one of the components during bonding of thermosetting pipe systems is recommended, such a manoeuvre with thermoplastic systems can lead to a leaking or failed joint.
- Use of incorrect cleaner or solvent cement.

7.7.3.2 Mechanical joints

These are available for all thermoplastic pipe systems, and although joint configuration and sealing materials may differ from one supplier to another, they all fall within three main groups, as detailed below.

7.7.3.3 Compression joints

Compression joints are limited in diameter and suitability of materials to be joined. They are generally limited to use on polyethylene systems, with a maximum diameter of 5 inches or 125 mm. They rely on the principal of compressing an elastomeric lip joint or O-ring between the pipe and an external compression fitting, the compression being obtain via threaded fittings on each side of the compression coupling. They provide a quick and efficient method of assembly, not only between pipe sections, but as a means of assembling a large range of fittings, such as reducers, tees, elbows and valves, as the fittings are themselves supplied with integral compression joints. A wide range of compression fittings within the same diameter limitations also exist as adapters for transferring from one material to another, for example from polyethylene to copper, lead or PVC pipe (see Fig. 22 above).

7.7.3.4 Spigot/socket with elastomeric seal

The most widely used jointing configuration for medium to large diameter PVC and ABS buried piping systems uses a specially formed socket which acts as a housing for an elastomeric O-ring or lip joint-type seal. When the spigot is inserted into the socket, the elastomeric seal deforms, creating a leak-tight seal acting on the inside of the socket and the outside of the pipe. Socket forms and joint dimensions and configurations vary from manufacturer to manufacturer and are generally not compatible. This type of joint is unable to

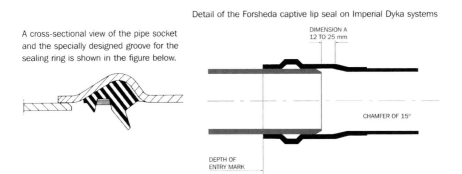

Figure 23. Spigot/socket joint. Courtesy of Dyka (UK) Ltd.

resist axial loadings and any such loading, depending upon the level of pressurisation of the system, must be transferred to sound undisturbed soil surrounding the excavated trench via concrete thrust blocks. As with all other non-restrained pipe systems, thrust blocks are required at each change in direction, deviation, valve and blanked off dead leg.

7.7.3.5 Flanged joints

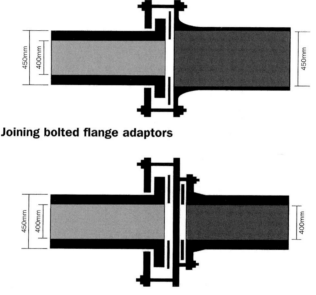

Figure 24. Flanged joint. Courtesy of Dyka (UK) Ltd.

*Dyka recommends the following guidelines should be observed when joining pipes using flange connections.

Flanged joints are the oldest and simplest form of jointing all types of pipe systems, and are retained on most thermoplastic pipe systems for connections to pumps, vessels, valves, hydrants and any in-line instrumentation. The most common practice is to use a thermoplastic collar, bonded or welded to the pipe, and equipped with a thermosetting or coated metallic loose flange (Van Stone-type flange). Due to the heavy wall thickness of many thermoplastic pipes, compared to metallic pipes for which flange dimensions were developed, it is often required to fabricate special loose flanges for which the OD and PCD correspond to one size or more above the nominal internal diameter of the thermoplastic pipe.

Thermoplastic stub flanges are relatively fragile when compared to metallic systems, and some precautions are required when bolting up, especially with crews more used to metallic systems.

- Check that flange faces are clean and have not been damaged during previous handling, especially that there are no significant scrape marks or deep scratches.
- Select the correct hardness grade of joint.
- Make sure that flanges and pipe work are correctly lined up, and that subsequent tightening will not induce any undue stress into the system.
- Bolt up in a diagonally opposed sequence, tightening each bolt in a multi-stage progressive manner, up to the final torque rating specified by the manufacturer.
- It is possible that a final torquing is required a few hours after the initial assembly, due to compression of both the gasket and the thermoplastic collar.

7.7.3.6 Threaded joints

Threaded joints are in general limited to small diameters, for assembling two different materials or some form of instrumentation. All thermoplastic threads are relatively soft and can be damaged by over-tightening. It is generally recommended to hand tighten only, or to use a strap wrench rather than serrated jaw wrenches.

There are many threaded pipe adaptors available either from the pipe companies or specialised fitting suppliers.

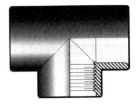
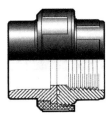

Figure 25. Threaded joint. Courtesy of Hydrodif.

7.7.3.7 Welding

Welding of thermoplastic pipe systems can be considered as the best possible solution for both long-term mechanical strength and sealing, and several different welding procedures can be used.

Although hot gas and hot gas extrusion welding can be used on most thermoplastic systems, we have reserved these two methods for the chapter dealing with dual laminate and thermoplastic tanks and vessels where they are perhaps more appropriate. It should, however, be noted that these two procedures are used for interventions on site where more common procedures cannot be used because of access or piping configuration. They are also used by specialised companies manufacturing non-standard fittings.

The following standards and guidelines concern welding of thermoplastic pipe systems:

ASTM D 3261. Standard specification for butt heat fusion polyethylene (PE) plastic fittings for polyethylene (PE) plastic piping and tubing.

ASTM F 1055. Standard specification for electrofusion-type polyethylene fittings for outside diameter controlled polyethylene pipe and tubing.

Draft BS EN 1042. Plastic piping systems. Fusion joints between polyolefin pipes and/or fittings. Determination of resistance to internal pressure at constant temperature.

WIS 4-32-08. Specification for site fusion of Pe 80 and Pe 100 pipe and fittings.

The first of the following three technologies, socket fusion welding, is used for assembling HdPe, PP and PVDF. The second method, butt fusion welding, is also used for HdPe, PP and PVDF as well as some of the more exotic fluoropolymers. The third procedure, electro-fusion socket welding, is reserved uniquely for polyethylene. Depending upon the procedure used, weld strength coefficients close to 1 can be obtained.

7.7.3.7.1 Socket fusion welding. Socket fusion welding can only be carried out with pipes and/or fittings made from the same material and to precise dimensional tolerances, within a range of diameters from 20 to 125 mm. The two items are welded together by an overlap of material without the additional of any extra material. To carry out the weld, the end of the pipe and the socket on the fitting are heated with a male/female heating tool, and once the correct temperature is obtained they are immediately pushed one inside the other.

Special tools are required, as supplied for example by Georg Fischer, and include calibration tools to clean and adjust the outer diameter of the pipe as well as the heating tool.

Typical welding temperatures for polyethylene are 250 to 270°C, which can be controlled automatically on the heating device, and typical heating times vary between 5 and 60 seconds depending upon the diameter of the pipe and the wall thickness (pressure rating) of the pipe.

Welding sequence

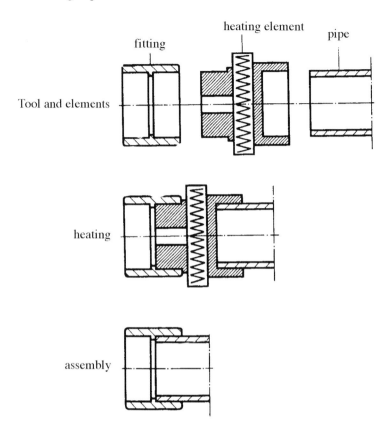

Figure 26. Socket fusion weld. Courtesy of George Fischer.

Once the correct heat and duration are obtained, the two items are withdrawn from the heating device, and are pushed together without any rotational movement until the pipe blocks on the shoulder inside the fitting. The two items are lined up and held together for a period at least as long as the heating time. A period of one hour should be allowed between welding and pressure testing.

7.7.3.7.2 Butt welded polyfusion. This is applicable as a welding technique to both flat sheet, for building thermoplastic tanks and vessels, and pipe, although different welding tools are required. A wide range of pipe butt fusion welding machines are available, each one operating over a specific range of diameters. The size range differs from supplier to supplier, but a typical selection of machine capacities would read as:

Figure 27. Butt fusion welding machine. Courtesy of KWH Tech.

- diameter 40 to 110 mm
- diameter 40 to 200 mm
- diameter 160 to 315 mm
- diameter 315 to 630 mm
- diameter 500 to 1000 mm
- diameter 800 to 1600 mm, with the two smaller machines being entirely hand operated.

The method requires the two sections to be welded to be held in clamps, which are designed not only to grip the pipe firmly but also redress any out of roundness. The opening and closing of the hydraulic ramps can be operated either by hand or by electric motor, and lateral clamp movements are either operator controlled or fully automated by a dedicated computerised control system. Prior to welding the pipe ends are trimmed, usually by an integral trimming device on the welding machine, except for small diameters where this operation is carried out with a hand tool. Once the pipe ends are trimmed, the two ends to be welded are hydraulically pushed against an electrically heated non-stick plate until molten, the pressure is withdrawn, the plate removed, and the two ends are then pushed together and held in place until the weld cools down. In actual fact the total welding process is broken down into five stages, during which timing, pressures applied and temperatures attained are all critical in achieving the best possible weld strength. All specific figures for the different diameters, materials and wall thickness are supplied with the machine.

> Stage 1 — the two pipes are pushed against the heated plate at a specific pressure until a bead of a certain dimension is formed.
> Stage 2 — whilst the heating plate remains in place the pressure applied to the two pipes is considerably reduced.
> Stage 3 — the heated plate is removed and the two ends are brought into contact immediately.
> Stage 4 — as the two ends are brought together the pressure is gradually increased until it reaches the level applied in stage 1.
> Stage 5 — as the joint cools to ambient temperature the pressure must remain applied at the same level as in stage 4. Should the pipe be required to be moved immediately, a cooling off period, equal to the initial cooling down period, should be respected between releasing the pressure and any substantial movement.

Simple basic machines require considerable operator input and experience in order to obtain consistently high results. The latest generation of computer-controlled machines will carry out all operations automatically once the pipes are correctly placed in the clamps, and will at the end of the welding cycle print out all relevant data concerning each particular weld, which not only forms part of the initial quality check, but is also the basis for long term traceability of material and welding performance.

No matter what the degree of automation a certain number of basic quality control checks should be made immediately the weld has cooled:

- use the bead gauge supplied with the machine to check that the dimensions of the external welding bead conforms to the welding specification;
- make an all round visual check for any excessive bead irregularity or mismatch of pipe diameters;
- cut off the external bead with the appropriate tool, and visually check the underside of the bead for signs of any contamination or slit defects; and
- check for cleanliness immediately around the joint area in order to ensure that no grease or dirt has entered the welding zone.

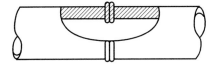

Figure 28. Butt fusion weld.

The pipe can be moved paying attention to limit movements and stressing around the joint once it is hand cool, and may be filled with water and tested 24 hours after completion of the final weld.

This procedure is also used for the manufacture of large-diameter fittings from pipe stock, often using specially designed welding tables, again available from most of the welding machine producers.

During butt fusion welding a bead is formed both inside and outside of the pipe, and as mentioned above, the external bead is removed and inspected as part of the quality control process. The bead on the inside of the pipe is inaccessible and remains in place. This is generally of little consequence except for some applications in the semiconductor industry, biotechnology, pharmaceutical and some foodstuff plants, where the bead is seen as a possible source of contamination due to bacterial growth within hidden or open cavities.

The Georg Fischer organisation has developed a high-purity range of PVDF pipe and fittings in diameters 20, 25, 32, 40, 50 and 63 mm, under the trademark SYGEF HP WNF, specifically for applications in the industries mentioned above, and has completed the product range with tooling and procedures which allow fusion welding with no internal bead, and by consequence no cavities in which microbiological organisms can develop.

The procedure consists of controlling precisely the heating of the extremities to be welded via the use of semicylindrical heating elements and the pressure applied to the two elements whilst at the same time blowing up a bladder placed inside the welding area, and which stops bead formation. The welding machine is completely automated and guarantees a high level of reproducibility, and is an inseparable part of the high-purity system, due to the utilisation of the external positioning ribs on all the fittings, and which permit an exact positioning of the fittings inside the jaws of the welding machine.

7.7.3.7.3 Electrically heated socket fusion welding. Also called electro-fusion welding, it is an automated version of the socket welding described in Section 7.7.3.7.1 and has been developed specifically for polyethylene, essentially due to the high volume utilisation of polyethylene pipe for the distribution of both gas and water. The jointing method is based upon special fittings which incorporate an electric heating element which, when current is applied, heats up via a time/temperature controlling device to the point that the two surfaces in contact melt and fuse. The fitting is allowed to cool down and the electrical connections are withdrawn. The assembled joint should not be tested immediately, and if possible 24 hours should be allowed before testing.

Different types of fittings exist for MdPe, HdPe and HPPe, and include a straight coupler for joining pipe-to-pipe, a reducing coupler for joining two different diameters, and an extremely wide range of elbows (22.5, 45 and 90°), Y pieces, equal and reduced tees, tapping tees, flanged and non-flanged branch saddles, stub flanges and hydrant branch saddles, for example. All fittings include the electric heating elements within their construction and serve as fitting and welding connection at the same time.

Diameters available run from 20 to 355 mm for straight couplers, tapping tees and all forms of branch saddles, whereas the maximum diameter for elbows, tees and stub flanges is 180 mm.

Control of the heating/fusion process is achieved via sophisticated controlling devices, the most advanced of which allow three main modes of operation.

- Automatic, whereby once the lead connecting the control box and the fitting is connected, the controller identifies the fitting and selects itself the correct fusion time. The machine operator has only to press the start button and the complete operation is carried out in an entirely automatic manner.
- Manual, in that the operator can programme the controller to the requirements of any specific fitting or environment.
- Barcode, when using fittings with a barcode identification. The operator scans the barcode on the fitting, the reading is automatically transferred to the controller which then programmes itself to the requirement of the fitting in use.

The most sophisticated systems also have facilities for input and output of selective data, full output voltage and output current monitoring, identification of potential overheating with automatic system abort and a temperature compensation facility for working environments at extreme temperatures.

The advantages of electro-fusion welding are:

- consistent quality of welding;
- there is no fusion bead on the inside of the joint;
- assembly and fusion are simplified and automated;
- a satisfactory level of fusion is automatically indicated by a monitoring system on each fitting; and
- complete complex pipework configurations can all be set up precisely and overall dimensions checked off prior to welding, and in many cases several or all of the joints can be made at one time.

168 *Thermoplastic Pipes*

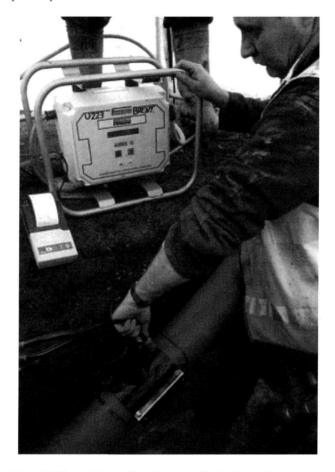

Figure 29. Electrofusion welding. Courtesy of Fusion Group Manufacturing.

Control of the finished weld is simple, with visual inspection of the fusion monitoring devices, a check that the assembly has not been moved during fusion, and a general inspection of the state and cleanliness of the weld and the surrounding area.

7.7.3.8 Qualification of welders

For manual welding see the section on manufacture of thermoplastic tanks and vessels.

The qualification for butt fusion and electro-fusion welding is in fact more a qualification to operate satisfactorily various machines, and manufacturers generally recommend that operators are qualified by testing trial welds against specific test requirements and procedures, as well as checking that correct welding procedures are established by the contractor and effectively followed by the machine operators. As an example, weld sample tests, procedures and criteria are specified in the UK in WIS 4-32-03, Appendix G.

7.7.4 Thermoplastic pipe fittings

Unlike the composite pipe market, the thermoplastic pipe market has standardised dimensions and configurations to a high level, and as such manufacturers have been able to achieve considerable success with the corresponding standardised fittings. There has tended to be a split between companies who manufacture pipe and others who manufacture fittings, although there are some exceptions to this rule.

A wide range of standard fittings are available in PVC, PVC-C, HdPe, PP, ABS and PVDF to meet dimensions and pressure ratings of the corresponding pipes. Bi-axial or otherwise strengthened PVC pipe systems may use standard PVC fittings as the outer diameter of both product lines are identical (the wall thickness of bi-axial pipes being thinner), or in some cases the pipe supplier also supplies a range of fittings specific to his own product range.

Non-standard fittings can be made for any of the thermoplastic pipe systems by cutting and welding of pipe and sheet.

7.7.5 Design of thermoplastic pipe systems

7.7.5.1 Product design

The basic design of a thermoplastic pipe is based on a long-term testing programme, much in line with the regression curve test to ASTM D 2992 used to establish the basic hydrostatic design basis of composite pipe systems. The burst strength of thermoplastic materials, as compared to the weeping pressure of composites, are, at defined pressure and temperature, time dependent. Test standards as defined in ISO 9080.2 generate design data based on a 50 year service life, through testing of pipe samples for up to 10 000 hours at 20°C, extrapolated to 50 years. A minimum safety factor of two is applied to the values extrapolated to 50 years.

7.7.5.2 System design

The classification group working on relevant CEN/ISO standards has recommended the application of an extra design factor, over and above the factor of two mentioned above, to cover 'unknown' loadings or environmental conditions which may occur, and, for example, the recommended supplementary design factor for polyethylene is 1.25.

There are many fundamental design criteria to all pipe systems, whether they be metallic, composite or thermoplastic. Thermoplastic materials do have such specific characteristics which must be taken into account when designing an underground or above ground pipe system.

The following points require more specific attention:

- The maximum calculated expansion and measures taken to accommodate or absorb this expansion.
- Pipe flexibility, whether it be bending radii on a buried line, or deflexion between supports on an above ground installation.
- Burial depth, acceptable soil or wheel loading, backfill material and compacting of the backfill.
- Pressure de-rating as a function of fluid or ambient temperatures.

All the major suppliers publish detailed in-depth guides concerning the choice, assembly, installation and testing of their products. Some of the companies, as well as some of the major distributors, have developed software packages which further facilitates the choice of materials, dimensions and pressure ratings, as well as assisting the engineer in the general consideration of the layout of the system, consideration of supports for above ground lines, and trenching and backfilling for buried lines.

Underground PVC pipe systems are usually installed with a spigot/O-ring seal joint, which will not resist the axial loading induced by even a low level of pressurisation. This loading is transferred to the surrounding undisturbed soil via concrete thrust blocks. Thrust blocks are designed and dimensioned taking into account the diameter of the pipe, the maximum internal pressure to which the system will be exposed and the load-bearing capacity of the surrounding soil. Thrust blocks are required at each change in direction (elbow or tee), each change of diameter and at each valve.

7.7.6 Pipe installation and testing

Thermoplastic pipes are much lighter than equivalent pipes in ductile iron or concrete, and as such lend themselves to handling and lifting by hand or light lifting equipment. This facility should not be abused to the extent that the pipe is damaged, even superficially, by handling with undue care and attention.

All major pipe manufacturers supply detailed information concerning the installation and testing of thermoplastic pipes. Much of the information is common to all piping materials, and we draw the reader's attention more closely to the specific consideration of thermoplastic materials.

Thermoplastic pipes can be considered as flexible or semi-flexible, and they will respond to flexing and backfilling on buried installations in a completely different manner to rigid pipe systems.

Low ambient temperatures will reduce the flexibility, and PVC pipe should receive particular attention when temperatures drop below 10°C, and all installation work should stop on PVC pipes when temperatures reach zero.

Trench widths should be kept to a strict minimum, but should also be wide enough to allow placing and compacting of bedding material against each quadrant of the pipe. For this reason a width, at the height of the crown of the pipe, of the pipe diameter plus approximately 300 to 350 mm is generally recommended.

Typical depths of coverage above the crown of the pipe are between 500 and 1000 mm, depending essentially upon the possible exposure to different traffic loadings. Trenches should be over excavated by approximately 100 mm in order that correct filling and compacting of the pipe bed can be achieved. The underbed should afford a continuous support along the total pipe length, with possible excavations around joints if required. The underbed should also be laid in such a manner that it guarantees the required fall and gradient.

Again, because thermoplastic pipes are relatively flexible, special care needs to be taken when backfilling the trench. In order to avoid deformation of the pipe due to the weight of the overlaying earth, and possible extra loading from traffic, crushing of the pipe must be limited by avoiding movement of the side walls of the pipe. This is achieved by compacting sidefill of the trench to at least 80% of optimum density, in successive layers of approximately 80 mm, up to a height of 100 mm above the height of the crown of the pipe. During this phase, care must be taken to avoid lateral displacement of the pipe and damage to the pipe. Constant checking is also required to ensure that the spaces under the lower half of the pipe are correctly filled and compacted on both sides. The remaining space in the trench can be filled with excavated earth in successive layers of approximately 300 mm, compacting each individual layer, and avoiding the use of heavy compacting equipment until the first layer of 300 mm has been placed and compacted.

Above ground pipe is especially sensitive to changes in ambient or fluid temperature, as well as the effect of direct sunlight on the pipe. Pipe systems must be correctly supported, taking into account the maximum temperature to which the pipe will be subjected. Continuous supporting may be required for smaller diameter pipe or for pipe operating at higher temperatures. Valves and other in-line regulation and instrumentation must be supported independently from the pipe, and such supports must be designed to avoid transmission of loads to the pipe when valves are manoeuvred. As thermoplastic pipe has coefficients of thermal expansion well above those of all other typical pipe materials, special attention must be given to the definition and the extent of thermal expansion and methods to be used to deal with it. In order to reduce significantly support and expansion problems with systems working at higher temperatures, it is often useful to examine the possibility of using dual laminate pipe systems.

Long pipelines should be pressure tested in relatively short lengths, especially at the beginning of a project in order to establish the conformity and reliability of welding procedures being used. Pipes should be backfilled sufficiently to avoid any movement, leaving joints exposed for detection of any eventual leaks. As is normal with all materials, air vents must be installed at all high points, and filling should be commenced slowly through the lowest point in the line. Once the line is full and all air has been evacuated the line should be left until the temperature has stabilised. Normal test pressures are in the range of 1.25 to 1.5 times the maximum design pressure of the system, and pressures are normally to be held for six hours or more.

7.7.7 Large-diameter fittings and access chambers for water and effluent mains

Corrugated double-wall polyethylene pipe is used to manufacture a very wide range of catch pits, valve or pump chambers, sampling or inspection chambers and simple access manholes. The same pipes, as well as solid-wall polyethylene pipe, is also used to manufacture a full range of non-standard fittings, both materials using various forms of thermoplastic welding as described above.

Figure 30. Prefabricated manhole. Courtesy of Bauku (www.bauku.de).

Polyethylene access chambers have considerable advantages over alternative concrete or brickwork constructions in that they are very light, easy to install, require little or no maintenance over their total service life and assure continuity of material with the pipe, avoiding all possibilities of leakage. The abrasion and corrosion resistance of the chamber is identical to that of the connecting pipework, and as such avoids the creation of any weak point. Each configuration can be equipped with single, double or multiple inlets, at single- or multi-levels, and should the chambers be subject to traffic loadings, light loadings can be handled with compacted granular surrounds, or concrete surroundings poured directly around the chamber for heavy or frequent traffic. All chambers can be equipped with the full range of standard access ladders or steps as well as many standard plastic steel or pre-cast covers and frames.

CHAPTER 8

Products and Applications — Composite and Thermoplastic Tanks, Silos and Other Vessels

Along with pipe systems, tanks, silos and a wide range of other similar vessels and constructions account for the largest utilisation of thermoplastic and composite materials in the water industry.

As can be noted from the previous details on national and international standards, the range of such documents applicable to tanks and vessels is extremely small when compared to those issued for pipe systems. This is perhaps understandable in that vessels are stand-alone equipment, assembled to other equipment only via flanges, the dimensions of which are standardised on a worldwide basis. A pipe system, made from a particular material, for a specific pressure rating, and in a specific dimension, is however, required to be generally available from a range of suppliers and for incorporation or assembly to existing equipment anywhere within a specific geographical area.

Tanks and other specific vessels such as CO_2 stripping columns, salt dissolvers and silos are made to individual fabrication standards by a large number of companies, offering very cost-competitive pricing due to the standardisation of shell diameters and end closures, standard nozzles and ancillary equipment. Many fabricators, as well as a wide range of jobbing shops or project oriented manufacturers offer specific custom-made vessels to suit the requirements of a specific application, process or dimensions.

In line with previous descriptions of pipe systems, we shall split our examination of vessels into two parts, on the one hand composite equipment, and on the other thermoplastics and dual laminates.

8.1 Cylindrical Composite Tanks and Other Vessels

Standard cylindrical composite tanks and vessels are available in all major resin systems described previously. We shall, however, consider epoxy resin-based equipment separately in Sections 8.6 and 8.7 as the production and curing procedures limit the dimensions and applications available.

Standard vessels are generally made by a combination of filament winding for the cylindrical shell, and contact or spray up moulding for the full range of end closures, although some integrally wound vessels incorporating cylindrical shells and flat, dished or conical end closures are also available. Centrifugally cast vessels are also manufactured by a few suppliers.

Using a basic vertical tank as an example, we can break down the fabrication and assembly as follows.

The shell diameter is taken from a fixed range of diameters for which each fabricator has made available standard mandrels and other tooling, generally in a range of 1000 mm up to 5000 mm (3 ft up to 16 ft), for a storage volume from 1 m^3 up to 250 m^3.

Flat, dished or conical bottoms and tops are moulded separately and assembled onto the cylindrical shell, often using a simple cemented socket joint with a hand lay up overwrap over the area of the joint, much as used when assembling pipe systems with a butt and strap overwrap joint. From this basic construction and assembly philosophy many variations have been developed, of which the following are of particular interest.

- **Cantilevered mandrels.** In most cases a mandrel is equipped with a central axis, and is supported by the filament winding machine at both ends. Once the winding is complete, the moulded shell is removed from the mandrel, each end cut to the required dimension, and the end closures are assembled as described above. A cantilevered mandrel is only supported at one end, and the free end of the mandrel is closed with the mould used to manufacture one end of the tank, generally the bottom of the tank in the case of a vertical vessel. This procedure allows the fabricator to manufacture the tank bottom and tank wall in one

Figure 31. FRP storage tanks. Courtesy of Forbes.

piece, which not only eliminates a costly joint, but also guarantees an improved structural continuity between the vessel wall and the end closure.

- **Integral winding of the shell and two end closures.** The fabricator moulds a reinforced inner liner of 2.5 to 3.2 mm thick, for both the shell and the two end closures. These three items are then assembled together with a butt and strap-type jointing, the fabricator having first placed metallic stiffening rings inside the shell construction. An axle for connection to the winding machine is attached to the two end closures, and the vessel 'skin' is placed on the winding machine. This skin is then slightly pressurised with air in order to develop sufficient rigidity for the winding process, which covers the cylindrical shell and the two end closures in a single integrated lamination. Upon completion the vessel is cured, the connections to the winding machine removed, an opening for the implantation of an inspection door is made, and the stiffening rings are dismantled and withdrawn from the vessel through this opening. Various finishing operations are carried out both inside and outside of

Figure 32. Dual laminate CO_2 degassers. Courtesy of Forbes.

the vessel in the areas of assembly of the three components and the area of attachment to the winding machine. This process ensures an integral, one-piece structural moulding, and at the same time avoids two expensive shop assemblies.

Both of the above mentioned procedures are valid for construction up to 4 or 5 m in diameter depending upon the structure of the winding machines.

A third procedure uses the continuous filament winding procedure developed by Owens Corning Engineered Pipe Systems, referred to above for the manufacture of medium to large diameter pipe. The 'pipe' is cut to the length of the required tank wall, and end closures are assembled on to it as for the basic system described above.

Nozzles or access doors, for example, are always moulded separately, either by contact or compression moulding, and are assembled to the tank with a hand lay up joint between the tank wall and the nozzle or access door.

8.1.1 Design

Using one of the typical design, production and testing standards, the UK national standard BS 4994 as a guideline, we can follow a typical process from the identification of initial requirements to handle a particular production or storage requirement up to the final tests and controls on the finished equipment. The initial section of this general procedure may usefully be followed by users of composite tanks and vessels, no matter what the actual standard eventually specified, for the original definition of a particular vessel.

Section one of BS 4994 covers the following.

Design and service conditions concerning

- Definition of fluids to be handled, their name, chemical composition, concentration and density.
- Design temperature.
- Design pressure (or vacuum) including any specific testing requirement.
- Service pressure.
- Process/loading cyclic conditions.
- Any other properties specific to the process or fluids being handled.

Site conditions of the installation

- Ambient and extreme temperatures.
- Superimposed loads such as wind, snow, pipework or other associated equipment.
- Any loads imposed during installation.
- Any seismic loading.

Any special conditions

- Agitation of the fluids.
- Vibrations generated by adjacent equipment.
- Pigmentation prior to inspection (see Chapter 12).

From a detailed examination of the above, and having defined the overall approximate dimensions and configuration of the vessel under consideration, the client and/or fabricator can establish:

- the resin system to be used;
- the type of reinforcing materials to be used and their physical properties; and
- type and thickness of the internal corrosion-resistant liner.

The final design details can then be defined, including:

- exact dimensions and permissible tolerances;
- design calculations and wall thickness;
- details of any local stiffening or reinforcement which may be required;
- bolting and flange materials;

- gasket materials and details;
- details of any external finish including any associated steelwork; and
- details for access and inspection openings.

Prior to completion of design and start of manufacture any specific requirements for a quality and inspection/testing system must be defined and agreed, including:

- Possibility to repair any laminate defects and the methods to be used to achieve such repair.
- Possible provision of special test laminates and the extent of mechanical testing to be carried out either on cut outs from the vessel or on prepared laminates (that is laminates prepared in parallel with the construction of the vessel, as being representative of the actual construction laminate). These laminates can then be submitted to various forms of destructive testing in order to control ultimate physical properties.
- The nature of tests and the required presence of any inspectors.

The vessel under study will then be placed into one of three categories depending upon the consideration of seven different criteria.

(1) Definition of the contents of the vessel as toxic, highly corrosive, corrosive, flammable and others.
(2) The chemical compatibility of the liner with the process fluid.
(3) The design temperature and the relationship to the HDT of the resin system.
(4) The design pressure and/or vacuum.
(5) The capacity of the vessel.
(6) The overall geometry and supports.
(7) Any other criteria which may be critical to the safety of the vessel.

The design documentation and drawing requirements are established as a result of the consideration of the previous criteria, and the subsequent definition of the category, and can vary considerably from category 1 to category 3. Category 1 requires design calculations as follows:

(1) Approval by an independent body.
(2) That the calculations cover hydrostatic loadings, applied pressure, applied vacuum, wind loads, lifting arrangements, supporting and seismic loading if applicable.

Category 1 also requires drawings as follows:

(1) Vessel/tank general arrangement drawings.
(2) Full fabrication drawings showing methods of manufacture.
(3) Installation procedures.

A category 3 vessel only requires design calculations as follows:

(1) Hydrostatic loadings.
(2) Lifting arrangements, and drawing requirements as (1) vessel/tank general arrangement drawings, and (2) installation procedures.

The quality system applied to the different categories of vessels also shows substantial variations depending upon the definition of category 1, 2 or 3. Category 1 requires the application of up to 17 different controls or phases, as follows.

Material records:
(1) Recording of resin type and quantity.
(2) Recording of glass type and quantity.
(3) Recording of names of personnel employed on the fabrication.
(4) Recording of the layers and types of glass*.
(5) Recording of the curing system*.
(6) Recording of any post-cure, if required.

Quality control tests:
(7) Spark test on thermoplastic liners if used*.
(8) Adequate documented information on the mechanical properties of the particular resin/glass laminate to be provided*.
(9) A production test coupon to be laminated with the vessel or obtained from nozzle cut-outs, and tested for ultimate shear strength, unit modulus and lap shear strength.
(10) A visual examination of nozzle cut-outs*
(11) Ash test on nozzle cut out–glass/resin ratio.
(12) Measurements of thickness.
(13) Acetone test (on polyester/vinyl ester-type resins)*.
(14) Barcol hardness measurements*.
(15) Residual styrene (on polyester/vinyl ester-type resins).
(16) In the case of thermoplastic lined vessels, measurement of weld and bond strength.
(17) An independent inspection.

For a vessel defined as category 3, only the seven controls marked * are required.

It is evident that the cost of two identical vessels, one designed to category 3, and the other to category 3, would show a considerable difference, and this type of price consideration, within or without the requirements of this particular standard, is of prime importance for the client in the decision-making process.

Section two of the standard covers materials and design loadings, under the following headings:

(1) Thermosetting resin systems, which include polyester, epoxy and furane.

(2) Reinforcing materials.
(3) Thermoplastic lining materials, if used.
(4) Mechanical properties, including consideration of the HDT of the resin system.
(5) Chemical or corrosion resistance.
(6) Effects on water quality.
(7) Construction of the chemical barrier (anti-corrosion liner).

It is worthwhile at this point to examine in more detail, on the basis of BS 4994, but applicable in general to all other standards and practices concerning corrosion resistant composite tanks and vessels, and also to some extent pipe systems, the basis construction of three types of anti-corrosion liners.

The text of BS 4994 defines the characteristics of the liners as follows for a thermoplastic lined or 'dual laminate' construction.

1. For thermoplastic linings the minimum bond strength of the structural laminate to the lining shall be 7 N/mm^2 in direct shear and 5 N/mm^2 width in peel when tested in accordance with procedures laid down by the standard. The standard includes that the bond strength may be achieved by the inclusion of a minimum of 0.45 kg/m^2 chopped strand mat laid up directly on the back of the thermoplastic lining.

2. The standard continues by offering two alternatives for thermosetting resin liners and specifies that in order to achieve the optimum properties the construction of the liner in contact with the fluid shall consist of the following:

A resin rich surface layer reinforced with a C-glass or synthetic veil, and having a total thickness between 0.025 and 0.50 mm.
A subsequent back-up laminate, to contain a minimum of 1.2 kg/m^2 of chopped strand mat, containing between 25 and 33% of glass by weight.
The total thickness of the two layers will be in the area of 2.5 to 3.0 mm.

The standard then proceeds to note that tanks and vessels constructed within the requirements of categories 2 and 3 may, under certain conditions, be manufactured with a back-up laminate limited to 0.6 kg/m^2 of chopped strand mat, which would result in a total liner thickness of between 1.5 and 2 mm.

To return to the overall examination of standard.

Item 8 considers the possibility of modifying the external surface layer of the vessel to reduce flammability and flame spread characteristics.

Item 9 examines allowable and design unit loadings, a point of significant interest and common to some extent to many of the standards for composite vessels.

The most important point considered is the K, or design factor, which is expressed as $K = 3 \times k_1 \times k_2 \times k_3 \times k_4 \times k_5$ where:

3 represents a constant to allow for long-term structural deterioration (ageing).

k_1 relates to the method of manufacture, with 1.5 for contact moulding, machine controlled filament winding and machine controlled spray up, and 3.0 for hand held spray up.

k_2 is a factor for the chemical environment and which may vary between 1.2 and 2, following criteria established by a guide and issued as an appendix to the standard.

k_3 is temperature related, and is expressed within a range of 1 to 1.25, again based upon guidelines issued within the standard.

k_4 relates to cyclic loading, and is expressed within a range of 1.1 to 2, based upon data supplied within the standard.

k_5 concerns the curing process, and varies from 1.1 for full cure plus post cure in the plant prior to shipment, up to 1.5 for vessels without a post cure, and designed for an operating temperature above $45°C$.

Calculation of minimum and maximum values from the above results in a design factor of 6.53 minimum and a theoretical maximum of 67.5 which is of course impractical.

In actual fact the standards defines a minimum design factor of 8, and good design practice will generally limit maximum design factors even for severe applications, to around 15.

Section three of the standard is concerned with the actual detailed design of the vessel, and covers in detail the following main points of interest.

- Design temperature and pressure.
- Laminate design and thickness.
- Consideration of cylindrical and spherical shells, subject to internal or external pressure and any local reinforcement or stiffening.
- Wind loadings.
- Rectangular tanks.
- Flat panels/square and rectangular with or without stiffening devices.
- Circular panels.
- Sandwich construction panels.
- Ends (hemispherical, semi-ellipsoidal, torispherical, conical, flat, bolted flat covers).

8.1.2 Applications

Typical configurations and dimensions of standard filament wound composite shop-built vessels are as follows:

- Vertical cylindrical flat bottom storage tanks from 1000 to 5000 mm diameter, for storage volumes of 1 to $250 \, m^3$.

- Vertical cylindrical tanks with conical or dished bottoms from 1 to 250 m^3.
- Horizontal cylindrical tanks with dished ends from 1000 to 4000 mm diameter.

From the conception of standard tanks many companies have developed a line of vessels for specific applications handling water and effluent, and a handful of companies located in the UK have played leading roles in this development, often with a complementary rather than competitive range of products, and by so doing have established the UK as a leader in the field of composite vessels for water treatment.

Klargester Environmental Engineering Ltd has used the inherent advantages of corrosion resistant composites to manufacture a wide range of specialised vessels including the following:

- Septic tanks — for non-mains drainage of wastewater, providing that the subsoil is capable of effectively dispersing the liquid effluent.
- Cesspools — or a simple holding tank used where other alternative equipment is unsuitable.
- Silage tanks — in some cases storage of silage and subsequent leakage and infiltration can have catastrophic effects on the surrounding environment. FRP silage tanks provide a long-term, cost-effective solution for the farmer.
- Grease traps and separators are designed to prevent grease from entering wastewater drainage systems which could result in pipe blockage.
- Full retention and silt separators are used in high-risk oil pollution areas, such as fuel handling and wash down areas. Single-chamber separators are generally recognised as being the most effective at separating imiscible particles and are available as standard, although multi-chamber versions can also be supplied. As contaminated water passes through the chamber(s), it is retained long enough for the hydrocarbon pollutants and sediments to separate out from the water and become entrapped within the separator.
- Bypass separators are used to collect water run off from areas such as roadways, car parks and other paved areas which may be contaminated by hydrocarbon pollutants, sediments and minor spillages. The bypass separator is designed to intercept the hydrocarbon pollutants and retain specifically sediments and spillages flowing during the initial part of a rain storm, when the majority of pollutants are swept up, allowing the remaining flow of rain water, which is relatively free of pollutants, to bypass the treatment process, and consequently maintaining a reasonable size and capacity for the total installation.
- Forecourt Enviroceptors are used in high-risk areas such as service station forecourts where they can reduce oil pollution levels of

rainwater to as little as 5 parts per million. They are fitted with shut off valves as standard to guard against fuel spillage, and can also be equipped with oil level alarms to continuously monitor the status of the separator and to ensure that the unit is correctly maintained.
- BioDisc treatment plants. A modular FRP effluent treatment-tank which can replace cesspools or septic tanks in order to meet the newer and more stringent requirements set by the local approving authorities. The unit involves a rotating biological contactor. Size and capacity can vary from single-house units up to installations for schools, hotels and restaurants, for example.

Forbes Plastics have developed another and extensive range of water and effluent-oriented FRP vessels, from small 200 l thermoplastic tanks up to 6 m diameter by 15 m high chop/hoop wound FRP and dual laminate vessels. The Forbes group initiated its entry into the tank market with thermoplastic products in 1961, and subsequently developed dual laminate constructions as their basic high volume vessel business. Recent developments have been oriented more specifically to a wide range of vessels for many forms of water treatment and environmental control, including the following.

- Large cost saving PUMA standard tanks, up to 200 m^3 for effluent treatment and storage, which exist in three standard diameters, 2400, 3050 and 4111 mm, and depending upon the resin system used are suited for storage of water, general and mildly corrosive effluent, sewage sludge and brine solutions, as well as for many industrial chemicals. PUMA tanks are designed for a minimum service life of 20 years, offer long-term resistance to water absorption and incorporate a specially developed npg (neo-pentol-glycol)-based isophthalic polyester resin which makes the internal surface of the laminate highly resistant to degradation and blistering. These tanks, when subjected to a full post cure are also suitable for the storage of potable water.
- Underground storage tanks for both potable and waste water or effluent, suitable for the collection of surface water, foul water, sewage, silage and general agricultural effluents. Its PUMA range of effluent tanks are made on automated chop/hoop winding machines, producing a sandwich construction, incorporating a resin structural system between FRP layers. This concept increases the strength of the vessel, and allows underground installation in high water table conditions without the more usual external reinforcement via external ribbing of the vessel.

Standard vessels are available in diameters of 2400 and 3000 mm, for capacities from 9 to 110 m^3.

(Note: the early days of FRP tank construction were known for some frequent and often spectacular failures. Design, production technology, raw materials and quality control systems have all made significant advances over the last 30 years, and these problems are essentially behind us. However, due care and attention is still required at all stages of tank design, production, testing, installation and use. Forbes Plastics has been a leading player in this development and we include in Chapter 12 a reprint of a previous article called *Tank Failure* written by Bryan Broadbent, sales director of Forbes Plastics.)

- A standard range of silos, all 2400 mm in diameter completes the basic range of storage vessels available from Forbes Plastics, with nominal capacities ranging from 6 to 38 m^3, although higher capacities are available on a project basis based on 3 and 4 m diameter shells. The PUMA range of silos can be manufactured in a wide range of resins to meet the requirements of the material stored. Typical applications include salt, flaked chemicals and fertilisers.

Forbes is the largest manufacturer of FRP silos in the UK, and its products offer some important features when compared to competition, including:

- An innovative filling system which allows fast uniform loading and complete filling to the specified capacity. This same design can also help in the reduction of particle separation and dust generation in some applications.
- The one-piece body offers no crevices allowing cross contamination of products, and the smooth internal finish assures complete emptying and facilitates cleaning and sterilisation.
- The translucent wall construction allows the level of contents to be checked at a glance against the integral calibration markings.

- A Minibulk integrally bonded safe storage system for hazardous chemicals, fabricated entirely from thermoplastic materials is also available and is described in the section concerning thermoplastic vessels.
- Salt saturators for the continuous production of saturated brine, with capacities from 7 to 140 tonnes per day. Brine is required for water treatment, as well as for the food and chemical industries, and salt saturators are used to provide a constant supply of brine solution. Forbes salt saturators are available in the three diameters already specified for standard storage tanks, for nominal continuous respective throughput of 2400, 3700 and 6800 l/hour, and with a possibility to increase this level by approximately 50% for a short-term, high-demand situation.

The saturator vessel contains layers of graded gravel on which rests a bed of salt covered by a constant head of water. The water dissolves the salt, and by penetrating the gravel, emerges as a clean saturated brine in an automatic process needing only a water supply and salt refilling from time to time.

A translucent vertical strip down the side of the vessel permits instant stock level inspection and the re-order level is marked. A permanent bed of salt at least 150 mm high must be maintained at all times to ensure that fully saturated brine is produced, and this bed also acts as an efficient filter removing dirt or dust which may have been introduced from the atmosphere or from the water supply.

- Aeration stripping towers for the elimination of dissolved CO_2 with water throughput of 6 to 200 m^3 per hour. The standard range of FD Degassers proposed by Forbes are in actual fact manufactured in thermoplastic materials, but special designs and larger capacity equipment is also manufactured in composite or composite/thermoplastic materials.

Capacities of standard equipment range from a water flow rate of 6 to 200 m^3 per hour, which corresponds with a nominal upper/lower diameter of 350/983 mm up to 1678/3050 mm, for overall heights of between 3240 and 6257 mm.

The FD Degasser operates on the principle of passing the water to be treated over a large surface area whilst blowing air, via a fan, against the flow. The resulting mass transfer of gas at the interface of the water and the air removes the acid forming carbon dioxide.

The units are designed to reduce the carbon dioxide content from 200 ppm to 5 ppm at 20°C. The unit comprises a packed tower with a base-mounted fan and integral sump.

Construction materials of the standard range of units are:

- tower, sump and cowl: black copolymer polypropylene containing 2% of carbon black;
- tower packing, pipe work, float, fan casing and impeller: polypropylene; and
- packing support grid: FRP.

- Stripping towers for the removal of organic solvents from groundwater. Solvent contamination of groundwater is an increasingly common problem for water authorities drawing water from wells and springs. Forbes stripping towers remove relatively high levels of aliphatic and aromatic organic volatiles, and many organic contaminants can be reduced to levels well below those established by the WHO. Forbes stripping towers rely on the corrosion resistance of both thermosetting and thermoplastic materials, and use specially designed packed towers with low energy–high efficiency packing.

Forbes also provide odour and fume scrubbing systems for wastewater treatment plants as well as emergency scrubbers for chlorine gas storage units in potable water plants.

The Balmoral Group is highly active in the fields of both thermoplastic and sectional composite tanks. A detailed review of sectional tanks follows in Section 8.4 and information on its rotationally moulded tanks is included in Section 8.5.1.

The company FRP, Fibre Reinforced Products, manufactures a wind range of filament wound composite vessels specifically for the area of water treatment and pollution control. These include the following.

- Septic and primary settlement tanks. A range of standard vessels, based on a diameter of 2600 mm is available for capacities from 18 000 to 54 000 l. Larger capacities, up to 240 000 l, and other diameters from 1.8 m up to 4 m are also available.
- Cesspools, following the configuration, standard and non-standard dimensions indicated above are also supplied.
- Silage effluent and settlement tanks for farmyard run off are used by farmers to avoid pollution of groundwater and rivers. Silage effluent tanks follow the dimensions and configurations indicated above, whereas the settlement tanks are manufactured in five sizes, 18, 27, 36, 46 and 54 000 l. The settlement tanks have internal separation plates to split the volume into three sections. They perform as a low-rate irrigation system, and the build-up of sludge inside the tank can be easily controlled and removed.
- Petrol interceptors as either bypass or full retention interceptors. Both types of interceptors meet the requirements of BS 8301 and are in line with current propositions for European standards. The full retention interceptor is specifically designed for high-risk areas such as fuel distribution depots, workshops and industrial process areas. Bypass interceptors are destined for areas of lower risk such as superstore car parks, industrial development zones and other areas involving large surfaces of hard standings. Both types of interceptors are available in capacities up to 240 000 litres.
- Silos — a wide range of silos, of diameters 2500, 3000 and 4000 mm complete the range of vessels supplied by FRP, with capacities up to 60 m^3.

Bio-Plus Environmental Systems Ltd is a specialist manufacturer of package sewage treatment plants and package pumping station, all based upon the physical and corrosion resistant properties of FRP.

Its range of products includes:

- Biological aerated filters, a three stage process available in a single FRP shell for 5 to 250 population equivalents, and which is fully described in Chapter 10.
- Standard or custom designed pumping stations with single or double pump options, to lift crude sewage, storm water and waste water in all residential, industrial and commercial applications.

- A wide range of grease separation systems and traps.
- Petrol, oil and silt interceptors, separators and traps designed to meet relevant legislation and standards.
- Septic tanks, cesspools, silage effluent and general storage tanks.

Other major suppliers of a range of standard composite tanks and vessels are to be found in most major industrialised countries, including Fiberdur-Vanck in Germany, who offers an extensive range of single- or double-wall FRP tanks in diameters from 600 up to 6000 mm in a wide range of standard configurations, and from 6500 mm up to 20 m diameter for vertical flat-bottomed tanks, Selip and FZ Fantoni in Italy, Casals Cardona in Spain, Carlier, Rousseau and SOVAP-Plasticon in France, as well as Tankinetics, and Justin Tanks in the USA. A more complete listing of fabricators worldwide is included in Chapter 14.

Centrifugally cast tanks and silos are available in the range from approximately 2 m in diameter up to a maximum of 4.2 m from one manufacturer in France.

For the manufacture of vessels having a diameter of more than 5 m or so, two main alternative construction methods exist, plus one novel technique developed by the company Tankinetics in the USA which is described below.

Many of the major end users of large composite structures are located close to the sea, or a navigable river or lake, often related to the requirement of a water intake for process or cooling purposes, and also for the rejection of treated effluent. A growing number of specialised fabricators are located on or close to a navigable river or canal. The combination of the two is allowing more and more vessels to be shipped by barge or boat, and dimensions up to approximately 10 m are now current in some parts of the world. Mandrel and machine dimensions and investment costs are substantial, and suitable suppliers are not available everywhere, and in this case, or should diameters above 10 m be required, the second alternative should be examined.

A handful of European and North American fabricators have developed machines and technologies for filament winding large cylindrical structures, up to almost 40 m in diameter, on site, or very close to the final location of the vessel. This process is dependent upon machines with mandrels which allow vertical filament winding, associated with contact moulding for inner liners and localised reinforcement or stiffening, and which can handle weights in excess of 50 tonnes (mandrel and laminate combined). Other production considerations specific to this style of manufacture are the huge material feed rates, which can reach more than 1000 kg/hour of resin-impregnated glass fibre roving, and which complicate both material handling and catalytic and curing procedures, all of this associated with local atmospheric conditions (temperature, humidity and sunlight) which can never be as controlled as within a fixed production workshop.

Figure 33. An oblated tank being transported to site. Courtesy of Tankinetics Inc.

The detailed laminate design is predominant among design considerations, with the need to specify the style and type of reinforcement, the number and orientation of the different layers, often incorporating simultaneous winding of direct and woven rovings. Other specific considerations include the possibility of handling and lifting both prefabricated units and the final structure, vacuum and pressure effects on large surfaces, expansion and contraction especially in the knuckle area of assembly of the vessel bottom to the sidewall, and also as related to nozzle connections to all associated pipework.

Case histories showing more than 10 years of continuous satisfactory service are included in Chapter 13, and include 12 off storage tanks of diameter 14.6 m by 12 m high (2000 m^3 each) and 2 off tanks 20 m in diameter by 12 m high (3700 m^3 each) all of which are installed in major industrial wastewater treatment plants in North America.

The third, and perhaps the most novel, technology has been developed by the US company Tankinetics Inc, a modular shipping and field erection concept designated Tankinetics' Oblation System. This technology facilitates shipping and erection of large-diameter cylindrical composite tanks.

Oblation enables the shop fabricated tank sections (up to 21 m in diameter) to be oblated according to precise mathematical formulae, nested together and shipped to site. Once on site, the sections are de-nested, de-oblated and bonded together. Tankinetics has more than 18 years of experience with the system, which it originally patented.

8.2 Rectangular or Square Composite Storage Tanks

Rectangular and square composite tanks can be split into two different categories, based both upon their own respective area of application and production technology.

The first concept of standard modular constructions is similar to the older modular steel structures, which in spite of various coating and protective procedures has limited corrosion resistance.

The flexural properties of composite materials are such that large flat surfaces are generally considered to be discouraged; however, recent moulding technology and innovative tank design has overcome this problem to some extent and allows large-scale manufacture of large rectangular or square storage tanks, especially for potable water. Panels are usually moulded by resin transfer or hot press technologies, which confer high and constant physical and dimensional qualities, as well as full resin cure, extremely important when the finished tank is to be used for the storage of potable water. The obvious interest of this type of construction is the possibility to ship large volume tanks in standard trucks or containers, and to build up the tank on site, often within a restricted space or with restricted access.

The concept involves bolted panels, usually with an external flange, although internally flanged constructions are also available, and we review two typical products from two different suppliers.

Dewey Waters supply both internally and externally flanged constructions. Externally flanged systems are based on two different panel dimensions, 1 m $\times$ 1 m and 1.22 m $\times$ 1.22 m, whereas their internally flanged construction uses the 1 m module only.

Five different base panels are available:

- Flat, with flange uppermost for tanks on concrete foundations.
- Base star, for internal or external bolting with half span foundations.
- 'Humphrey', for internal bolting and full span foundations.
- Dome base, for external bolting fully drained base on full span or half span foundations.

192 *Rectangular or Square Composite Storage Tanks*

Figure 34. An externally flanged modular tank. Courtesy of Dewey Waters.

- Sump, equipped to fit any size outlet, with flush and full draining fit.

Sidewalls are built up with standard Star sidewalls, in two different thicknesses to suit both shallow and deep tank construction. Covers are built up with:

- Standard dome panels, shaped to shed rainwater but with walkways at the edges for safety.
- Standard flat panels for use when headroom is limited.
- Insulated flat panels for temperature control and containment.
- Standard manway panels for access.
- Raised ball valve chamber panel.
- Quick access inspection hatch.

Figure 35. An internally-flanged modular tank. Courtesy of Brinar Plastic Fabrications Ltd.

Lid supports (vertical columns) in PVC are fitted to give support to lids and are stressed to accept live, wind and snow loadings. Internal stainless steel tie rods are used to cancel out sidewall stress.

Tank sizes vary from 1 m×1 m×1 m up to 9 m×6 m×4 m deep, with a maximum capacity of 216 m^3, and can be supplied in a fully insulated version if required.

Brimar Plastic Fabrications also supplies both internally and externally flanged systems, and we have retained its internally flanged system for a more detailed examination.

Its totally internally flanged (TIF) sectional tanks are based upon panel widths of 300, 500 and 1000 mm, with any height up to 4 m, which means sides of 2 m long and above can be designed with increments of 100 mm.

This concept also calls for internal columns and horizontal tie bars in stainless steel.

These internally flanged tanks are fully insulated, and have all the benefits of other types of sectional tanks. However, as all the flanges are fastened together from the inside, a theoretical minimum space of only 25 mm is required around the outside of the tank. The base panels, being internally flanged, allow the tanks to be installed directly on to a flat and level surface if required.

Both of the systems described above are WRc approved for the storage of potable water.

The custom-designed concept of rectangular tanks allows the manufacturer to build an exact storage volume within the smallest possible surface area, and this concept can offer the most economical storage system within a limited space or for a specific process. This advantage, is, however, easily overtaken by the high cost per cubic metre of volume when compared to cylindrical sections. Wall thickness of rectangular tanks need not necessarily be very thick, but they do require a high level of stiffening by composite or metallic/composite beams and/or ribs. This can lead both to excessive weight, significantly increased external dimensions and subsequent excessive cost. This somewhat unbalanced cost/performance ratio does not however rule out the manufacture of some very large rectangular storage tanks, as for example the tank supplied by Balmoral Composites Unit, concerning a $10.5 \times 4.9 \times 2.8$ m high (140 m^3) potable water storage tank for an offshore oil production platform.

8.3 Thermoplastic and Dual Laminate Tanks and Vessels

Thermoplastic tanks and vessels can be made by various techniques, including rotational moulding, using solid or hollow section spiral winding or by cutting, forming and welding thermoplastic sheets.

8.3.1 Rotationally moulded thermoplastic tanks

The Balmoral Group, located in Aberdeen, UK, offers one of the largest ranges of tanks and vessels available in rotationally moulded polyethylene, producing a range of vessels with capacities from 70 to 10 000 l.

The range of products include:

- Bulk liquid storage tanks, primarily designed for water storage, in capacities ranging from 1135 to 10 000 l, approved by the WRc. All tanks have a 500 mm access and lockable cover, and are guaranteed for ten years.
- Septic tanks, with capacities from 2700 to 6000 l (corresponding to a population of 4 to 22) and available for drain invert depths of 1 or 1.5 m.
- Sewage treatment plants, as the patented Sequential Batch Reactor (SBR) which combines the efficiency of activated sludge aeration with batch treatment to provide high efficiency under extreme loading conditions. The manufacturer claims a reduction of main pollutants including ammonia by 96%, phosphate reduction by 88%, and a batching system which eliminates peak surges. Capacity of served population varies from 6 to 48, with total tank volumes, primary settlement and reactor space combined, from 4000 to 16 000 l.
- Interceptors. Balmoral produces a range of full retention and bypass interceptors, suitable for a drainage area of between 220 and 14 200 m^2. The interceptors are designed to stop oil and other hydrocarbons from entering the drainage system, thereby reducing water pollution.

All the above vessels are made in polyethylene by using rotational moulding. Polyethylene is an extremely tough and robust material, with high impact resistance, and is ideally suited to these smaller vessels which are often stocked in builders' or distributors' yards, are handled frequently with a wide range of equipment, and are required to survive collisions with hard and sharp objects. The rotational moulding process offers ideal economic conditions for long production runs of standard vessels, and vessel bodies are usually made in a one-piece moulding.

8.3.2 Spiral wound and welded shell construction

Sheet welding is not the only technology used to make medium to large diameter thermoplastic tank shells. Spiral wound and welded sections, as described in the section concerning spiral wound and welded pipe, are also used for the fabrication of vessels. Solid wall or hollow wall extruded sections are wound onto a heated cylindrical mandrel of the required diameter at a temperature of approximately 220°C (for HdPe), and are automatically welded together as each subsequent rotation of the mandrel is made. The wall thickness of the extruded section is increased as the strength requirements of the tank shell increase. Bottoms, tops or covers, nozzles and manways are then welded to the basic shell as for all other thermoplastic tanks, often using highly automated welding equipment to weld both inside and outside of the tank.

Typical applications in the water-related industries are of course a wide range of both standard and custom designed tanks, limited in general to 10 to 15 m^3 capacity for fully welded vessels, and 50 m^3 for spiral wound constructions.

Two materials account for practically all thermoplastic tanks, black or natural polypropylene and high-density polyethylene, with U-PVC, PVDF, E-CTFE and FEP reserved for a few specialised applications. In all areas except for spiral wound fabrications, HdPe does show signs of being replaced by polypropylene, except where low ambient temperatures can be expected, due to the relative fragility of polypropylene at low temperatures. Polypropylene has a wide range of chemical resistance, and can resist temperatures from roughly zero to 80°C depending upon the fluid being stored, whereas HdPe covers a range from approximately −20°C up to 45°C.

Companies such as Allibert Manutention have specialised in the construction of a wide range of standard tanks and vessels in HdPe and PP, for storage volumes of 250 to 50 000 litres, for temperatures of −30 to 45°C for HdPe, and −10 to plus 90°C for constructions in PP. Allibert also offers constructions in PVDF if required. Standard diameters are 700, 1000, 1200, 1600, 2000, 2500, 2900 and 3100 mm, all compatible with transport by road and rail.

Starting from its basic range of storage tanks, Allibert has developed vessels specific to water treatment such as reactors fully equipped with mixers, internal separations and baffles, heating coils or steam injectors, for example, as well as flocculant storage and preparation vessels for products such as aluminium sulphate, ferric chloride, ferric sulphate and ferric chlorosulphate.

8.3.3 Thermoplastic welded tank and vessel construction

Thermoplastic sheets, essentially U-PVC, PVC-C, PP, HdPe and PVDF, can be formed under heat and pressure to make curved wall sections, dished ends, cones and similar forms common to the tank and vessel market. They can then be welded together by three different welding techniques, fusion, hot gas or extrusion welding, as described below.

8.4 Thermoplastic Welding Technology

8.4.1 Fusion welding

This is applicable as a welding technique to flat sheet for building thermoplastic tanks and vessels, and different fusion welding machines are available, each one operating up to a specific range of width and thickness.

The principle requires the two sheets to be welded to be held in clamps, the opening and closing of which are controlled by hydraulic rams. Prior to welding, the sheets are cut to size and the ends trimmed. The two ends to be welded are hydraulically pushed against an electrically heated non-stick plate until molten, the pressure is withdrawn, the plate removed, and the two ends are then pushed together and held in place until the weld cools down. In actual fact the total welding process is broken down into five stages, during which timing, pressures applied and temperatures attained are all critical in achieving the best possible weld strength. All specific figures for the different diameters, materials and wall thickness are supplied with the machine.

Stage 1 — the two sheets are pushed against the heated plate at a specific pressure until a bead of a certain dimension is formed.

Stage 2 — whilst the heating plate remains in place the pressure applied to the two sheets is considerably reduced.

Stage 3 — the heated plates is removed and the two ends are brought into contact immediately.

Stage 4 — as the two ends are brought together the pressure is gradually increased until it reaches the level applied in Stage 1.

Stage 5 — as the joint cools to ambient temperature the pressure must remain applied at the same level as in Stage 4. Should the sheet be required to be moved immediately, a cooling off period equal to the initial cooling down period, should be respected between releasing the pressure and any substantial movement.

Simple, basic machines require considerable operator input and experience in order to obtain consistently high results. The latest generation of computer-controlled machines will carry out all operations automatically once the sheets are correctly placed in the clamps, and will at the end of the welding cycle print out all relevant data concerning each particular weld, which not only forms part of the initial quality check, but is also the basis for long-term traceability of material and welding performance.

No matter what the degree of automation a certain number of basic quality control checks should be made immediately the weld has cooled.

- Use the bead gauge supplied with the machine to check that the dimensions of the welding beads conform to the welding specification.
- Make an all round visual check for any excessive bead irregularity or mismatch of the two ends of the sheets.
- Cut off the beads with the appropriate tool, and visually check the underside of the bead for signs of any contamination or slit defects.
- Check for cleanliness immediately around the joint area in order to ensure that no grease or dirt has entered the welding zone.

8.4.2 Hot gas welding

In this process a gas, usually dried and de-oiled air, is heated via an electric device, and is then directed to heat the two surfaces to be welded and the welding rod used to weld the two items together. The thickness of the materials to be welded is generally between 1 and 10 mm.

The two materials to be welded and the welding rod must be of the same nature, and for high-quality welding exactly the same materials. Welding rods are available in various dimensions and shapes, but essentially 2 to 4 mm, and round or triangular.

A welding device heats the welding rod and the materials to be welded, at the same time allowing the operator to apply pressure onto the welding rod, and to pull the welding rod along the line of the weld.

Typical welding parameters for HdPe and PP are as follows:

Loading onto the welding rod
Diameter 3 mm — 10 to 16 N
Diameter 4 mm — 25 to 35 N
Air temperature: HdPe — 300 to 350°C/PP — 280 to 330°C
Air volume — 40 to 60 l per minute.

The welding rod is generally laid into a triangular space, achieved by shaving one or more of the materials to be welded. The thickness and strength of the weld can be built up by applying successive welding rods, uniformly disposed around and on the initial pass.

8.4.3 Extrusion welding

This is a similar technology to that described above, but used for relatively thick material (10 mm and above). In this case the welding machine is equipped with a mini extruder, which extrudes a molten welding rod into the weld, which has itself been pre-heated up to the correct temperature by hot air supplied by the welding machine. A variant of this process replaces the extruder by a device which pulls several welding rods at the same time and builds up a thick weld in one pass.

8.5 Dual Laminate Constructions

A thermoplastic inner liner and a supporting external composite structure are used instead of pure thermoplastic constructions for three often interdependent reasons:

- When dimensional or other considerations mean that the thermoplastic materials themselves cannot support the loading applied to the vessel.
- When cost, especially when dealing with the more high performance (and high cost) thermoplastics, would mean that the overall cost of a pure thermoplastic construction would be prohibitive.
- When service or design temperatures are too close to the maximum operating temperature specified by the manufacturer of the thermoplastic pipe or sheet.

When selecting a dual laminate rather than straight composite materials the reasons may be:

- Improved corrosion resistance of the thermoplastic.
- Improved abrasion resistance of the thermoplastic, although it should be noted that the addition of certain fillers to thermosetting resins may achieve abrasion resistance levels equal to, or in some cases better than, thermoplastics, especially at higher temperatures.
- Ease and cost of manufacture, especially for non-standard dimensions or configurations where standard mandrels and moulds are available, and where the use of the thermoplastic liner as the basic form excludes the requirement to build one-off moulds.

There is little or no limitation in shape, form or configuration of vessels in all the material variations, and one can say that any vessel configuration made in mild or stainless steel can also be made in thermoplastic, dual laminate or composite material.

There is no hard and fast rule which defines comparative cost levels of dual laminates and straight thermosets, as different resin systems and different liner materials themselves induce a high variation in the cost of raw materials. Different companies may have specialised in one or other of the techniques, and specific and related investment may influence considerably the final cost. Perhaps the only basic but over-generalised rules are as follows:

- Standard tanks and similar vessels of up to about 15 m^3 capacity are probably cheapest in straight polyethylene or polypropylene, especially, but not only, if manufactured by rotational moulding.
- Standard tanks above 15 m^3 are generally most economic in a basis isophthalic-based composite structure.

Dual laminates rely generally on a high-strength bond between the thermoplastic sheet or pipe used as the inner liner, and the supporting composite structure. The bond between PVC and PVC-C and a composite material is achieved by using a specially formulated resin to promote a chemical bond between the two materials after cleaning, and in some cases roughening, the surface of the thermoplastic material.

Non-structural internal equipment inside various process vessels can often be made more economically using solid thermoplastic materials inside a thermoplastic or dual laminate vessel, using normal thermoplastic welding techniques, as opposed to composite internal structures inside a composite vessel, which are generally more difficult and expensive to fabricate, and may be subject to deficiencies in the strength and corrosion of secondary bonding of laminates to the original structure of the vessel.

8.6 High-Pressure Filters

Under the heading of high-pressure filters we shall limit ourselves to what is commonly known as reverse osmosis pressure vessels, although this designation can be somewhat misleading, and the filtration membranes used within these vessels. The above mentioned designation stems from the original and expensive concept of obtaining potable water by reverse osmosis of seawater at high pressure (70 bar, 1000 psi). Today, vessels are used at pressures as low as 3 bar (45 psi), for four main filtration applications, including the original reverse osmosis (RO), and investment and running costs have decreased considerably.

Dimensions of these vessels have been determined by the major suppliers of filtration membranes, which have three nominal diameters, 8 inch, 4 inch and 2.5 inch, and a standard length of 14, 21 and 40 or 60 inches. Membranes of 14 and 21 inch long are available in 2.5 and 4 inch diameter, and are used in small single-vessel applications as found in small boats and laboratories, for example. 40 inch (1016 mm) and 60 inch (1524 mm) membranes are used as single units in all diameters, but in most cases, especially for the 8 inch diameter, 2 to 8 membranes are installed in one long single vessel, and in most cases banks or trains of vessels are built and linked up together to form a filtration unit. The worldwide dimensional standardisation of the membranes has allowed the overall business to develop within an environment of healthy competition between both the membrane and vessel manufacturers, and an end user can move from one supplier to another with comparative ease.

As mentioned above the initial market area for this type of membrane filtration was reverse osmosis for the production of potable water from seawater or brackish water. Due to initial high costs of this technology, reverse osmosis sea water desalination was essentially reserved for Middle East countries having abundant sources of energy combined with a chronic shortage of fresh water.

Subsequent membrane developments have improved their performance, lowered costs and have allowed the industry to benefit from an expanding RO market, and at the same time to diversify into the nano- and ultrafiltration markets, with, for example, the production of high volumes of potable water from river water, using nanofiltration membranes.

Typical examples of membrane filtration outside of the desalination business include:

- Process water for power plants.
- Process water for a wide range of chemical plants, including water recovery from municipal effluent plants.
- Ultra-pure water for electronic components and semiconductor plants.
- Oily waste water treatment plants.
- Treatment of water and beer in breweries.
- Aseptic water in hospitals, clinics and laboratories.
- Removal of colour from the effluent of dye plants in the textile industry.
- Concentration and clarification of beverages, such as grape juice or orange juice.
- Concentration of amine acids and sugar in the food industry.
- Recovery and concentration of spent acids in the chemical industry.

The actual filtration process uses composite spiral wound filtration membranes of various configurations and a range of thermoplastic materials, sometimes enclosed within an FRP skin. These membranes are then packed within a composite pressure vessel as detailed in Fig. 36 below.

Note that when referring to the actual membrane used inside the FRP or composite vessel, the meaning of composite as applied to the membrane means simply a construction made from the combination of two or more materials and not the use of fibre reinforced plastic as we generally use the word composite.

This type of filtration can be broken down into four general categories, microfiltration, ultrafiltration, nanofiltration and reverse osmosis. Details of these four categories are given below in the section concerning the design and construction of the actual membranes.

Most pressure vessels supplied for this type of filtration are made from composite materials, although stainless steel vessels are also available, especially in areas where high-temperature sterilisation with high-pressure steam is required, and thermoplastic vessels or casings are dominant in small low-pressure units. Due to the specialised nature of construction there are only a few significant manufacturers of composite pressure vessels in the world and these are listed in the directory of companies.

There is one dominant standard for the design, manufacture and testing of theses vessels, that is the ASME Code, Section X, which is recognised all over the world. Further details on this code are given in Section 5.

This code specifies requirements of raw materials, production technologies, two different design procedures and criteria for both prototype and on-line pressure testing, all contained within a quality system specific to the scope of the code itself.

This code, within the limits of Class 1 procedures as used by all the major suppliers of RO pressure vessels, requires that under specific conditions the vessel shall resist 100 000 pressure cycles from zero to the design pressure of the vessel, followed by a hydrostatic burst test during which the vessel shall resist a progressive pressurisation from 0 bar to six times the design pressure over a period of not less than 60 seconds. For example, an 8 inch vessel, designed for 70 bar shall be pressurised from 0 to 70 bar 100 000 times, and shall then be pressurised to 420 bar over a period of not less than 60 seconds. The code requires that this test be carried out on each prototype vessel and subsequent re-qualification pressure vessels at fixed intervals. All production vessels are tested at 1.5 times design pressure during 1 minute.

ASME Section X was not written specifically for the type of vessel under consideration, and many other features of these vessels are of significant importance.

End closures, made from thermoplastic or composite materials, sometimes with a metallic backing plate, are sealed with O-rings, lip rings or quad rings. The efficiency of these types of seals at the full range of pressures to which the vessel will be submitted throughout its service life are not covered by the code, and each type of seal presents both advantages and disadvantages. For example, a lip ring may not seal correctly at low or zero pressure, which could present problems during prolonged periods of shut down, whereas O-rings may show slight leakage at high pressures, i.e. during actual service. Lip rings may make the end closure more easy to remove after a long period of service, but any difficulty met at this stage may also be the result of the shape and configuration of the ends of the vessels, for example if they are cylindrical or conical. Sealing of all types of joints is, however, dependent upon the dimensional tolerance and surface qualities of the inner surface of the vessel in contact with the seal. Most manufacturers used highly polished chromium plated steel mandrels, machined to very precise tolerances and problems with sealing should not arise with vessels purchased from reputable manufacturers.

Resin systems used are generally amine-cured epoxies, but vessels using polyester resins and thermoplastic inner liners are also available. The choice of resin system has been historically based upon mechanical properties in order to withstand the pressure tests indicated above, with the thinnest possible laminate. Actual legislation concerning the suitability of certain materials to be in contact with potable water may influence future developments, and vessels with dual resin systems may appear, with, for example a cycloaliphatic epoxy system as an inner liner for compatibility with potable water, and an amine-cured structure for strength and cost. We may also see further developments of the dual laminate structure, with thermoplastic inner liners reinforced with an external composite layer.

All vessels, other than those using an opaque inner liner must be painted in order that no light shall penetrate into the vessel. Should this occur, microbiological growth can develop which will rapidly cause a deterioration of the performance of the membrane, and subsequent deterioration of the quality of water produced.

The actual membranes placed inside these vessels are all made from combinations of thermoplastic materials, sometimes including some composite material, either for the permeate pipe in the centre of the membrane, or the cladding shield on the outside of the membrane.

The FilmTec Corp, a leader in the field of spiral wound membranes, supplies a historical introduction concerning the origins and early development of this type of membrane, followed by some practical information concerning the construction and use of the membranes.

Since their development as practical unit operations in the late 1950s and early 1960s, RO and UF have been continually expanding the scope of their applications. Initially, as mentioned above, RO was applied to the desalination of seawater and brackish water. Increasing demands on the industry to conserve water, reduce energy consumption, control pollution and reclaim useful materials from waste streams have made new applications economically attractive. In addition, advances in the fields of biotechnology and pharmaceuticals, coupled with advances in membrane development, are making membranes an important separation step, which compared to distillation, offers energy savings and does not lead to thermal degradation of the products.

Basic membrane research with respect to the foundation of the FilmTec Corp and the development of the Filmtec FT30 membrane started in 1963 at the North Star Research Institute in Minneapolis, Minnesota, USA.

Since then, products and product improvements have been developed in the North Star Laboratories, such as:

- The first composite reverse osmosis membrane;
- Microporous polysulphone membrane;
- The first non-cellulose thin-film composite membrane (NS-100); and
- Ultra-thin cellulose hemodialysis membrane.

An important development in the thin-film composite approach was the discovery of microporous polysulphone support films in 1967 by John Cadotte. Not only did this result in twice the flux of the ultra-thin cellulose acetate composite membrane, but it subsequently led to the invention of four important non-cellulosic thin-film composite membranes by Cadotte: NS-100, NS-200, NS-300 and finally the FT30 membrane while at Filmtec Corp.

To finish this historical introduction, the FilmTec Corp was founded in 1977 and became a wholly owned subsidiary of the Dow Chemical Co in 1985.

The various filtration technologies which currently exist can be categorised on the basis of the size of the particle removed from a feed stream.

- Microfiltration (MF), used to remove particles in the range of 0.1 to 1 μm. Working pressure is very low, generally in the range of one bar

(15 psi), and the pressure vessels are generally supplied in thermoplastic materials.
- Ultrafiltration (UF), used to remove particles in the range of 20 to 1000 Å, at pressures up to approximately 10 bar (150 psi).
- Nanofiltration (NF), used to remove particles in the range of 1 nm (10 Å) at typical service pressures of 6 to 16 bar (90 to 150 psi).
- Reverse osmosis (RO), used particularly, but not uniquely, in the field of desalination, using both seawater and brackish water. RO is also used widely in all types of industry, including food and beverage preparations, industrial process water, ultra-pure water for the semiconductor industry. Service pressures can range from 16 to 100 bar (450 to 1500 psi).

Two basic types of membrane design are in current use in water treatment, that is spiral wound and hollow fibre modules. Hollow fibre modules vary in dimension from one supplier to another, and pressure vessels to contain such membranes are available in a wide range of materials. A typical application of composite housings with hollow fibres can be seen in the highly successful hollow fibre module described below and supplied by Aquasource. The design of the vessel is specific to the Aquasource concept and is not available for membranes from an alternative supplier.

We return to the FilmTec Corp for further information on typical spiral wound membranes, which account for most composite pressure vessels supplied today in the RO and associated markets.

FilmTec membranes are thin-film composite membranes in a spiral wound configuration. Spiral wound designs offer many advantages compared to other module designs, such as tubular, plate and frame and hollow fibre module design for most of the reverse osmosis applications in water treatment. Typically, spirally wound configurations offer significantly lower replacement costs, simpler plumbing systems, easier maintenance and greater design freedom than other configurations.

The construction of a spiral wound FilmTec RO membrane element as well as its installation in a pressure vessel is schematically shown in Figure 36 below.

The FilmTec FT30 thin film composite reverse osmosis membrane is made from an aromatic amine, 1,3-benzene diamine. It is defined as a thin-film composite membrane consisting of three layers, a polyester support web, a microporous polysulphone interlayer, and an ultra-thin barrier layer on the top surface, as shown in Figure 37 below.

The major structural support is provided by the non-woven web, which has been calendered to produce a hard, smooth surface free from loose fibres. Since the polyester web is too irregular and porous to provide a proper substrate for the salt barrier layer, a microporous layer of engineering plastic (polysulphone) is cast onto the surface of the web.

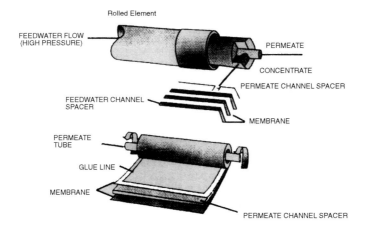

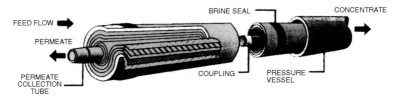

Figure 36. Exploded view of a FilmTec membrane. Courtesy of FilmTec Corp.

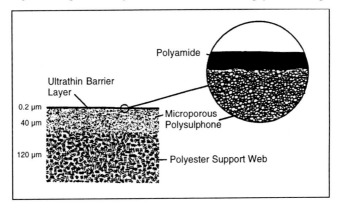

Figure 37. Structure of a membrane. Courtesy of FilmTec Corp.

The polysulphone coating is remarkable in that it has surface pores controlled to a diameter of approximately 15 Å. The FT30 barrier layer, about 2000 Å thick, can withstand high pressures because of the support provided by the polysulphone layer.

Membranes of similar construction are available from several multinational companies, offering a range of materials and construction to handle a wide range of fluids at temperatures from as low as 1°C up to 95°C or more. For more details concerning suppliers see Chapter 14.

Aquasource, a subsidiary of the Dégremont group of companies has, as mentioned above, developed a highly successful alternative to spiral wound membrane filtration for the production of potable water. Dégremont has retained the ultrafiltration process from the various processes available, because the size of the membrane pores corresponds to the needs of clarification and disinfection of a wide range of raw water destined for potable applications. This ultrafiltration process also has the added advantage of preserving all essential salts needed to maintain the quality of potable water.

The process calls for membranes consisting of long hollow fibres whose walls act as a filter to remove all particles greater than 0.01 µm, including pollens, algae, parasites, bacteria, viruses, germs and cysts (giardia and cryptosporidium). These membranes are encapsulated within a composite pressure vessel, and contrary to the spiral wound process, feed water is injected into the membranes and the permeate is collected on the outer skin of the cylindrical vessel via a flanged composite stub. Due to the generally low pressure, all associated feed, reject and permeate piping is generally fabricated from thermoplastic materials.

The Aquasource process uses both dead end and cross-flow filtration, that is when the water has low turbidity or suspended solids content, the water, under the effect of pressure, is distributed along the fibre and flows 'head on' through the porous wall. This process is designated as dead end filtration. When, however, the feed water has increased turbidity or a higher level of solids then the trans-membrane pressure is increased, the re-circulation pump included within the system being brought into play. The water flows at an increased velocity along the fibres sweeping away the accumulated cake, the clogging is controlled and a constant flow rate is maintained. This process is designated as cross-flow filtration.

When operating in the cross-flow mode, powder activated carbon can be injected into the re-circulation system, and can be used to treat chemical pollution such as that caused by organic matter, pesticides and pigments.

Dégremont commercialise the Aquasource units as standard self-contained skids, with capacities of 20, 40, 80 and 160 m^3/hour, often destined for villages or small urban areas, as part of a general upgrading of existing units, for hospitals and holiday resorts and for pre-treatment of feed water destined for reverse osmosis. The system is also meeting with considerable success as a self-contained, easily transported potable water production unit for use in areas suffering from major environmental catastrophes both in Europe and in many developing countries.

8.7 Small Composite Low Pressure Filters, Containers and Storage Tanks

Within this section we intend to cover a specific product range and market sector, dominated by a handful of manufactures located essentially in North America and Europe, but with recent extensions into the far East and the Pacific area.

The basic configuration of this type of vessel is a cylindrical body with two hemispherical end closures, generally equipped with polar boss openings or nozzles which are required to facilitate high-volume production methods, although they may also be equipped with other nozzles on the cylindrical shell. Such vessels, with capacities ranging from 5 l up to 30 m^3, from diameters as low as 125 mm up to 2400 mm, for pressure ratings up to 10 bar, are used for a wide range of applications:

- Hot water storage heaters
- Reverse osmosis permeate or pre-filtration storage tanks
- Residential water softeners
- Sand or carbon filters
- Brine tanks
- Settling tanks
- Retention tanks in chlorine or ozone disinfectant units
- Degasifiers

Depending upon the size and pressure rating of the vessel, several different manufacturing techniques are available. Based upon the three different production methods used by one of the leading producers of these types of vessels, The Structural Group, it is possible to demonstrate the specificity which can be built into a standard composite production process, to adapt it to the manufacture of specific vessels for particular applications.

The Structural Group, with production plants in Ohio, USA, Europe and India, and since a recent take over now part of the Pentair Water Treatment group, has developed its own technology and associated trademarks:

- COMPOSIT — commercial and industrial vessels, in diameters 18 to 96 inch (450 to 2440 mm) for volumes up to 7000 gallons (32 m^3).

Products and Applications — Composite and Thermoplastic Tanks, Silos and Other Vessels   211

- POLYGLASS — pressure vessels, in diameters 6 to 16 inch (150 to 400 mm) for volumes up to 50 gallons (220 l).
- FRP tanks, in diameters 6 to 36 inch (150 to 900 mm) for volumes up to 270 gallons (1.2 m³).

We examine first the COMPOSIT process for the construction of large vessels using a thermoplastic inner shell and an overwound filament wound structural laminate.

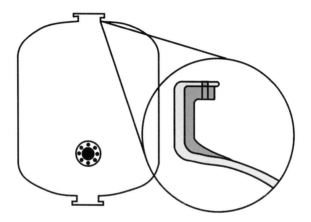

Figure 38. COMPOSIT vessel — stage 1. Courtesy of Structural NV.

Stage 1 — the inner shell is moulded using a rotomoulding process to form a thermoplastic liner, generally in polyethylene, but other materials such as polypropylene, PVDF or ECTFE are also available to meet specific requirements. The features required in the vessel are integrally moulded into the shell as either flanged or threaded openings, providing a smooth, seamless and weld-free inner surface.

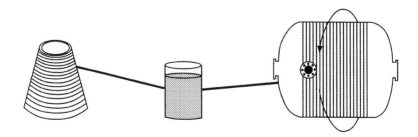

Figure 39. COMPOSIT vessel — stage 2. Courtesy of Structural NV.

Stage 2 — continuous epoxy impregnated glass fibre rovings are overwrapped around the inner liner in an initial polar winding pattern incorporating both the cylindrical shell and the two end closures. This winding is followed by a circumferential reinforcement of the cylindrical shell. The epoxy resin is pigmented to the client's colour requirements, and all winding is carried out on a three-axis computerised filament winding machine.

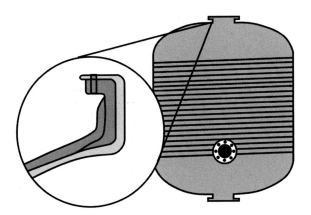

Figure 40. COMPOSIT vessel — stage 3. Courtesy of Structural NV.

Stage 3 — the filament wound vessel is then transferred to a curing oven in order that the laminate develops the full mechanical properties required to ensure maximum long-term reliability. This type of vessel, with pressure ratings up to 150 psi (10 bar), can be installed on various supporting structures in both a vertical and horizontal position. An interesting application is a 1200 mm diameter vessel of this type, subjected to cyclic pressurisation up to 10 bar at 55°C, and used as a brominator for industrial water systems.

The POLYGLASS system for the production of pressure vessels is similar to the previous process, but offers a more economical solution for smaller, lower rated pressure vessels with a high-density polyethylene inner liner. Ten different diameters are available, from 5 to 16 inch, for operating pressures of 150 psi (10 bar) at 120°F for threaded connections and 150°F for the flanged alternative (approximately 50 and 65°C respectively) and all vessels are certified to the NSF standard 44 for contact with potable water.

The FRP process developed and exploited by Structural uses neither the thermoplastic inner liner concept nor the filament winding production method, and instead uses a glass fibre preform/resin injection method.

Figure 41. Polyglass vessel — stage 1. Courtesy of Structural NV.

Stage 1 — a glass fibre preform is obtained by controlled spray up technology according to the size and shape of the vessel, to a specific weight and wall thickness.

Stage 2 — the glass fibre preform is inserted into a steel mould, and a removable internal bladder is placed within the preform. The mould is then closed and sealed.

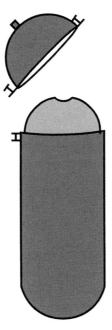

Figure 42. Polyglass vessel — stage 2. Courtesy of Structural NV.

214 *Small Composite Low Pressure Filters, Containers and Storage Tanks*

Stage 3 — the internal bladder is pressurised in order to push the preform hard up against the outer steel mould. A controlled amount of catalysed resin system is then injected into the annular space between the bladder and the mould, producing a laminate with a controlled thickness, weight and glass/resin content.

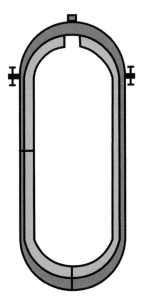

Figure 43. Polyglass vessel — stage 3. Courtesy of Structural NV.

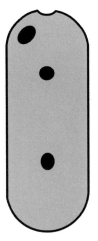

Figure 44. Polyglass vessel — stage 4. Courtesy of Structural NV.

Stage 4 — the laminate is heat cured, the bladder is removed and the one-piece seamless moulding is removed from the steel mould. Post-cure operations include the tapping of threaded openings located according to the client's requirements.

The maximum pressure ratings are 150 psi (10 bar) at 120°F (50°C) for vessels made with polyester resins and 150°F (65°C) when using vinyl esters. All vessels are certified to NSF Standard 44.

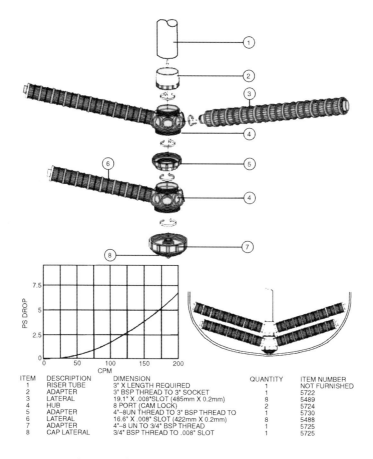

Figure 45. A typical Bajonet installation. Courtesy of Structural NV.

The total range of pressure vessels is completed by a modular internal distribution system, under the trade name Bajonet. Bajonet systems are designed for practically all forms of water and wastewater treatment and filtration, and will fit pressure vessels from 12 to 96 inch (300 to 2400 mm). The modular components are easily installed because lateral connections only require a one-quarter turn to lock into place. The design of the slots assures a uniform vertical distribution of flow throughout the entire bed without channelling. The slot openings are moulded uniformly with a self-cleaning 'V' contour, assuring continuous maximum flow with a minimum pressure drop. Polypropylene distributor components will operate at continuous temperatures up to 50°C or alternatives in PVDF will extend the operating range up to 65°C.

CHAPTER 9

Products and Applications — Miscellaneous Equipment

9.1 Tank Covers and Odour Control Equipment in Wastewater Treatment Plants

Odour control, that is either neutralisation or capture of odours produced within various sections of both municipal or industrial wastewater and effluent treatment plants, is becoming an ever increasing consideration in the design of new plants and the upgrading of existing ones. This is especially important in two major cases, firstly new high-volume plants capable of treating effluent equivalent to one million inhabitants or more, from which such volumes of unpleasant if not dangerous emanations would spread over large areas of the surrounding environment, affecting large numbers of local inhabitants. The second case may concern much smaller stations, but their location within built-up urban areas once again exposes a large number of inhabitants to some significant discomfort. Whilst various promotions of methods to reduce the extent of odour emission are under way, via the introduction of specific chemicals into the treatment process, and actually a group of water utilities in the UK are co-operating in a project involving a number of test methods to provide basic information on the development of odours from sewers and sewage treatment plants and how they could eventually be treated, the principle of confining odours within a limited space, and neutralising any localised emission from this zone is by far the most frequently used technology.

Typical areas within wastewater treatment plants which can release significant levels and intensities of odours to the atmosphere are clarification tanks and decanters, aeration tanks and channels, both trickling and high rate filters and sludge digesters and sludge storage tanks. Other smaller but significant zones which may require coverage are grit detritors, inlet channels and archimedian screw pumps, as well as access openings in general transfer channels in the larger plants.

Circular self-supporting composite tank covers up to 35 m in diameter have been in service for many years, and rectangular tank covers up to 100 m long or more are now commonplace. Covers are designed around two main concepts, a beam and in-fill arrangement for rectangular and small circular tanks, or a self-supporting assembly of modular sections based on a conical or domed form for larger circular tanks.

Figure 46. FRP tank cover and polypropylene odour control unit.

Whilst most storage and treatment tanks and basins are built from concrete or coated steel, composite materials have many significant advantages over both these materials and aluminum when covers are concerned.

- Weight — with a specific density of 1.8 to 1.9, combined with a judicial choice of moulded or pultruded profiles, it is often possible to assemble complete covers with little or no specialized lifting equipment or cranes. Large diameter circular covers can be assembled on the ground, close to the tank to be covered, and lifted on in one piece.
- Corrosion resistance — the level of corrosion resistance of either isophthalic or vinyl ester resins, to both internal and external environments, is assured practically indefinitely, and little or no maintenance should be expected over the normal 20 to 25 year life span of the treatment plant. The incorporation of pigments in the structural laminate as well as in the surface gelcoat eliminates even any requirement to repaint the covers, although some fading of colours will occur over an extended time period due to natural ultraviolet radiation.
- Access and removal of sections of the covers — due to the low weight of the material it is possible to remove one or more panels, usually by hand, to gain access to the tank or equipment in the tank, or in the case of a regular requirement, simple to operate access doors or panels are specifically built in to the construction.
- Architectural considerations — the facility with which a composite moulding process can be adapted to meet the requirements of even the most imaginative industrial architect is well known, and has permitted the construction of many spectacular projects at a reasonable cost, impossible to imagine in any other suitable material.

- Cost effectiveness and flexibility in design — the use of composite materials offers both the opportunity to use relatively inexpensive standard moulds thereby limiting investment costs in each new project, but also allows the designer to build in modifications, localised reinforcements and stiffening to suit the requirements of each specific project.

Ducts and the associated scrubbers or absorption towers to neutralise odours in air drawn off from tanks are in the main manufactured from polyethylene or polypropylene, both of which offer complete corrosion resistance, light weight and sufficient strength to deal with the very low pressures, negative and positive, regularly encountered in this type of installation. Absorption towers as large as 4 m in diameter and 12 m high have been built in polyethylene, as installed inside buildings they are protected from wind loadings, potentially a source of more stress than those due to the functioning of the absorber itself.

One of the most important companies involved in the odour control cover market is A-Form Projects Ltd in the UK, having more than 25 years' experience in the design, manufacture and installation of a wide range of composite covers in Europe and the Middle East. The company has introduced many novel features on covers supplied over the last few years, and is responsible for the design and manufacture of many of the largest composite covers ever constructed anywhere in the world. Two particular projects are used as case histories in Chapter 13, one involves a series of eight folding tank covers, each 6 m wide and 13 m long, associated with a stringent specification and the second example, probably the most spectacular tank cover installation ever, covering 8 off primary settling tanks, each 65 m long and 25 m wide, using a combination of rigid composite beams and flexible reinforced PVC membranes.

Design of tank covers is generally based upon or drawn up around local national standards for composite vessels, but due to the specialised nature of this equipment, it is possible to draw up a specific general specification prior to addressing the actual engineering problems.

The list below, developed from a design process used by A-Form Projects, suggests a number of points which may be considered, and which should form a specification for the design, pricing and construction of a cover:

- Define the shape of the tank, and the tolerances on the stated dimensions. This is of great significance for concrete tanks and older tanks in a refit project. A site visit and a rigourous dimension control prior to finalizing design drawings are to be recommended.
- Establish a description of any equipment which interfaces with the cover, for example, a bridge, pipework, electric cables, supports for equipment, hand railings and ladders.
- Define all requirements for access, for penetration into the tank, for inspection of tank contents, for tank cleaning or equipment within the

tank, or for servicing and adjustment of this equipment, together with a consideration of the frequency of these requirements. Other points of consideration could be hose entries for cleaning, ventilation of the space under the cover, type, size and location of walkways, non-slip requirements and hand railing.
- Relative translucency of the cover that can be specified to 80%.
- If the cover, or parts of the cover, are to be removed, how frequently and what means of handling are either required or already available.

Having completed the functional specification of the cover, the structural parameters can be examined and defined. Design loadings on this type of equipment can be split into sections:

- Wind loadings as defined by local legislation and/or current practice.
- Snow or other exceptional and similar loadings (such as sand in areas subject to significant wind-driven displacement of sand).
- Dead loadings due to the weight of any equipment supported by the tank, either permanently or occasionally, and man loads, which can be split into three types:
- Occasional access for maintenance and cleaning;
- frequent access for operational purposes; and
- areas defined as general walkways.

Occasional access for maintenance and cleaning is often calculated on the basis that the area or cover under consideration should be strong enough to support two people and their normal equipment (a toolbox for example).

Frequent access considers the same level of loading, but due to the frequency on intervention, requires that deflection of the cover is limited in order to install a high level of security, which is both real and sensed by the persons concerned. Deflection is generally limited to the smaller of 1/200 of span or 15 mm, which ever is the smallest.

Areas considered as general walkways are usually subject to the deflection of 1/200 of span, when the whole area would occasionally have people standing on it.

Unlike pipe systems or storage tanks which are generally subject to continuous loadings close to their design limit, tank covers tend to be subject to full design loadings on a very occasional basis, i.e. there is no snow or wind loading and no maintenance staff on the covers for perhaps more than 90% of the time, and practically no possibility of all loadings operating concurrently, and designs based on a minimum safety factor of three, as compared to ultimate properties, are common.

When designing a cover one must not forget the function of the tank and the consideration of the effect of this function on the cover or vice versa. Two developments from A-Form Projects demonstrate the advantages to be gained by considering the two aspects together.

First of all, the requirement to cover tanks with rotating full or half bridges permitted the company to design a concept by which the tank cover is attached to the bridge and rotates with it, supported on wheels running on the periphery of the tank. The relatively low inertia of the tank requires little or zero extra energy from the motor driving the tank bridge, and adds the extra advantage of protecting the total structure within the cover. A second development, actually a further development of the preceding concept, involves the design of a clarifier which integrates a sludge scraper and an odour control cover. The combination of two well-proven technologies has produced a design that is hydraulically more efficient, as well as being cheaper and easier to install than conventional equipment. The complete cover revolves, as mentioned in the previous example, driven by a simple static motor. The scrapers, built entirely in composite materials, are attached to the roof and are arranged in a scroll configuration. This gives an extremely efficient scraping action, allows a much slower rotation, and consequently less turbulence. An initial project involving 8 off 20 m diameter covers was under construction at the time this book went to press. Should general access on covers not be required, or should the installation be of a less permanent nature, then various forms of thermoplastic covers are available, based upon a reinforced thermoplastic membrane tensioned over the surface of the tank. This concept can be taken one step further, removing the tensioning device and adding some form of flotation equipment, which allows the cover to rise and fall with the level of fluid in the tank. By maintaining the cover in close contact with the surface of the fluid, little space is created allowing the emission and concentration of odours.

A recent and typical example of this latest method, using material supplied by GSE Lining Technology was installed on a $5000\ m^2$ buffer pond in a wastewater treatment plant, the odours from which were upsetting local residents. This application uses high-density polyethylene floating covers that provided a seal tight cover that moves, via a number of suitably positioned floats, with the level of the effluent in the pond. All gases emanating from the effluent are extracted and cleansed by passing through a series of biofilters prior to release to the atmosphere.

9.2 Well and Borehole Equipment

Both thermoplastic and composite materials are used as casing and/or tubing for shallow depth water pumping, whereas high performance epoxy resin-based composite materials are generally retained for deep wells such as low-temperature geothermal water production, water flood enhanced oil recovery schemes in semi-depleted oilfields and general water and effluent injection or disposal.

Well casing is defined as a tubular product inserted into a borehole to retain the borehole walls and provide guidance for any required tubing or immersed pump. Casing sections at the bottom of the well are generally slotted or drilled to allow infiltration of the water from the aquifer, and are generally referred to as screens. Casings are generally designed upon the basis of their resistance to external pressure and may be grouted or cemented into the borehole.

Tubing, again a tubular product is designed to be inserted within a casing, and depending upon the pressure available within the well, may be required to support an immersed pump or a packer (device to maintain the tubing in place within the casing at the bottom of the well). For large-volume, low-pressure pumping, a casing may be used as a tubing, within an even larger casing.

9.2.1 Thermoplastic well casings

Economic considerations generally limit the depth of boreholes for the production of potable water, and most wells can be equipped with relatively inexpensive PVC casing. This type of material is generally supplied in lengths of approximately 3 and 6 metres, commensurate with the dimensional requirements of transport by container and possibilities to handle and install on site.

The main applicable standard for PVC casings are:

- **ASTM F 480-99** Standard specifications for thermoplastic well casing pipe and couplings made in standard dimension ratios (SDR), schedule 40 and schedule 80.
- **DIN 4925-1, DIN 4925-2, DIN 4925-3,** which cover respectively diameter 35 through 100 mm, using a whitworth thread, diameter 100 through 200 mm and 250 through 400 mm, both of which use trapezoidal thread connections.

Composite casings, a relatively small market, tend to be designed, constructed and tested around ASTM standards for composite pipe.

Using a leading US producer of thermoplastic casing and associated material, Certain Teed, as an example, we can examine the range of material typically available.

Certa-Lok PVC Well Casing is designed for a large range of applications, including domestic, irrigation, municipal, industrial, geothermal, reverse osmosis supply wells and monitoring wells.

All Certa-Lok products use Certain Teed's coupling and spine locking device to form an instantaneous full strength joint, with no solvent adhesion or threading. The casing is supplied in lengths of 20 feet (6096 mm), in a variety of wall thicknesses, for installation in wells with depths of approximately 350 m.

The advantages of PVC casing over traditional metallic materials are numerous, and include:

- Light weight. The weight is approximately 20% of a comparable metallic casing, leading to easier and cheaper transport and on-site handling.
- Corrosion resistance. The PVC material is resistant to most chemicals, soft, brackish and saline water, and is immune to electrolytic and galvanic corrosion.
- Non-toxic. The PVC casings are NSF approved, will not impart any taste to the water, nor will it support bacterial growth.
- Cost efficient. The joint is instantaneous, and achieves full strength immediately upon assembly, is solvent free, and two twenty-foot sections can generally be jointed in around two minutes. The corrosion resistance and mechanical strength results in an extremely long service life, and should the casing require removal it can be quickly disassembled and removed from the borehole without having to cut out joints. The undamaged material can also be re-installed quickly and efficiently.

Certa-Lok PVC well casing is available in eight different diameters (6.9 to 17.4 inch (175 to 442 mm)) in various SDR to suit different well depths.

Certa-Lok Integral Bell Well Casing, that is casing with an integral bell rather than a separate coupling, offers all the advantages of the material described above, and is available in 4, 4.5, 5 and 6 inch diameter, in several different SDR. Due to the integral joint, assembly and installation times are further reduced.

To complete the range of casings, Certain Teed offer an extensive line of slotted casings and accessories.

It should be noted that contrary to practice with metallic casing, thermoplastic casing must never be driven but should only be installed in an oversized borehole.

Other points of interest when considering the installation of thermoplastic casings are:

- The internal diameter of the casing, taking into consideration the required flow rate, and the subsequent dimension of the pump and tubing.
- The ground structure and the overall depth of the well.
- The resultant installation, backfilling and/or grouting techniques.
- The temperature and nature of the fluid.

Certa-Lok PVC Drop Pipe. A PVC tubing or drop pipe for attachment to a submersible pipe, generally suspended within a well casing, using the same joint configurations as described above for casings. With this type of joint the submersible pump can be withdrawn for servicing and re-installed with minimum cost and down time. A Certa-Lok PVC drop pipe weighs 107 pounds per 20 ft length compared to 400 pounds for 6 inch Sch 40 steel (47 kg for 6096 mm/ 178 kg for steel).

Certa-Lok PVC drop pipe is available in sizes 1 to 8 inch, for installations up to 170 m deep, is designed and manufactured to meet or exceed ASTM specification D1785 requirements (Sch 80), and can be installed and used under the conditions given in Table 6.

The product line includes a range of fittings, including couplings, adaptor couplings — female by solvent weld socket, adapter couplings — Certa-Lok male by male thread NPT and Certa-Lok stainless steel male by male thread NPT.

An alternative approach has been made by another supplier of thermoplastic casings, this time situated in the Middle East. The National Plastic and Building Materials Industries LLC, based in Sharjah in the UAE,

Table 6. Quick selection guide for Certa-Lok PVC drop pipe for submersible pipe. *Courtesy of CertainTeed Corp.*

Pipe size (inch)	Pump weight (kg)	Flow (l/min)	Maximum discharge pressure at well head (bar)								
			0	1.7	3.4	5.2	6.9	8.6	10.3	12.1	13.8
			Maximum setting depth (m)								
1.5	38	95	137	125	112	100	87	75	63	50	38
2	90	170	129	116	103	90	77	64	52	39	26
3	135	378	170	155	142	127	113	100	85	71	57
4	180	662	150	135	120	106	91	77	63	48	34
5	270	1040	143	118	114	100	85	71	56	42	27
6	360	1514	150	136	121	106	92	77	62	48	33
8	450	2650	145	131	113	100	86	71	56	40	22

This information is taken from a quick guide published by CertainTeed, with most units transcribed into metric dimensions and volumes. CertainTeed point out these data were intended only for the purposes of an initial evaluation, based upon typical parameters of weight and flow. For a given surface discharge pressure, setting depth must be limited to the maximum values indicated. For definitive calculations, or to engineer any installations which do not fit the table data, CertainTeed require the use of the latest edition of its Design Worksheet (literature code #40-37-42).

produces a range of PVC casings and screens in accordance with the previously mentioned metric standard DIN 4925-1 / 4925-2 and 4925-3. There are two basic ranges, one for shallow and medium depth wells, and a second with a thicker heavy duty wall thickness for wells of greater depth. The casings are supplied with a male pipe thread on the spigot end, and a female thread at the socket end.

Screen sections can be supplied with either a plane or ribbed surface. Ribbed surfaces increase water permeability and help prevent clogging of intake openings. Slots are arranged horizontally to improve the mechanical strength of the screens and are designed to give open areas ranging from 6 to 12%. Slot widths vary between 0.2 and 3.0 mm.

Johnson Screens, part of the USF Filtration and separations organisation has introduced two more interesting applications for a thermoplastic alternative, under the designations PVC Vee-Wire and Schumasoil.

Technologies developed for actually making slots in thermoplastic casing are generally based upon perforation or cutting. Johnson Screens have adopted a wire wrap / welding rod technology, whereby continuous PVC "wire" is wrapped around and welded to PVC rods.

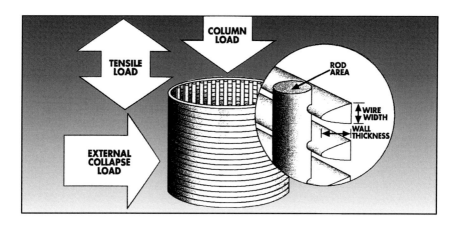

Figure 47. PVC Vee-Wire screen. Courtesy of USF Filtration and Separation.

The vee shaped wire, which narrows inwardly, tends to prevent any clogging by sand. When combined with the possibility offered by the manufacturer to vary the space between the wires, screens can be designed and manufactured to offer the maximum open area and dimensions suitable for any size and type of formation material.

Hydraulic tests comparing vee-wire screen with slotted screen, using an identical slot width, have shown that draw down with vee wire is 20% less than with the slotted alternative. PVC offers also the advantage that it may be treated repeatedly with acid to remove encrustation. Screens to 6-inch diameter are made with a rod base (fig 47), whereas 8-inch diameter screens have a channel-rod base for enhanced strength.

The second material, Schumasoil, is a porous polyethylene well screen, originally developed by the Schumacher Company in co-operation with the University of Karlsruhe, well known for its involvement in many aspects of handling and treating water. Schumasoil screens are made by sintering high-density polyethylene beads without glues or solvents, and are designed for many difficult operations where conventional screens are not suitable.

Figure 48. Schumasoil porous polyethylene well screens. Courtesy of USF Filtration and Separation.

As the entire screen is made of a porous material, it can be installed without the requirement of gravel packing. This makes it suitable for use in horizontal wells or in vertical wells completed in heaving sands, two conditions where gravel packing would be costly or impossible. The level of porosity is determined by the sizes of the particles used in the sintering process, and can range from 29 to 45% making the screens suitable for use in even fine grained or silt materials. The screens are currently available in six standard diameters (3 inch through 12 inch) and with five levels of average pore size (40 through 500 microns) and are suitable for continuous service a temperatures up to $80°$.

Because of their original properties, Schumasoil screens have been used in many different and sometimes difficult applications, including;

- Soil vapour extraction, where the screens open area allows high efficiency laminar flow of vapours.
- Air spacing where high volumes of air distributed evenly along the entire screen increases cost-effectiveness of remediation programs.
- Bioremediation, where the non micro-porous surface resists biological fouling.
- Groundwater extraction, when downhole conditions prohibit conventional completion of ground water wells, or monitoring, remediation and infiltration wells.

9.2.2 Composite tubing and well casing/screens

Composite tubing and casing tend to fall into two categories, polyester-based systems for shallow to medium depth water wells, and epoxy-based systems for greater depths, higher temperatures, or for oilfield-related applications.

The first type of product is offered by companies such as Sarplast Initiative Industriali in Italy or the Burgess Well Company located in Nebraska, USA. The second type of product is well represented by Fiberglass Systems, based in San Antonio, Texas, USA.

Sarplast proposes a standard range of casing from 100 to 500 diameter, in three different wall thicknesses corresponding to three different maximum installation depths of 100, 200 and 300 m. They also indicate that designs to greater depths are available, as well as an extension of the range of diameters up to as high as 3000 mm.

Screens are available in two different formats, that is monoskin screens, with cut slots, suitable for medium depth applications. Open areas of 12% and above are standard, which would indicate fewer problems with clogging compared to lower values with PVC casing. A second higher performance configuration, designated pipe base screens, with 'V' shaped slots is also available. This configuration, see Figure 49, has, it is claimed by Sarplast, several advantages:

- They provide the same performances of the continuous slots screen in stainless steel.
- The composite material gives the screen higher tensile strength and collapse resistance, and as such is more suitable for deeper wells.
- The 'V' shaped slots eliminate clogging.
- The relatively low speed of inlet flow and the high transmitting capacity, as well as the limited pressure drop across the screen allow an economical water supply because of the reduction in consumption of energy by the pump.

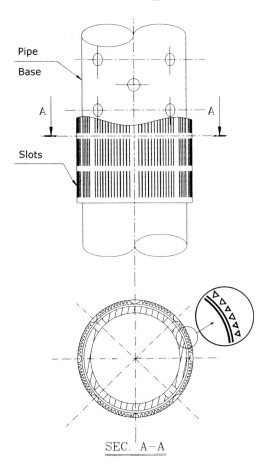

Figure 49. Screen with 'V' shaped slots. Courtesy of Iniziative Industriali SpA.

- The vertical slots reduce the friction between the gravel pack and the screen. The 'crushing' of the gravel pack, and the subsequent consequences on screen clogging, are much lower.

Sarplast use integral bell and spigot joints on all casings. The hydraulic sealing is obtained by single or double O-rings, and the axial resistance is taken by a mechanical locking device that is inserted into prepared groves in both the spigot and socket via an opening in the socket. Associated well head piping using the same material and jointing configurations can also be supplied.

The Burgess Well Company proposes a range of casings and associated column pipe (tubing), based upon a polyester resin system, with an alternative using a vinyl ester. All casings and screens are assembled with proprietary 'V' threads and Teflon tape associated with a Teflon-based pipe dope.

Casing is available in 2 to 16 inch diameter in three different wall thicknesses and ratings, whereas tubing is proposed in 4 to 12 inch diameter.

Higher performance composite tubing and casing is available from Fiberglass Systems under the trade name Star who supply anhydride-cured epoxy resin based tubing in diameters of $1\frac{1}{2}-4\frac{1}{2}$ inch, with pressure ratings up to 4000 psi, and casing in diameters $4\frac{1}{2}$, $5\frac{1}{2}$, 7 and $8\frac{5}{8}$ inch, in 1500 and 2000 psi ratings, all for service temperatures up to 93°C. Although mainly developed for oilfield applications, Star Fiberglass tubing and casing have been used successfully in low temperature geothermal wells, to depths as low as 1750 m. Smith Fiberglass proposes similar materials, including anhydride-cured epoxy casings in $4\frac{1}{2}$ to $13\frac{3}{8}$ inch and Ameron, who offers a similar range of materials via its acquisition of Centron, complete the trio of the most significant producers in this market.

Depending upon the diameter and pressure rating, each pipe joint may be supplied with an integral threaded female socket at one end, and a matched male threaded spigot, or pin, at the other, or as an alternative with a female threaded coupling bonded to one end of the pipe joint instead of the integral coupling. Assembly between the different lengths is achieved via API 8RD threaded joints which use a wide range of proprietary sealing materials.

9.3 Coatings and Linings

9.3.1 Initial protection of new equipment

9.3.1.1 Pipe systems

The family of thermoplastics and thermosetting composite materials have proven over the last 10 years their technical and commercial dominance in the field of in-situ pipe renovation. Many thermoplastic materials are also used to line steel pipe, under the designation of plastic lined pipe, essentially in the field of aggressive chemicals, although some applications are possible when handling chemicals for water treatment or in some ultra-pure water applications.

Typical liner materials are polypropylene, polyvinylidene chloride (PVDC), PVDF and PTFE.

9.3.1.2 Tanks, vessels and other constructions

Linings of storage tanks, and even road tankers with composite or thermoplastic materials is a relatively common practice within the chemical and petrochemical markets, and a great deal of formulations and application technologies exist worldwide for the water treatment and handling activities discussed in this book. The range and scope of these materials is such that an entire handbook could be dedicated to this activity. A brief introduction is, however, attempted in Section 9.3.2.2.

Most FRP jobbing shops can offer laminated liners inside storage or process vessels, and many of the manufacturers of dual laminate tanks and vessels can offer either bonded or loose thermoplastic liners within a steel envelope.

9.3.2 In-situ lining of existing equipment

9.3.2.1 Pipe systems

By far the largest actual market, and without doubt the largest market for many years to come for pipeline renovation by lining, is the potable and wastewater sector. Due to the deterioration of water and sewer lines installed essentially in densely built up urban areas over the last 100 years or so, many companies have developed highly effective renovation and rehabilitation techniques, using various and wide ranging no-dig technologies. The essence of these technologies is the ability to resolve problems with buried pipe systems such as corrosion, deposits, leakage and damage whilst causing minimum disturbance to surrounding surface traffic, commerce and residents.

Several manufacturers of both thermoplastic and composite pipes have adapted their standard products to the requirements of this activity, whilst other companies have devoted their total business to the in-situ repair of pipe systems.

Using the wide range of technologies presented by one of the world's leading specialists in trenchless pipeline rehabilitation and renovation, the UK company Subterra, it is possible to explore most of the available solutions for relining existing pipe systems, all of which avail themselves of thermoplastic or thermosetting resins.

Before considering the actual lining technologies themselves, it is important to carry out an in-situ examination of the pipe to be lined via a CCTV survey, or as with Subterra technology, an enhanced level of information using ground probing radar (GPR) or in-sewer ground probing radar (ISGPR).

GPR and ISGPR are two effective surveying methods that offer a no-dig opportunity to accurately identify areas of pipe damage, leaks and other anomalies, such as wet or dry voids around the pipeline, soil types and other solid masses in the surrounding ground.

GPR is used in the location of services and pipes, and to survey a wide area from the surface, and can accurately survey and assess the substrata, both prior to excavation and following rehabilitation. It can also be used to obtain a rapid profile, as an aid in leak detection by mapping out ground moisture in the vicinity of suspected leaks.

ISGPR, a newly developed technology based upon experience drawn from previous use of the GPR system, is a system that can 'see' both into the wall of the pipe and the surrounding soil to give a picture in real time of the actual conditions underground.

Once complete, both technologies allow the presentation of the survey as a simple coded layout with an optional full written report complete with scaled CAD drawings portraying the interpreted information.

Proceeding now with a review of actual lining and coating technologies, and based upon the designation adopted by Subterra for each technique, it is possible to identify 10 or more significant procedures for the in-situ renovation and repair of buried pipe systems.

Epoxy spray lining — ELC 257/91 is a two-part solvent free second generation epoxy resin lining material, offering advantages over the first generation system from Subterra (ELC 173/90), particularly with respect to slump resistance and durability. The application of a thin epoxy resin internal coating by in-situ spraying can improve flow, ease pressure problems and address a variety of water quality issues. Lengths of up to 200 m and thicknesses of 1 mm can be sprayed in one pass. Computerised monitoring of both pumping and winching rates ensures a high degree of control, ensuring optimum performance, and provides a corrosion-resistant, cost-effective lining for water mains. The material can be applied using standard spraying equipment incorporating heated materials storage tanks and umbilical hoses.

The ELG 257/91 epoxy resin formulation has been approved for use in contact with potable water in the UK, USA and Canada, Belgium, France, Hong Kong, Norway and Singapore.

Epoxy resin linings can be applied to a wide range of materials, including cast and ductile iron, steel, asbestos cement, non-reinforced, reinforced and pre-stressed concrete and composite materials. Typical areas of application are potable water mains, raw water pipes, fire mains and industrial process water systems.

Epoxy resin linings in general are solvent free formulations, have excellent adhesive properties, even on moist surfaces, and offer minimal shrinkage. ELC 257/91 is touch dry in 4 hours and the coated line can be returned to service after curing. In the UK, the cure period for all in-situ epoxy resin linings for water supply applications is stipulated by the Drinking Water Inspectorate as a minimum of 16 hours from completion of the spray-up.

ELC 257/91 epoxy resin provides a hard, durable corrosion resistant barrier layer over the material to be protected, and the application does not block off customer service connections.

Sprayed epoxy resin linings are also available from a wide range of suppliers some of which are listed in Chapter 14.

Another company highly active in the non-structural epoxy lining market has also developed a fast-setting, spray-on lining system based upon a two-part urethane formulation. Pipeway Ltd, together with E Wood Ltd have developed the E Wood/Pipeway Copon Hycote 169 lining system, which has recently been approved by the DWI.

Copon Hycote 169 is a two-part urethane system, light grey in colour, with a similar but glossier appearance to epoxy coatings when cured. After mixing and application to the pipe walls, initial set is achieved in approximately 40 seconds, with the coating hard and unmarkable after approximately two minutes. With such physical properties, DWI approval allows CCTV inspection to be undertaken 30 minutes after completion of the lining. A total cure period of two hours is required from completion of the lining to commencement of return to chlorination procedures. These comprise, as with epoxy lining, a 30 minute chlorination contact time followed by one hour of flushing.

This same company is also developing a product designated Pipesaver Pipeline Renovation, a new method for the structural renovation of existing pipelines. The process involves the enhancement of cement mortar lining by the use of an overlaying polyethylene membrane to provide leakage prevention and improvement of water quality. The Pipesaver process is at the present time being actively developed in conjunction with the WRc. DWI approval for the material under Reg 25(1)(a) for use in potable water mains has already been granted.

Slip lining — or insertion of a new undersize pipe, usually in polyethylene, within an existing pipe is perhaps the simplest form of pipeline rehabilitation. The technique results in a reduced bore diameter as substantial space is taken up by the new pipe. Depending upon the clearance of the new pipe within the existing one, some cleaning or removal of obstructions may be required. In many cases, it has been found that a maximum annular clearance of as little as 5% of the mains diameter, even less for diameters above 300 mm, has been more than satisfactory.

The basic lining operation entails the insertion of a winch cable through the existing line, which is then attached to the front end of the liner. The liner is then pulled into the bore of the existing pipe, and is connected at both ends to the existing system. Durapipe-S&LP publishes maximum pulling loads for both MDPE and its grade of HPPE, Excel, for diameters 20 to 1000 mm, in SDR 11 to 26. Depending upon the diameter, these vary between 0.10 and 163 tons, giving some idea of possibilities with this technology. In normal conditions, welded strings of up to 1000 m of HPPE can be pulled without imposing exceptional axial strain, although it is recommended to maintain a period of 24 hours for pipe relaxation prior to grouting or tying in.

The void between the new and existing pipe may be filled with grouting, depending on the particular application. The Water Research Centre publishes the *WRc Sewage Rehabilitation Manual*, which contains all relevant information, including information on allowable external pressures at grout curing temperatures.

This method allows for connection to service connections along the way, and offers an economic and quick solution for long lengths, in diameters up to 1600 mm, including the possibility to line large-radius bends. This particular solution can repair both leakage and structural decay by a judicious choice of diameter and SDR.

Rolldown lining — or the insertion of a close fit liner is a step forward from the previously described slip lining. Rolldown is presently available for pipelines in the diameter range of 100 to 500 mm. The Rolldown system uses standard polyethylene pipe in a way which creates a close fit liner within an existing pipe. The diameter of the pipe is reduced concentrically by pushing it through sets of rollers onsite. The processed liner then retains its reduced size indefinitely until it is reverted back towards its original size by pressurization with cold water. The reduced diameter liner pipe is then pulled through the host pipe, and once in place is pressurised to revert the diameter to form a close fit within the existing pipe. The Rolldown technology uses standard pressure piping (up to 16 bar) in the SDR range 11 to 33, depending on the operating pressure and structural condition of the host pipe. Rolldown linings can operate either as a stand-alone pressure pipes or as a thin-wall liner to eliminate leakage in an otherwise structurally sound pipe. In both cases the smooth bore of the polyethylene will maximise flow capacity, and in the case of thin-wall lining, may even increase the flow capacity of the original pipe.

Lengths of up to 1000 m can be installed at one time, including passage through bends up to 11.25°, assisted by the fact that the reduced diameter are held until final pressurisation of the line.

The process requires no heating, low winching forces and can be carried out on a stop/start basis should, for example, any unforeseen problems be encountered. Liner end termination fittings, service tappings and branch connectors are commercially available to complete the system.

The total process is demonstrated by Figures 50–52.

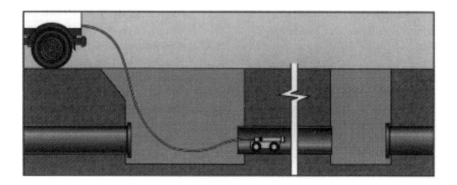

Figure 50. Inspection with CCTV and cleaning. Courtesy of Subterra.

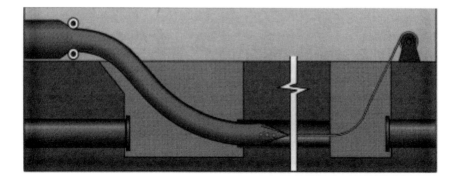

Figure 51. Reduction in diameter and insertion of the liner. Courtesy of Subterra.

236 Coatings and Linings

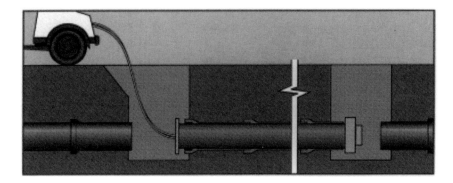

Figure 52. Reversion and sealing. Courtesy of Subterra.

Subline system — this process can be used to install close fit, thin-wall PE liners in host pipes in the diameter range 75 to 1600 mm. The Subline system is similar to the above-mentioned Rolldown system, in that it relies on deforming a standard polyethylene pipe on site to the extent that it can be pulled through an existing pipe, and then regains its original form via pressurisation with cold water. The Subline system is specifically designed for thin-wall lining from 75 to 1600 mm. The polyethylene liner is pushed

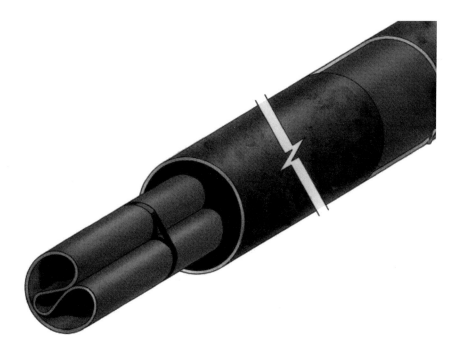

Figure 53. Folded liner inserted into the host pipe. Courtesy of Subterra.

through formers which fold it into a 'C' shape, and which is maintained by temporary straps. This deformation which reduces the cross sectional area of the liner by up to 40% of its orginal diameter allows sufficient clearance for insertion of lengths up to 1000 m or more in the existing pipe, including negotiation of bends up to 45°.

Once in place the folded liner pipe is pressurised with cold water, which snaps the temporary straps and allows the liner to revert to a close fit inside the existing pipe. The main advantages compared to the Rolldown process are the availability of larger sizes and an improved passage through existing pipeline bends.

Installation procedures are generally in line with those described for the Rolldown product.

Similar systems are available from Wavin, for diameters of 100 to 400 mm, under the designation 'Compact Pipe' who also supply associated electrofusion couplers, as well as NuPipe from Insituform, which uses PVC pipe as a liner, for applications up to 300 mm in diameter. Yet another version, from Uponor Anger, under the designation Omega-Liner, uses a modified U-PVC (PVC Alloy) which is deformed and wound onto a drum, and by reducing its size by 30 to 40% facilitates the insertion into the host pipe. Steam and pressure are applied to induce these liners to regain the original shape and size, and thereby moulding them to the profile of the host pipe.

Pipe bursting — and the associated lining is the only in-situ process which allows the replacement of an existing pipe with a pipe of equal or even larger diameter. The existing pipe is split by a tool, which as it progresses through the pipe expands to crack it, and at the same time creates an expanded borehole. The same tool also pulls into place a new structural polyethylene pipe to replace the old one.

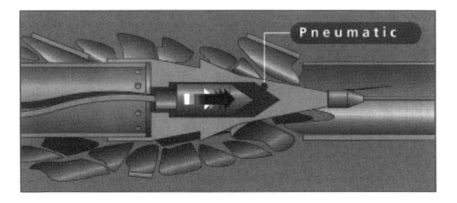

Figure 54. Pipe bursting — pneumatic. Courtesy of Subterra.

238 Coatings and Linings

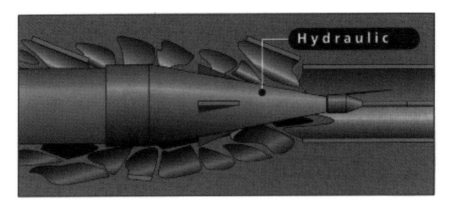

Figure 55. Pipe bursting — hydraulic. Courtesy of Subterra.

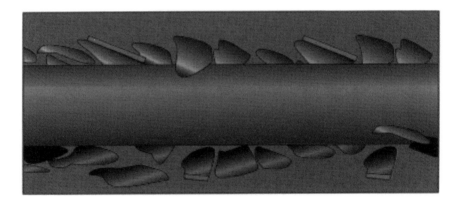

Figure 56. Pipe bursting — insertion of the new liner. Courtesy of Subterra.

The most modern tools can crack and open up most unserviceable pipelines, even if local reinforcement and repairs have already been carried out, and risk of damage to adjacent pipework is minimised with the latest generation of hydraulic tools.

In this field, Durapipe-S&LP has developed a new product called Safeguard, which is a normal polyethylene pipe with an abrasion-resistant external coating. The usual practice in the past has been to draw a sacrificial polyethylene sleeve immediately behind the cracking tool, and then a new polyethylene main could be inserted into the sleeve, avoiding any risk that the new main had received any significant damage due to its passage through the debris of the cracked mains. The elimination of the sacrificial sleeve reduces both the cost and the duration of the total operation.

It is perhaps interesting to note that similar technologies are used on no-dig installations for installing new pipelines, when using impact moling to pass, for example, a pipe under a road or rail track without opening a trench, or when using directional drilling to pass under larger structures or rivers. In both cases, a polyethylene pipe or sleeve can be drawn through the opening, on the back of the mole or drill.

Remote Line — a composite repair system for sewer lines. All the previous relining systems based on thermoplastic materials could be used in systems which required access to existing service connections and pipe line branches. The Remote Line and similar systems rely on thermosetting composite materials and can be used in systems with service connections and side branches. Whereas thermoplastic systems can be used with both potable water and sewage, subject to specific design assessments, the composite systems at the present time tend to be used predominantly for applications of a non-potable nature.

The Remote Line process offers a structural, full-length, manhole-to-manhole lining system, accessed from existing manholes, in diameters of 150 to 600 mm, for lengths of 100 m or more depending upon diameter and existing pipeline configurations. Once installation is complete, any lateral connections are re-opened with a remote controlled cutting machine.

The glass fibre reinforced liner is prepared to specific dimensions in the factory and is winched into the line to be repaired. Once in place, the section is sealed off and the liner is expanded to fill the pipe with compressed air. Steam is injected to cure the resin system, and when curing is complete the ends of the liner are trimmed, and the annulus ends sealed with an epoxy mortar. Any lateral joints are re-opened by using a tractor-mounted remote controlled cutter.

The close fit and reduced wall thickness (from 3 mm upwards) minimises bore reduction whilst adding the required structural capability, and the reduced installation/cure cycle times often mean that the repaired line can be restored to service within the same day.

A similar process is used in the 'Remoteline patch' system to repair small lengths of damaged pipe, using a system of packers to block off the two ends of the 'patched' area during cure. The cured liner is cut and trimmed with the same remote controlled cutting devices used to re-open laterals on larger systems.

Inpipe — is an advanced, close-fit, cured-in-place, manhole-to-manhole sewer lining system, similar to the above-mentioned Remote Line system. It differs from the above in that the liner is inserted in an inversion operation, using compressed air to inflate and insert the liner. Once inverted the liner is cured with ultraviolet radiation rather than steam. The UV curing system is not sensitive to cold spots in the host pipe, and as such a higher degree of uniform polymerisation can be guaranteed, and inner and outer foils compatible with the UV system minimise styrene emissions and also prevent migration of resin up laterals. Cutting and finishing of the liner is carried out as for the previous systems.

In-situ composite linings are also available from Insituform Technologies using similar but somewhat different technologies. The Insituform pressure pipe liner is manufactured from a felt tube, coated with a permanently bonded polyethylene layer, and loaded with resin by a vacuum impregnation process. On site, this pipe is inverted into the deteriorated pipe, and when total inversion is complete, the resin is cured by circulating hot water inside the pipe. Once cure is complete, service laterals are restored using robot-controlled cutting devices. The associated Insituform Point Repair process is used to repair localised problems in an otherwise sound pipe.

The Insituform process has been used in pipes with a wide range of shapes and diameters up to 2750 mm (9 ft), and is carried out in conformity with ASTM F 1216-98 — Standard Practice for the Rehabilitation of Existing Pipelines and Conduits by the Inversion and Curing of a Resin-Impregnated Tube.

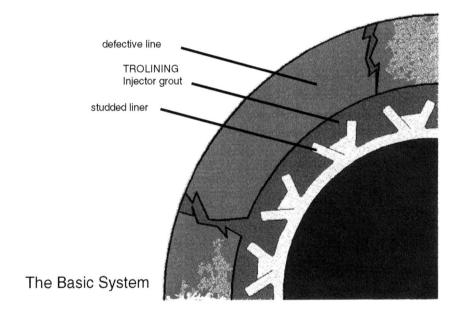

Figure 57. Troliner — basic system. Courtesy of TROLINING GmbH.

TROLINING systems — a somewhat different approach to lining deteriorated sewer systems has been taken by the German company TROLINING GmbH, based on polyethylene, but incorporating polyethylene studs fixed to the back of the liner, and specially developed mortars and annular grouting technology.

The basic system uses a single inner liner with anchors on the outside of the HdPe sheet. The inner liner is pulled into the sewer and the TROLINING-Injector grout is poured into the annular space created by the anchors. The grouting will fill adjacent voids in the existing sewer pipe walls, thereby consolidating and repairing any damaged portions of existing sewer pipe.

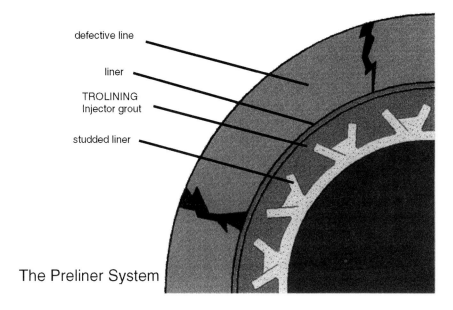

Figure 58. Troliner — pre-liner system. Courtesy of TROLINING GmbH.

The TROLINING-Preliner system uses an additional HdPe sheet to create a defined annular space for grouting. The preliner system is essential in areas with a high water table.

Even more sophisticated systems are available such as the Monitoring System which will detect any subsequent infiltration and the Double System wherein two or more layers of lining/grouting add increased strength to the liner in case of high external loading. A further possibility exists to integrate cable ducting behind the liner, inside the grouting, up to a diameter of 20 mm without reducing the line cross-section.

A new TROLINING Preliner System for large cross-sections allows rehabilitation of lines up to 3500 mm (140 inches) diameter, or oval and teardrop sections of similar dimensions. The sections of liner are pulled into the sewer line and joined by extrusion welding (in accordance with DVS 2207, part 4). The studs and spacers create the annular void for the grouting between the inner liner and the pre-liner, and steel bars or mesh can be included to increase the load-bearing capacity of the overall structure.

The liner system is supported by a wood or steel formwork before a high-strength grouting is injected into the annular void. This grouting provides the load bearing capability of the composite system, and the sandwich construction between the liner and inner liner protects the grout from internal and external aggressive fluids. The manual nature of this system lends itself to construction of difficult areas such as elbows and transitional sections.

The Finish-based organisation KWH Pipe has developed over almost 20 years its own specific short module relining systems, under the name VipLiner. The VipLiner renovation method is a slip-lining system whereby a string of short modules are placed into an existing carrier pipe. The KWH method does not require excavation and will therefore minimise traffic and street disruption. Moreover, in most cases, the sewage system could be kept functioning throughout the renovation works, and no de-watering is generally required. Laterals are connected from inside the system, using a patented process and special equipment, that once again saves on time- and money-consuming excavation costs.

VipLiner units are made from HdPe 100, have an effective length of 500 mm and are assembled via a patented clipped system with an elastomeric seal. They have smooth internal and external surfaces, a ring stiffness class of SN 16 to ISO 9969 and a leak tightness to EN 1277. They are available in diameters 110, 125, 160, 180, 200, 225, 250, 280, 315, 355, 400, 450 and 500 mm.

Installation is effected via existing inspection or access chambers, and all installation equipment for pushing in the liners, injecting grouting (low-density foam concrete) and sealing lateral connections are available on a hire or purchase basis. One of the largest sewer rehabilitation projects in the world was completed in Kuwait, using mostly VipLiners, and is described as a case history in Chapter 11.

Rehabilitation of lead pipe systems — a new requirement imposed by many governmental bodies as well as the WHO has set new and very low levels concerning the maximum amount of lead in potable water.

Most lead pipe still in use is relatively small in diameter, relatively complex, and is typically used inside buildings or for pipework linking buildings to the mains. Several new products are appearing on the market to resolve this problem, which could involve millions of kilometres; all of them are based on thermoplastics.

Wavin has introduced its NEOFIT product, a pipe made from PET (polyethylene terephtalate), smaller in diameter than the lead pipe into which it is inserted. The liner pipe is pushed and/or pulled into the pipe to be lined, and once inserted, the liner is inflated using a Wavin custom-built, semi-automatic expansion unit. The result is a tightly fitting liner, installed in approximately one hour from start to finish for an average-size installation.

Renovation and relining of existing underground pipe systems will be one of the major activities in the water industry over the coming years as existing and new technologies provide economic and technically sound solutions for a growing number of products. In line with the development of this market, specific standards and codes of practice are being developed. We list some of these documents below.

ASTM F 1216. Standard practice for rehabilitation of existing pipelines and conduits by the inversion and curing of a resin impregnated tube.

ASTM F 1504. Standard specification for folded polyvinyl chloride (PVC) pipe for existing sewer and conduit rehabilitation.

ASTM F 1743. Standard practice for rehabilitation of existing pipelines and conduits by pulled-in-place installation of cured-in-place thermosetting resin pipe.

ASTM F 1867. Standard practice for installation of folded/formed polyvinyl chloride (PVC) pipe type A for existing sewer and conduit rehabilitation.

ASTM F 1871. Standard specification for folded/formed polyvinyl chloride (PVC) pipe type A for existing sewer and conduit rehabilitation.

ASTM F 1947. Standard practice for installation of folded polyvinyl chloride (PVC) pipe into existing sewers.

JIS K 6940. Glass flakes vinyl ester resin lining films.

We also include a listing of WIS standards, that is Water Industry Specifications issued in the UK by the Water Research Centre, and available via their publications department in Swindon (see bibliography).

WIS 4-34-02. Glass fibre reinforced plastics (GRP) sewer linings.

WIS 4-34-04. Renovation of gravity sewers by lining with cured-in-place pipes.

WIS 4-34-05. Polyester resin concrete (PRC) sewer linings.

It is also interesting to note that NACE International (National Association of Corrosion Engineers) has, within its standards on Material Requirements, Recommended Practices and Test Methods, developed some standards which can be applied to pipe lining and coating, as well as other coating and lining technologies discussed below, such as:

RP0288-94 — Inspection of linings on steel and concrete.

RP0394-94 — Application, performance and quality control of plant applied fusion bonded epoxy external coating.

RP0490-95 — Holiday detection of fusion bonded epoxy external pipe coating of 250 to 760 µm (10 to 30 thousandths of an inch).

TM0183-93 — Evaluation of internal plastic coatings for corrosion control of tubular goods in an aqueous flowing environment.

TM0185-93 — Evaluation of internal plastic coatings for corrosion control of tubular goods by autoclave testing.

9.3.2.2 Protection of other constructions for fluid containment

Protective coatings are so diverse and numerous as to justify a handbook of their own, and this particular work can only serve as a very basic introduction. Under the heading of protective coating of constructions for fluid containment we can include applications on walls and floors, tanker loading areas, concrete or steel structures for a wide range of applications, bunds and sumps, large tanks and filters, for example.

Lining materials can vary from spray up epoxy or polyurethane, glass flake filled polyester, vinyl ester or epoxy resin formulations, structural laminates using the total range of thermosetting resins, to rigid or flexible thermoplastic sheeting or lining. The choice of materials is a function of the type and condition of the supporting structure, the environmental conditions in which the coating can be applied, and the nature of the fluid in contact with the coating.

The success or not of a protective lining is most often decided by the state of the supporting structure immediately before the application of the coating. The structures should generally be dry and clean, free from all contamination, and as far as possible all friable or corroded material should have been removed. Steam or other high-pressure cleaning often followed by shot or sand blasting is generally recommended.

Using a small selection of companies involved in the coating and protection business we can look more closely at some of the technology and materials used.

Spray up epoxy and polyurethane coatings. E Wood Ltd, one of the world's leading suppliers of coating formulations, has, in its Copon range of high-performance coatings, a wide range of spray-on epoxy and polyurethane coatings. This market has evolved over the last few years in that not only do materials destined to be in contact with potable water need to meet stringent health and safety rules, but also need to meet new environmental protection rules when the application is in process, notably concerning the use of volatile solvents.

Typical properties of the Copon range of coatings are:

- Solvent-free and waterborne technology.
- Single-coat applications for most situations.

- Approval to UK and European legislation for materials used in contact with potable water.
- Excellent protection of both metallic and mineral surfaces.
- High level of abrasion, impact and chemical resistance.
- Suitable for application in the factory, on-site and in-situ.
- Minimum downtime of storage tanks and process vessels.
- Long service life with minimal maintenance.

Typical applications include internal and external coatings of carbon steel and ductile iron pipes, concrete pipes, and mild steel and concrete tanks and vessels. The latest product developments which meet with the most recent legislation concerning contact with potable water include:

- Copon Hycote 162PW and 162PWX, both solvent-free epoxy systems;
- Copon Hycote 165, a solvent free polyurethane; and
- Copon Hycote 169, an instant setting solvent-free polyurethane.

Other products, including Copon Hycote EA9WB, a water-based epoxy, are available, and many of them can also be used, in certain conditions, in contact with potable water.

Glass flake linings. ITW Irathane International, as well as being one of the leading specialists in urethane and epoxy coatings have also developed, in co-operation with Fuji Resin, various glass flake-based lining systems. In glass flake lining technology thin glass flakes are incorporated into a resin, which when applied in liquid form, cures to create tough, glass like products which form a powerful bond with the supporting substrate. Irathane glass flake linings are constructed from multi-layer barriers, which remain stable for long periods in corrosive environments. This cohesive barrier is formed by a pattern of flakes lying parallel to the substrate within the resin matrix, preventing the ingress and permeation of corrosive elements. The excellent bond strength developed between the lining and the substrate contributes to the final high level of performance.

Irathane stress the importance of being able to select the correct product for each application by:

- choosing the best resin matrix for each application;
- changing the type of glass flake to suit the application, using small flakes for less demanding environments and larger flakes to resist the most severe chemical attack; and
- incorporating additional reinforcement for combating arduous environments, including heat, stress and highly corrosive chemicals.

Irathane offer two main systems, the 800 and 900 ranges. The final choice is to be made taking into account chemical resistance, temperature resistance, stress areas, life expectancy, costs and logistics. By controlling the formulation and product characteristics, coatings can be manufactured with minimal residual stress, a coefficient of expansion similar to metal and excellent adhesion to metal and concrete. The 800 range can be applied by either spray or trowel.

Sprayable products in the 800 range contain the smallest size (0.5 mm) glass flakes and can be easily applied as thin coatings using conventional airless spray equipment. Their temperature capability can be enhanced by increasing coating thickness up to 1 mm, after which a larger diameter glass is necessary.

The 900 range of products is claimed by Irathane to be the most advanced glass flake lining available, offering maximum protection even when applied in thin coatings. Optimum resistance is provided to thermal shock, chemical concentration and very high temperatures. Both spray- and trowel-applied grades incorporate large flake sizes to ensure maximum resistance to permeation.

When using sprayed systems, and in order to achieve the thinnest linings with the maximum number of flake layers, glass flakes are also included in the topcoat. In addition, these flakes are chemically treated in order to achieve maximum parallel alignment. Due to the size of the flakes (1.2 mm), Irathane propose using only spray equipment supplied by itself.

900 series trowel-applied systems are formulated to produce thinner linings than those normally associated with trowel-applied systems, providing excellent resistance to thermal shock and stress.

The 810 system uses an isophthalic polyester, and is recommended for seawater cooling systems, structural steelwork and external surfaces of pipes and equipment in corrosive environments.

The 820 and 920 series are based on a bisphenol A vinyl ester and are recommended for chemical storage, effluent containment and cooling water systems, especially above ambient temperature.

Higher temperature applications, and those requiring enhanced corrosion resistance, are covered by the 830/930 systems based on novolac vinyl esters, the 940 system which uses a brominated vinyl ester, and the 950 system, for which a chlorendic acid polyester is used.

The performance of Irathane glass flake linings can be demonstrated by a series of tests and experiences.

First of all the adhesion of the lining to a steel substrate. In a fatigue test, conforming to method B of ASTM D671-63-T, where dual amplitude stress was applied to carbon steel coated with 2 mm of Irathane T920 AR grade, no abnormalities were observed after 10 million cycles at 196 MPa. The test was then run at 245 MPa, and whilst the carbon steel substrate cracked at 570 000 cycles, no abnormalities were observed in the Irathane lining.

When considering heat resistance, Irathane glass flake systems are stable over long periods of time, and as a function of the characteristics of the resin system retained.

Elongation may be of concern in some applications, and in order to protect concrete structures against ground movements, Irathane flake systems can be modified to allow up to 1.5% elongation at rupture. Where seasonal crack movements make more extreme demands on the lining, Irathane flake flex formulations will be required. This product incorporates a unique layer, which enables the lining to remain securely bonded without disruption. This performance is due to the internal structure, which undergoes cohesive reorientation when subjected to extension and contraction forces.

AC (abrasion- and corrosion-resistant) grades of Irathane glass flake linings can also offer outstanding resistance to abrasion and erosion in even highly corrosive environments.

ITW Irathane International also offers via both its UK and worldwide network of authorised applicators, the following coating systems:

- Irathane, an abrasion, corrosion and chemically resistant elastomeric polyurethane coating, lining and casting system.
- Online, an acid-resistant and ceramic-filled epoxy system.
- Aqualine, a liquid-applied, two-part, 100% solids polyurethane system which combines elasticity with tensile strength and abrasion resistance is designed to protect concrete and accommodate structural movements in large water-containing structures. Developed specifically for waste and potable water containment in large concrete structures, Aqualine is of special interest in reservoirs, water towers, settlement and effluent tanks, secondary containment areas and sewage treatment plants in general. Aqualine is approved for use in contact with potable water in many European countries.

Structural laminates. Colvic plc, located in the UK, has developed two aspects of the lining business using glass fibre reinforced thermosetting resin laminates to protect and/or seal existing structures.

Its first activity, defined as Hygienic Fibreglass Wall Lining involves the direct moulding or lamination of an FRP structure onto walls and floors of buildings, often in the food processing industry, in order to obtain a hygienic, seamless, impact-resistant surface. The process fills all voids, cracks and cavities, helping to obtain bacteria-free surfaces, which are easy to clean with high-pressure hoses or even chemical cleaners. The lining is impervious to moisture penetration and fungicidal growth.

The lining process requires that all walls and floors are cleaned to remove impurities and loose material. A polyester-based primer is then sprayed onto the substrate and brushed over to ensure complete coverage, with any voids being filled with a polyester-based mortar.

The FRP laminate is then built up and rolled out over the total surface, with the lining being moulded around any obstacle, such as electric sockets, cables and electrical boxes, and the laminating process is repeated until the required thickness has been built up. The laminate is allowed to cure, and then an abrasion-resistant polyester surface coating which has a ceramic-like finish is applied. The final lining can be guaranteed for a period up to 20 years.

The second type of application involves the construction of a tank within a tank, and we use a typical case history to illustrate the process.

The application involved a steel modular tank, approximately $11 \times 6 \times 5$ m installed in a food processing plant. The tank was badly corroded inside and was contaminating the water it contained. The tank was leaking badly and required painting on the outside. The tank interior was first grit blasted which removed previous coatings of paint and corroded material. After priming with a polyester resin, the tank was lined with a 3 mm thick laminate using chopped strand mat. A final abrasion-resistant polyester surface coating was then applied. During the total operation, particular attention was paid to the interior reinforcing bars and gusset plates. The total operation was completed within five days.

One of the most effective features of this type of lining is the ability of the material to seal tanks with existing serious leakage. A tank can be lined with a laminate thickness up to 50 mm in areas where serious damage has occurred, and lining can be a economic alternative to the replacement of seriously corroded tanks. The resin system can be adapted to deal with many fluids, from water to strong acids, industrial effluent and general wastewater, for example.

Thermoplastic linings. We look here at processes as developed by Steuler Surface Protective Systems for the manufacture of mechanically anchored thermoplastic Bekaplast lining systems for sewage/ effluent piping, basins and tanks.

The Bekaplast concept developed in Germany by Steuler relies upon the creation of a mechanical bond between concrete, a relatively cheap, high-strength material with little chemical resistance, and a thermoplastic material which is relatively expensive, has rather low mechanical properties, but offers high levels of corrosion resistance.

The concept is used in flat and curved forms to line sewage pipes, circular or rectangular tanks, as an integral part of a new construction, or as repair or rehabilitation of existing and corroded structures.

The standard grade of Bekaplast material is a 5 or 8 mm thick HdPe sheet, although special materials such as polypropylene or PVDF can also be supplied. One side of the thermoplastic sheet is provided with a defined number of conical anchor studs, which ensure a permanent bond with the concrete. This form of positive and frictional interlock is able to withstand the stresses caused by the restraint of the movement engendered by the different thermal expansion rates of the two materials.

For the production of new sewage pipes, the Bekaplast material, rolled and welded into a tubular form, is placed on the pipe mandrel like an internal

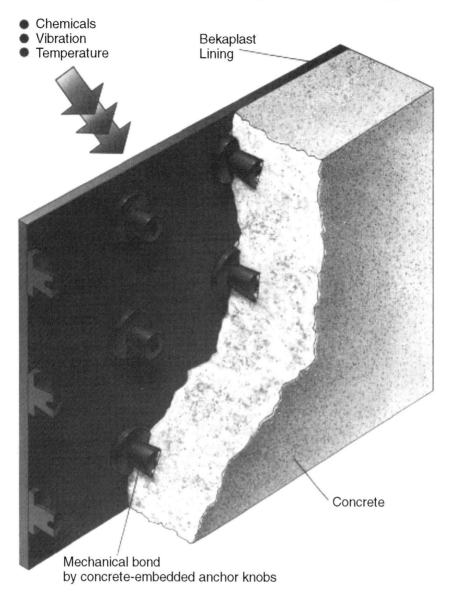

Figure 59. The form of the positive interlock. Courtesy of Steuler Industriewerke GmbH-Surface protection systems

formwork, and the concrete pipe is then produced in the normal manner. The finished pipes are laid in the normal manner, with pipes being inserted into each other by means of welded sockets, and the joints between each pipe section are then extrusion welded to produce a gas- and watertight seal. Standard pipes are available in a diameter range of 800 to 3200 mm.

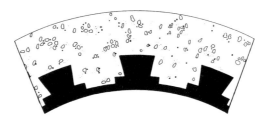

Figure 60. The mechanical bond between the concrete and the liner. Coutesy of Steuler Industriewerke GmbH-Surface protection systems.

Figure 61. A new lined concrete pipe. Coutesy of Steuler Industriewerke GmbH-Surface protection systems.

When the material is used to reline existing sewers or pipes, individual liner lengths are introduced into the system via existing manholes. The annular space between the liner and the old pipe is then grouted with a special low-viscosity quick-setting mortar. Once the grout is set, the radial liner seams are sealed with an extruded weld.

When used as lining in square or rectangular tanks the Bekaplast sheets are used as formwork for both floors and vertical walls using techniques and specific ancillary fittings and joints developed by Steuler. The advantages of the system are the same for all applications:

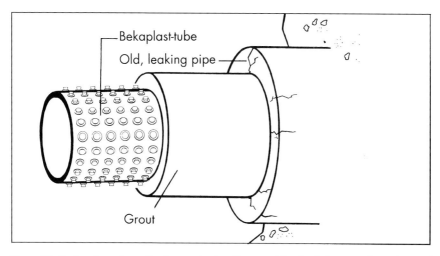

Figure 62. Section through a relined concrete pipe. Coutesy of Steuler Industriewerke GmbH-Surface protection systems.

- A gas- and watertight welded system.
- A uniform mechanical bond between the concrete structure and the thermoplastic lining, which avoids differential expansion of the two materials, and absorbs the stress created.
- The high corrosion resistance of thermoplastic materials.
- Smooth non-stick liner surfaces that can increase flow rates and decrease head loss.
- A high-impact, tough, abrasion-resistant material, which in case of any exceptional damage can be repaired easily.

A further development by Steuler, designated Bekaplast–DWS, is a double-wall lining system for concrete structures containing water-polluting liquids. As a result of increased environmental concerns and related legislation, permanent and defined monitoring of lined underground reinforced concrete structures is becoming more frequent, and in some countries such as Germany, is already mandatory. The German Water Act requires that leakages be detected and stopped before pollution of the surrounding area arises. The Bekaplast–DWS system consists of an outer thermoplastic sheet, which is, welded in the factory to an inner sheet in a manner, which forms a complete, double-wall panel. This double-walled lining becomes an integral part of the concrete structure since the anchoring knobs provide a mechanical lock to the concrete. After removal of the concrete forms, the joints in both the outer and inner sheets are sealed by fusion welding. This provides an integral liquid-tight

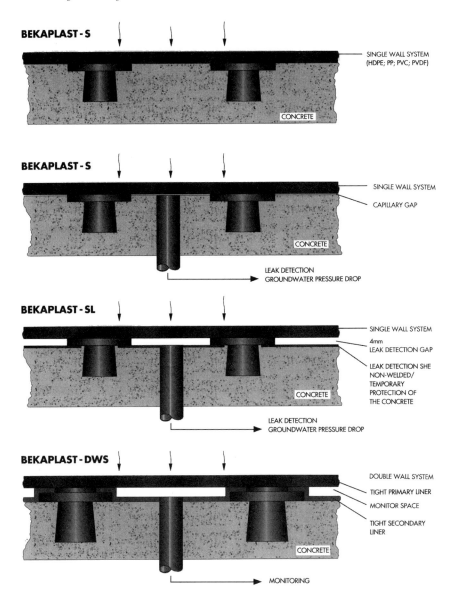

Figure 63. The range of Beckaplast wall linings. Coutesy of Steuler Industriewerke GmbH-Surface protection systems.

and pressure-resistant leak detection inner space. Monitoring devices are connected to this inter space in order to detect immediately any leak.

Figure 63 illustrates the different forms and arrangements possible with the Bekaplast sheeting.

9.4 District Heating and Cooling Processes

Buried pipework for the distribution of hot or cold water, essentially located in high-density, built-up areas is one of the most constraining applications for pipe systems in any material. Many systems are required to operate 24 hours per day, 365 days per year, with shutdowns limited to a few hours, and any maintenance or repair to the actual system often requires major and costly work, with disruption of surface traffic. It is understandable that operators and owners of such systems look to high-performance material which offers every chance of continuous service for at least 20 years, often more, and which offers a good resistance to not only the fluid being handled, but also to aggressive soil conditions and possible galvanic corrosion.

Figure 64. Installation and bonding of insulated epoxy pipe and fittings.

Many major district heating schemes operate at temperatures well above the possibility of composite materials, but a substantial market does exist for heating schemes limited to 110 to 120°C, and for cooling schemes handling chilled water at 3 to 5°C.

Composite pipes for heating are made from both amine- and anhydride-cured epoxy, and chilling water can be handled by both polyester and epoxy-based composite systems, as well as thermoplastics, especially polyethylene. Specialised contractors offer pre-insulated composite pipe systems using rigid polyurethane foam as the insulation material, injected into an annular space formed by the composite pipe and a thermoplastic, generally polyethylene, outer casing. The definition of a long-term design basis for the total system, pipe, fitting and assembly, generally a bonded joint, at the operating temperature of the system is of paramount importance, and long-term regression curves can be extrapolated to 20 or 40 years to calculate a valid hydrostatic design basis.

Installation of this type of pipe system requires special attention in several areas, notably:

- Definition of the minimum space between adjacent pipes in the trench.
- Minimum bending radius of pipe.
- The quality and materials of backfilling.
- The anchoring of high-temperature systems at all changes in direction and at all derivations, generally via concrete thrust blocks poured around the fitting to be restrained.
- Containment of any forces generated by expansion, especially when joining composite materials to steel pipework or within inspection or valve chambers.

The only standard identified as applicable to this service is the US Military Specification MIL-P-28584B for pipes and fittings conveying water at pressures to 125 psi and temperatures to 250°F, for which US producers of epoxy pipe supply a specific product range.

Another related application is for the construction of condensate returns on all heating networks where pressures and temperatures of the condensate return are low but extremely aggressive for mild steel and cast iron. One other related application is covered as a case history in Chapter 11, and concerns the use of epoxy pipe systems in low-temperature geothermal heating plants.

9.5 Industrial Process, Demineralised and Deionised Water

Industrial process water, including cooling water, is often water which has been used previously within the plant and which is recuperated rather than being discharged to waste, or in many cases seawater is used as cooling water.

In the first case the water may contain a progressive build up of mildly corrosive components, most of which can be handled efficiently by both composite and thermoplastic pipe systems, the choice between the two being made generally after consideration of temperature and pipe dimensions. Pipe systems handling process water may have to operate within areas in which the external environment maybe as corrosive as the fluid being handled and once again the inherent corrosion-resistant properties of plastic materials impose their selection.

The requirement for demineralised and/or deionised water is common to a wide range of industrial processes, and installations of this type are common in all forms of water treatment plants. Plastics are a natural choice for all pipe systems and vessels handling these fluids, with PVC, PP, HdPe and PVDF showing practically unlimited service lives at temperatures commensurate with the properties of each material. Pressure vessels in both straight FRP or with thermoplastic inner liners as described in Chapter 8, are specified as standard for all types of applications. FRP, essentially epoxy-based pipe systems come into their own on larger installations where dimensions and pressure ratings may make thermoplastic systems impracticable.

9.6 Ultra-Pure Water and Similar Applications

The storage and transport of ultra-pure water and other fluids requiring maximum levels of purity and zero contamination is the domain of PVDF and PFA, depending upon the degree of purity to be maintained, whether considering pipework, tanks or other storage vessels.

PVDF-HP (high purity) and PFA are produced under extremely controlled conditions, using specially selected raw materials. The companies Symalit AG and Georg Fischer have co-operated to produce and market a range of pipes and fittings under the designation PVDF-HP of diameter 16 to 225 mm, and Georg Fischer has developed an associated welding system SYGEF HP WNF which allows the elaboration of welds with no hollows or excess material on the inside of the weld.

PVDF and PVDF-HP offer excellent chemical resistance, to pH 11, a high degree of purity, excellent surface quality with high mechanical properties and abrasion resistance. As well as offering an excellent resistance to ageing, PVDF is also resistant to UV, β- and γ-radiation. PVDF is dimensionally stable from -40 to $150°C$. As well as the pipe system mentioned above, Symalit also produce a range of semi-finished materials: lining laminates and sheets up to 1.25 m wide, profiles and welding rods.

PFA extends the limits of use of thermoplastics due to its almost unlimited chemical resistance and a dimensional stability from 190 up to $260°C$. Symalit produces PFA pipes of diameter 16 to 32 mm, lining laminates and films, as well as welding rods, for example.

9.7 Composite Structures and Gratings

When considering a composite structure we are essentially concerned with the application of pultruded sections and profiles, the definition of their relevant mechanical properties, and corrosion and fire resistance. Composite gratings can also be fabricated from pultruded sections, but can also be manufactured by other moulding processes.

To return to the pultrusion process, described briefly in Chapter 4, it is important to note that the mechanical properties of a pultruded profile of a given shape and size may vary from one producer to another, this being one of the problems often met when dealing with these materials, and which should be addressed by the various standards under preparation. The mechanical properties are a function of the choice of first of all the resin matrix, polyester for most basic applications, vinyl ester or epoxy for generally higher performances, or phenolic if the finished product is obliged to meet standards concerning non-flammable properties. The type and format of the reinforcement is equally important. E-glass is the usual material, in the form of direct or woven roving of various weights, or even chopped strand mat. The percentage of glass to resin, as well as the amount of any fillers mixed into the resin, will also influence significantly the final physical properties of the profile.

Most pultrusion companies supply a large range of standard profiles, as I- and U-beams, equal and unequal angles, T-bars, square and round tubes and bars, strips and flat profiles, and, depending upon the actual profile, dimensions as large as 400 mm are now no longer exceptional. Depending upon tooling costs and the volume of the production run, most companies will also manufacture profiles to the specific requirements of a company or project. This sector of the market is responsible for an extremely wide range of profiles, from a few millimetres wide up to a width of 1500 mm, for example as a standard profile sheet for the fabrication of covers for water treatment tanks.

Pultruded profiles offer many interesting properties for applications in the water treatment and general industrial market:

- Corrosion resistance, providing almost limitless service in aggressive environments, with no requirements for painting or repair. All pultruded profiles are pigmented, but can be painted if required.

- High strength-to-weight ratios, with a higher strength-to-weight ratio than steel, and a density less than one-third of that of steel, making transport, handling and assembly easier and more economic.
- Excellent electrical and thermal insulation.
- Anti-magnetic and spark free.
- Can be supplied with non-slip finishes for areas of access.
- Cutting, machining and assembly can be made easily and quickly in plant or on site. Assembly is generally made with stainless steel bolts, although composite bolting is available for specific applications.

Typical applications include the construction of:

- Walkways and bridges
- Ladders and handrails
- Filter screens
- Maintenance platforms
- Tank covers and supports
- Pipe racks and supports
- Cable trays

Gratings can be manufactured by either assembly of pultruded profiles, which offer the best overall physical properties, or by various moulding processes, which can result in a wide range of properties. Gratings obtained by assembling pultruded profiles are preferred for all heavy-duty services, wider span applications and where special configurations and shapes are required. They are generally formed by interlocking two different profiles, locked up mechanically, and the typical glass/resin ratio of 60/40 obtained with the profiles offer long-term durability combined with high loading capacities. Typical weights vary between 8 and 25 kg/m^2 depending upon the height of the profile used and the percentage open area (from 15 to 70%).

High-grade moulded grating is made by laying up an interlaced glass fibre structure within a mould, then injecting the resin and curing the finished product in an oven. Typical glass/resin ratios maybe as low as 30/70 and the resin may contain fillers. Typical weights vary between 12 and 20 kg/m^2 and open areas are usually around 70%. This type of grating meets the requirements of many applications and offers considerable savings when compared to pultruded constructions.

A second moulding technique, involving the use of short fibres and moulding compounds offers a very economic solution, but with extremely limited physical properties, and is generally used on continuous or close supports.

All glass fibre gratings are lightweight, easy to cut and can be fixed together or to the steel or composite supporting structure with a wide range of standard fittings, generally supplied in 316L stainless steel. High-grade grating is extremely resistant to impact, and grating will deflect considerably under heavy impact loading before returning to its original form. Corrosion resistance can vary from good to excellent depending upon the resin system, and all gratings can be supplied with slip-resistant finishes.

9.8 Water Treatment Associated with Air Pollution Control

Air pollution control, often achieved by passing gaseous effluent through a scrubber, tends to transfer the pollutants from the air to water. Much of the air pollution control equipment for temperatures below 150°C are manufactured from thermoplastic or composite materials, including scrubbers and absorber towers, ducts and stacks, together with pipework handling both process water and effluent. The same range of equipment and materials can be used to fabricate effluent treatment tanks, settling tanks and pipework for all phases of handling and treatment of the aqueous effluent.

9.9 Miscellaneous Equipment for Water and Effluent Treatment

A wide range of products and materials has been developed specifically for the aggressive and constant attack by industrial and municipal effluent, which caused considerable deterioration of conventional materials throughout the twentieth century. Both simple and sophisticated engineering and production technologies have lead to huge savings and improved performance in a wide range of equipment, and we have attempted to group a number of these products and processes together in Chapter 10 under the heading of novel applications.

9.10 Onshore and Offshore Oilfield Applications

Water is inseparable from oil in many of the production technologies used to produce oil and/or gas in both onshore and offshore applications, either for water flood and water injection schemes to enhance the production of partially depleted fields or for potable water production, seawater cooling, and fire-fighting mains on offshore production rigs.

Water injection to enhance recovery of oil and gas in partially depleted fields has become common in all parts of the world, often using relatively corrosive water from saline or brackish sources. Pressures may be as high as 2000 psi and composite epoxy-based pipe systems built to specification API 15 HR are required. Here we enter the world of the oil business, the reliance on API/AINSI standards and practices, which for composite pipe imposes specific dimensions, thread patterns for the assembly of pipes and fittings, in many cases compatible with metallic products, and an emphasis on physical properties of the products rather than chemical resistance which is basically assumed as being more than sufficient for all but a few procedures and applications.

Offshore oil and gas production rigs are equipped with water makers, a reverse osmosis unit for producing fresh water from seawater using composite pressure vessels as described in Chapter 8. These water makers are often piped up with a mixture of composite and super duplex pipe, the super duplex being retained because of pressure ratings outside of the possibility of composite materials.

Although composite materials are generally more or less flammable, at least to the extent that in a sustained fire condition, the pipe will eventually fail, due to the progressive consumption of the different layers of the filament wound structure, they are often retained for the fabrication of fire mains and deluge systems. In normal fire conditions, with water circulating within the pipe, tests have shown that in many cases the composite pipe system will maintain sufficient physical integrity to the extent that the material will outlast the survival time allotted to the rig in case of a general fire. This basic level of performance can be improved to the level that the system can be used in hydrocarbon fires. Ameron supply its 2000M-FP range, on which both pipes and fittings receive a thick intumescent coating in the factory prior to shipment. When exposed to fire, the intumescent coating expands to form an incombustible foam char that both protects and insulates the pipe.

Bondstrand series 2000M-FP has been demonstrated to be capable of maintaining service pressure following exposure to a hydrocarbon fire (temperature approximately 1000°C) of no less than 10 minutes' duration in the dry (empty) condition, and 3 to 6 hours' duration in full flow condition with the pipe wall remaining below 120°C at all times.

9.11 Buried Fire Mains

With constraints similar to those evoked concerning pipe systems for district heating schemes, underground fire mains have become an application for which composite pipe systems are readily and often specified, in a wide range of plants from carbonated beverages to refineries and petrochemical plants, from automobile plants to LNG terminals. All rely on the long-term internal and external corrosion resistance of composite epoxy-based pipe systems, unaffected by the constant contact with untreated, often saline or brackish water, or with foaming agents as well as all types of aggressive soil conditions, galvanic corrosion and fluctuating levels of ground water.

One particular specification and approval system applies to composite pipe systems destined for this application, the Factory Mutual System, which imposes a specification, quality control system, and above all a product qualification pressure testing requirement for every diameter, configuration and pressure rating of pipe, fitting and joint. Pipes and fittings of all configurations are assembled together and subject to a hydrostatic pressure test at four times the rated pressure of the components of the system. Once the pressurisation has been achieved, the test pressure is held for 2 minutes. Any weeping or leakage during this period should be cause for refusal of the Factory Mutual (FM) qualification.

FM publishes and imposes rules for all aspects of the design and installation of FM-approved pipe systems, contained within their Installation and Maintenance data sheets. FM-approved pipe systems are available from all the major producers, Ameron, Fibercast, Future Pipe Industry and Smith Fiberglass. Approved pressure ratings, diameter ranges and jointing systems vary from one producer to another.

Pipe systems may also be listed by Underwriters' Laboratories and Underwriters' Laboratories Canada (UL and ULC) for the same service.

9.12 Jetties and Locks

With composite materials having been adopted for the manufacture of the major part of the small boat market it is not surprising that many of these small boats, together with many larger ones tie up on a jetty which may be partially built or more often protected by composite materials.

Growing concern around the effects on the environment of wood-preserving agents used in marine fenders has opened the door for alternative materials, and applications in both thermoplastic and composite materials have already been developed and tried on an industrial scale.

The design, manufacture and installation can be relatively simple for low load-bearing applications, or very complex, involving the combination of thermoplastics, composites and concrete for large heavy-duty installations.

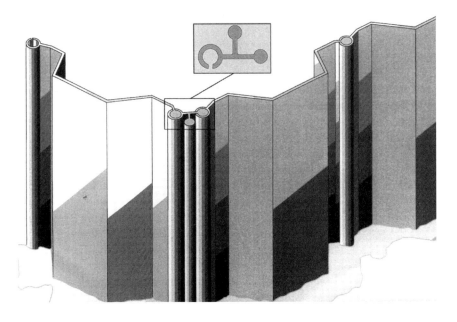

Figure 65. Composite Korlok profile. Courtesy of International Gratings Inc.

We have identified four significant manufacturers, all located in North America, who manufacture a range of products to suit different conditions and requirements.

International Grating Inc (IGi) produces a pultruded sheet pile for use as a simple marine retaining wall, under the trademark Korlok, as a deep C-beam 16 inch (400 mm) wide and 4 inch (100 mm) deep, together with a pultruded T-connector to form corners and to provide a point of departure for further extensions.

IGi claims that the combination of the interlocking design and wide panels simplifies installation, using any of several methods: jack hammer, water jet, or vibratory driver or compactor. The finished sheet pipe wall can be capped with wood, concrete or composite materials.

Hardcore Composites has developed the SCRIMP (Seemann Composite Resin Infusion Process) technology for the manufacture of composite fender piles and associated panels which offers an advantage over pultrusion technology in that different wall thickness and distribution of reinforcing materials are possible. This technology allows the manufacturer to adapt a design to each specific project or even different parts of a single project. When required, local stiffening can be obtained via the inclusion of balsa or three-dimensional cores built also with the SCRIMP process.

Lancaster Composite Inc has developed an 'FRP Precast Composite Pile' based upon a filament wound epoxy resin-based shell, wound at specific angles in order to enhance the longitudinal properties, combined with the injection inside the shell of cementatious foam. The expanded foam transfers a permanent stress into the composite shell, and the combination of tension in the reinforced shell combined with the compression in the cementatious foam results in a very stable structure.

More than 21 000 ft (6400 m) of 12.7 inch (323 mm) diameter composite piling has been used by US Army Corps of Engineers Waterways Experiment Station in association with the Greater New Orleans Express Commission to replace deteriorated timber piling on the Lake Ponchartrain Causeway near New Orleans.

Seaward International Inc, a leader in all forms of waterfront infrastructures has developed products based upon a combination of both thermoplastic and composite materials. Its 'Seapile' fender system is based upon a polyethylene extrusion, reinforced with composite bars in the axial direction. These composite bars are introduced into the extruded thermoplastic as it is being produced, giving the thermoplastic structure the axial strength required to handle the hammering used during installation. Standard dimensions of the Seapile system are 13 and 16 inches (330 and 406 mm), with unit length being limited to the possibility of transport. To date, the longest piles delivered are 105 ft (32 m).

In France, Les Voies Navigable de France (VNF) is restoring France's network of canals, which are now increasingly used for tourist-related rather than commercial activities, including lock gates, built in wood and steel, and which have now corroded, rotted or warped to the extent that they are both difficult and potentially dangerous to operate.

VNF approached the Direction des Construction Naval (DCN) for the construction and maintenance of military vessels in steel and composite in Lorient in the west of France, in order to develop and manufacture a new generation of lighter and corrosion-resistant lock gates, using the knowledge and production technology which the DCN has acquired in the construction of composite waterfront structures, minesweepers and other light vessels.

DCN retained an isophthalic resin system due to the good ageing characteristics in an aqueous environment, combined with a contact moulding process in open moulds. This process allowed fibre orientation and concentration to be adapted to meet the requirements of the different parts of the lock gates, including the inclusion of horizontal stiffeners. The resulting arc-shaped composite panels are installed in a stainless steel frame, the combination of which reduces both maintenance costs and the overall weight to half that of the original gates. This lighter weight facilitates handling and installation, as well as the actual opening and closing of the gates, and the extra buoyancy reduces stress on the mechanical assembly during these operations.

9.13 Irrigation

Thermoplastic pipe systems have become a natural choice for the large scale irrigation systems now in place in both arid and non-arid parts of the world, with the various grades of polyethylene (HdPe, MdPe and LdPe) having the largest share of the market.

The reasons for this success are clear:

- Low weight and flexibility facilitates both handling and coiling of pipe.
- The high-impact resistance of polyethylene avoids damage during the constant handling and transport of the material.
- The excellent corrosion resistance of polyethylene means no coating or painting, and an excellent resistance to both internal and external aggression.

The success of thermoplastics in the irrigation market is not restricted to the properties and performances of the pipes, but is also due to the facility to mould high-volume and relatively 'high-tech' fittings and accessories which ensure both the assembly of the different pipes and fittings and a steady uniform regulated water feed.

Several national and international product standards and codes have been developed for these materials, and we have chosen to include them in this section rather than in Chapter 5 which is dedicated to the most important sets of standards.

Typical standards are:

North America.

ASTM F 690. Standard practice for underground installation of thermoplastic pressure piping irrigation systems.

ASTM F 771. Standard specification for polyethylene (Pe) thermoplastic high-pressure irrigation systems.

ASTM F 878. Standard specification for polybutylene thin-wall drip irrigation tubing.

ASTM F 1176. Standard practice for design and installation of thermoplastic irrigation systems with maximum working pressure of 63°F (17°C).

France.

NF U 51-432. Irrigation equipment. Polyethylene tubes for micro-irrigation equipment.

NF U 54-086. Plastics. Unplasticised polyvinyl chloride (PVC-U) for underground irrigation. Specifications.

Germany.

DIN 19658-2. Polyethylene (PE) coiling pipe and hoses for irrigation systems. Part 2: Hoses with fabric inlay, rigid foam; dimensions and technical delivery conditions.

DIN 19658-3. Polyethylene (PE) coiling pipe and hoses for irrigation systems. Part 3: Hoses of non-rigid material with woven inner layer, dimensions and technical delivery conditions.

Cross-border international standards

Europe

EN DIN 12324-2. Irrigation techniques — reel machine systems — Part 2: Specifications of polyethylene tubes for reel machines.

PrEN 12734. Irrigation techniques. Quick coupling pipes for movable irrigation supply. Technical terms of delivery and testing.

ISO

ISO 8779. Polyethylene (PE) pipes for irrigation laterals; specifications.

ISO 8796. Polyethylene (PE) 25 pipes for irrigation laterals; susceptibility to environmental stress cracking induced by insert-type fittings; test method and specification.

ISO DIS 8796. Polyethylene (PE) 32 pipes for irrigation laterals; susceptibility to environmental stress cracking induced by insert-type fittings; test method and specification.

ISO 9625. Mechanical joint fittings for use with polyethylene pipe for irrigation purposes.

ISO DIS 13460. Agricultural irrigation equipment — plastic saddles for polyethylene pressure pipes.

CHAPTER 10

Novel Applications

In this chapter we attempt to identify and describe briefly some of the newer (and in some cases not so new but perhaps more imaginative applications) of thermoplastics and composite materials in the water treatment business, although the particular application may not be unique to the identified suppliers. Details concerning addresses and contact numbers of the companies mentioned are to be found in the directory of suppliers in Chapter 14.

10.1 Composite Modular Settling Tanks

F Z Fantoni, founded in 1966, is located in northern Italy, operates out of a modern 8000 m² plant and is specialised in the design and manufacture of tanks for oenological, general industrial and water treatment markets. We describe here its patented Comby settling tank.

The Comby 10 settling tank is constructed entirely from composite materials, except for PVC spacers and the external support structure in galvanised steel. It works on the combined principles of centrifugation and gravitational laminar decantation, using multiple inclined planes, resulting

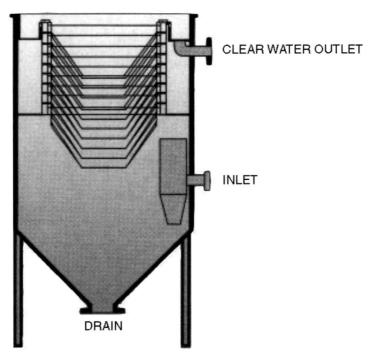

Figure 66. Comby composite settling tank. Courtesy of FZ Fantoni SpA.

in a settling area 10 times greater than the diameter of the tank. It has no moving parts, and the modular construction allows rapid assembly and dismantling for maintenance, cleaning or eventual extension to the filter pack.

The concept involves a silo-type shell, an effluent intake that incorporates a centrifugal device that immediately separates out the heavier particles, and a modular pack of superimposed truncated filter planes.

The initial sedimentation is induced by the centrifugal effect at the entry into the tank, whilst remaining solids in suspension are obliged to ascend through the set of inclined planes, undergoing laminar decantation. Once the decanted particles reach a certain thickness on the conical surfaces, they slip off and fall to the bottom of the tank, where sludge gathers and is drawn off.

The tank is normally self-cleaning, and it is generally sufficient to reduce the level of the liquid in the tank under the level of the modular pack to remove all sludge deposited on the conical surfaces.

Ten different standard sizes are available, from 600 to 3000 mm diameter, corresponding to a nominal flow of 0.8 to 40 m^3/hour, and the main dimensional characteristics and performance data are given in Table 7. The manufacturer claims that with a few modifications the vessels can also be used as oil separators.

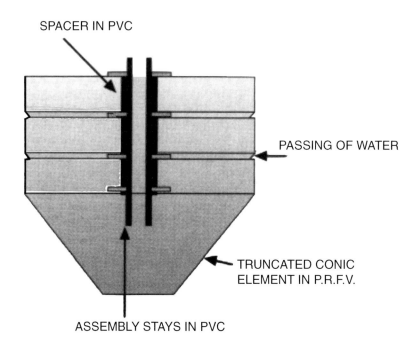

Figure 67. Spacer arrangement. Courtesy of FZ Fantoni SpA.

Table 7. Dimensions and capacities of the Comby 10 standard range of settling tanks

Diameter (cm)	Cylinder height (cm)	Cone height (cm)	Capacity (litres)	Useful area (m^2)	Minimum retention (min)	Spec Q (m^3/m^2)	Useful Q (m^3/hour)
60	90	45	300	1.0	32	0.8	0.8
75	115	56	550	1.5	42	0.8	1.2
90	135	67	1000	2.5	43	0.8	2.0
120	200	90	2600	6.0	47	0.8	4.8
140	300	105	5000	9.5	61	0.8	7.6
160	280	120	6500	12.0	59	0.8	9.6
180	280	135	8200	15.5	57	0.8	12.4
200	280	150	10000	22.5	49	0.8	18.0
230	300	172	15000	30.0	52	0.8	24.0
250	300	187	17500	40.0	46	0.8	32.0
300	300	225	26500	50.0	53	0.8	40.0

Data is based upon a speed of sedimentation of 1.3 and a minimum density of 1.10.
Courtesy of FZ Fantoni SpA.

10.2 Package Sewage Treatment Plants

These are pre-fabricated self-standing modular units to handle residential effluent that have developed considerably over the last ten years as legislation imposes new constraints on sewage treatment and disposal, especially for units or communities unable to obtain access to mains drainage. These new units are now widely available to provide effective and economic treatment of sewage for communities of 30 to 250 persons.

We examine here two slightly different concepts of these small units, which have one major point in common, that is the use of composite and thermoplastic materials.

Bio-plus Environmental Systems Ltd, located in the north of England, are a specialist manufacturer of package sewage treatment plants, package pumping stations and a large range of pollution control products manufactured with composite materials. They offer a range of 16 package sewage plants under its designation BS3/1250RB to BS18/90HC, for population equivalents of 30 to 250 persons.

WPL Ltd, located in the south of England, specialise in the provision of wastewater treatment for developments unable to gain access to mains drainage, and offer 12 versions of its HiPAF — high-performance aerated filter, for population equivalents of 60 to 200 persons.

Both companies offer a design based on a three-stage/multi-compartment rectangular or cylindrical FRP tank, all of which are buried so that only the roof or hatch-covers are visible above ground.

The standard BS range of Bio-plus plants are designed to achieve the following effluent discharge standard:

- Biological oxygen demand (BOD) of 20 mg/l;
- Suspended solids at 30 mg/l; and
- Ammoniacal nitrogen at 20 mg/l.

The unit works on the basis of three successive stages of treatment:

- Stage 1 — a primary settlement zone, based on the principles used in septic tanks, is designed to reduce the BOD of the influent sewage by 30

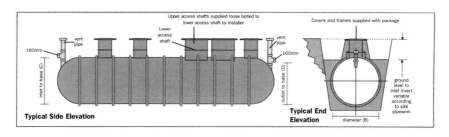

Figure 68. Overall configuration of a Bio-plus unit. Courtesy of Bio-Plus Environmental Systems Ltd.

Table 8. Dimensions and capacities of the BS horizontal cylindrical models

Model reference	Population equivalent	Design hydraulic flow (1/day)	Design organic load (kg BOD/day)	Length (A)	Diameter (B)	Inlet to base (C)	Outlet to Base (D)
BS6/90HC	30	6000	1.8	3700	1850	1900	1700
BS7/90HC	40	8000	2.4	4700	1850	1900	1700
BS8/90HC	50	10 000	3.0	5520	1850	1900	1700
BS9/90HC	60	12 000	3.6	6820	1850	1900	1700
B10/90HC	70	14 000	4.2	4600	2500	2600	2350
B11/90HC	80	16 000	4.8	5500	2500	2600	2350
BS12/90HC	90	18 000	5.4	6450	2500	2600	2350
BS13/90HC	100	20 000	6.0	6450	2500	2600	2350
BS14/90HC	120	24 000	7.2	8310	2500	2600	2350
BS15/90HC	150	30 000	9.0	10 170	2500	2600	2350
BS16/90HC	175	35 000	10.5	10 170	2500	2600	2350
BS17/90HC	200	40 000	12.0	12 290	2500	2600	2350
BS18/90HC two-tank system	250	50 000	15.0	8310	2500	2600	2350

All dimensions are in mm. For the BS18/90HC two-tank system, the dimensions of each tank are detailed. Courtesy of Bio-Plus Environmental Systems Ltd.

to 40%. The settled sewage flows on to the secondary treatment in Stage 2, and surplus sludge is drawn off at intervals by a suction tanker.

- Stage 2 — a biological aerated filter (BAF) with air supplied from an adjacent compressor.
- Stage 3 — the final settlement zone ensures that any carryover of particles is settled and not carried over into the final effluent. The settled sludge which accumulates at the bottom of this zone is returned automatically by airlift periodically. The overall configuration and typical dimensions and capacities are detailed in Table 8.

Smaller package plants to serve population equivalents of 5 to 10 persons are available in a Bio-cell range from Bio-plus.

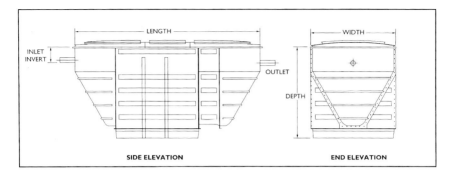

Figure 69. Overall configuration of a HIPAF unit. Courtesy of WPL Ltd.

Table 9. Dimensions and capacities of the HiPAF Midi Range of effluent treatment stations

Model	HP6	HP7	HP8	HP9	HP10	HP11	HP12	HP13	HP14	HP15	HP17.5	HP20
Population equivalent	60	70	80	90	100	110	120	130	140	150	175	200
DWF (m^3/d)	12	14	16	18	20	22	24	26	28	30	32	34
Length (m)	3.55	3.95	5.15	5.4	6.15	5.9	6.3	7.1	7.1	7.6	8.4	9.6
Width (m)	2.7	2.7	2.7	2.7	2.7	2.7	2.7	2.7	2.7	2.7	2.7	2.7
Depth (m)	3	3	3	3	3	3	3	3	3	3	3	3

Design parameters: dry weather flow (DWF)=200 l/head/day; biological oxygen demand (BOD)=60 g/head/day; maximum flow=3×DWF; invert depth=0.5 to 1.5 m; discharge standards BOD=20 mg/l; suspended solids=30 mg/l; ammoniacal nitrogen=10 mg/l. Other discharge and design parameters can also be accommodated.
Courtesy of WPL Ltd.

The overall configuration, design parameters and overall dimensions of the alternative system available from WPL Ltd are detailed in Table 9.

The HiPAF system from WPL works by treating the effluent in three stages as indicated above. Raw sewage enters the first primary settlement chamber where much of the solid matter settles to the bottom to form sludge. Displaced and partially settled sludge passes through a baffle into the second primary chamber where further settlement takes place. This baffle is designed to prevent floating solids entering the bio-filter compartment. The first airlift forward feeds the settled primary liquor into the bio-filter, which is where the settled sewage undergoes treatment by a bacterial process. This two-stage compartment contains high-void thermoplastic media where the oxidising bacteria collect in a jelly-like substance known as biomass. The liquor passes through the bio filter by displacement. A flow of air is introduced below the plastic media bed via a series of fine bubble diffusers to provide the oxygen supply, which enables the growth of the biomass.

The treated effluent then flows through a notched weir into the humus settlement chamber where final settlement takes place prior to discharge. At frequent intervals, an air-lift return transports the settled humus sludge back to the primary chamber for co-settlement with the incoming sewage. This recirculated effluent introduces dissolved oxygen into the primary section, which further assists the effectiveness of the treatment plant.

WPL also manufacture a smaller, circular unit, designated Compact, for installations serving up to 50 persons, and, by juxtaposition of two or more Midi units, can supply its Modular range for populations up to 2000.

Other products available from the same company include rapid gravity tertiary sand filters, conical settlement tanks up to 6 m in diameter, complete with launders, weirs and bridges all made from composite materials.

10.3 Prefabricated Pumping Stations

Series Fäkalex FPS KP manufactured by Feluwa Pumpen GmbH combine traditional pumps and fittings with a thick-wall HdPe tank, for the collection and evacuation of wastewater and sewage from domestic properties,

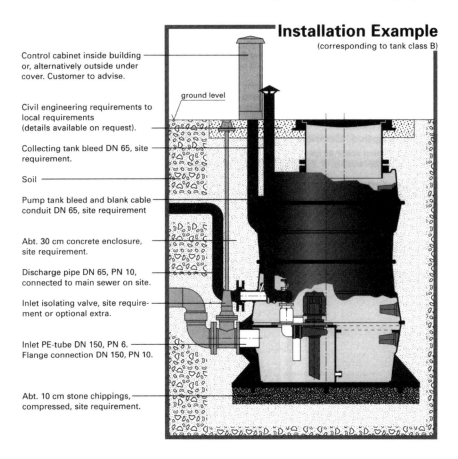

Figure 70. Overall arrangement of a pumping station. Courtesy of Feluwa Pumpen GmbH.

industrial sites and holiday camp sites where effluent needs to be collected and then evacuated towards existing mains or treatment plants. The specific advantages of this type of construction are:

- Delivery to site of a fully equipped pumping station, ready to be fitted up to inlet and outlet pipework and the electrical supply.
- Corrosion resistant heavy duty HdPe one-piece moulded tank, complete with access steps, with a smooth internal surface which helps to prevent build-up of solids.
- Lightweight and easy to handle and install.
- Facility to adjust the overall height on site, within a range of 500 mm. The vertical tolerance is 5°.

Feluwa Pumpen GmbH is a specialist manufacturer of a large range of pumps, used in all areas of water and effluent treatment, submersible and centrifugal pumps, pumps for special purposes and irrigation plants.

10.4 Modular Biomedia

These are for water and wastewater treatment, as manufactured by:

- Brentwood Industries Inc under the trade name ACCU-PAC Biological Media.
- 2H Kunststoff GmbH.
- Munters Euroform GmbH under the trade name BIOdek.
- Rauschert Verfahrenstechnik for structured blocks, random carriers and suspended carriers under the trade name Biolox 10 and Bioflow 30.
- Marley Cooling Tower under the trade name MARPAK.

Thermoplastic media for both wastewater and potable water have developed over the last 25 years or so, from basic simple random packed shapes to today's highly engineered, complex, and in the main structured elements, usually available in block form. They have found a wide range of applications, including trickling filters, aerobic biological filters, degassing of potable water and even as packing in scrubbers on deodorization plants. They are available in various forms and materials, including PVC, PP and Pe, and in many cases the sheet thickness can be varied to suit different loading conditions.

Trickling filters using traditional rock filling have been used with some considerable success for at least 100 years, due to a wide range of advantages when compared to processes based on activated sludge. These advantages include:

- Low levels of supervision.
- Being more resistant to hydraulic or toxic shocks they recover more quickly.
- Significant energy savings.

At the same time, these traditional filters had serious negative aspects, including:

- High risks of clogging.
- Limited height and high construction costs.

Figure 71. Typical filter media in polypropylene. Courtesy of 2H Kunststoff GmbH.

The use of modular thermoplastic structured media, such as BIOdek supplied by Munters Euroform GmbH allows the engineer to maintain and enhance the advantages and at the same time remove some of the more negative aspects. The high void content, typically around 95% considerably reduces the clogging problem associated with traditional fill and the high surface-to-volume ratios, variable between 100 and 400 m^2/m^3 means enhanced contact between the effluent and the microorganisms used to breakdown the organic waste. This same high void content and specific geometry of the modular systems improves the natural draft of air, which again makes it possible to work with higher BOD loadings.

The unit weight and structural properties of this type of media, together with the possibility to supply modular blocks with different load bearing capacities, make it possible to design taller units, reducing the ground surface occupied, and consequently offering a much more economic solution. Due to the type of construction, and the low dry weight of between 20 and 100 kg/m^3, installation and eventual removal and replacement is greatly facilitated. Several manufacturers offer, for the larger projects, the possibility to assemble the blocks on site, using mobile production or gluing equipment, thereby reducing the volume and cost of transport by as much as 85%.

Media can be manufactured with 60° cross-flow or vertical flow patterns, depending upon the requirements of the application, and either incorporate stiffened areas or can be supplied in different thicknesses to obtain structural properties such that media depths of up to 10 m can be obtained.

Figure 72. BIOdek reinforced media. Courtesy of Munters Euroform GmbH.

All materials used are resistant to practically all concentrations of municipal or industrial effluent, as well as fungi, bacterial growth, as well as being resistant to ultraviolet radiation.

Typical areas of application for plastic media are in:

- Trickling filters for BOD removal, nitrification and/or denitrification.
- Submerged biofilm processes for BOD removal, nitrification, denitrification and anaerobic treatment.
- Potable water degasification (CO_2 removal).
- Odour scrubbing.

Munters Euroform also manufactures a plastic lamella product, TUBEdek, using similar materials and production technology to that used for the media described above.

TUBEdek is a tubular system of inclined channels creating optimum conditions for parallel plate sedimentation. The developed surface area is increased by a factor of 5 to 10 compared to the base area, and with the availability of different channel shapes, variable height and slope allow a full selection of possibilities enabling the engineer to achieve optimum conditions.

TUBEdek can be used in:

- Sedimentation after flocculation;

- pre-sedimentation and final sedimentation of municipal wastewater; and
- industrial waste and surface water treatment.

Brentwood Industries offers a wind range of products for bio-modular media for water and wastewater treatment under the trade names ACCU6Pac VF-3800 and VF-5000 in a vertical flow configuration, ACCU-PAC CFS-3000, CF-1900, CF-1200 and CF-650 in cross-flow configurations for applications such as aerobic and anaerobic wastewater treatment, nitrification, odour control, air stripping and oil/water separation. The company also supplies fill media for cooling towers under the designation VF 19 PLUS and TURBOsplash PAC.

Accu-Pac tube settlers, under the reference IFR-6000, for potable and non-potable water applications complete the range. This product line is designed to achieve enhanced settling capacities, by providing multiple tubular channels, sloped at an angle of 60°. Individual tubes are continuous and smooth, with the configuration and shape of each tube designed to obtain a low Reynolds number and laminar flow conditions for rapid accumulation and settlement of solids through the tubes.

10.5 Advanced Polymeric Membrane Systems for Potable and Wastewater Treatment

Zenon Environmental Inc, based in Burlington, Ontario, Canada, with subsidiaries throughout the Americas, Europe and the Middle East, is a leading company in the field of purification of potable water and the treatment of wastewater, and has developed a range of membrane technologies for microfiltration, ultrafiltration, nanofiltration and reverse osmosis.

One of these technologies, the ZeeWeed membrane process, relies on the combination of chemical inertness and the high level of physical properties of hollow fibre polymeric materials, similar in performance to those used in the more conventional use of hollow fibres inside pressurised vessels for ultra- and micro-filtration. The level of porosity of theses membranes is situated between the level of micro- and ultrafiltration, with a nominal 0.035 µm, and an absolute 0.10 µm pore size, ensuring that, for potable water applications, no

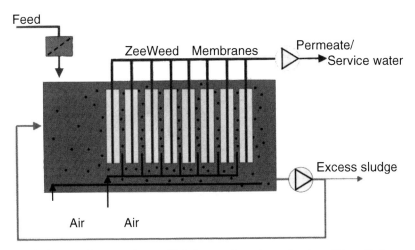

Figure 73. Operational concept of an outside-in immersed membrane. Courtesy of Zenon Environmental Systems Inc.

particle matter including Cryptosporidium oocysts, Giardia cysts, suspended solids or pin flocked pollutants will escape to the treated water stream.

The ZeeWeed process makes use of a simple aeration basin or tank, often manufactured in epoxy lined steel, into which are placed the membrane modules in a vertical position. Each module is made up of a frame, from which are freely suspended the membranes, rather like packets of spaghetti.

The membranes operate under a partial vacuum within the hollow fibres by way of a centrifugal pump. The treated water passes through the membrane, enters the hollow fibres and is pumped to distribution by the permeate pumps. Air is introduced at the bottom of the membrane module to create turbulence which scrubs and cleans the outside of the membrane fibres, allowing them to function at high flux. This air will also oxidise iron and other organic compounds, generating better quality water than provided by microfiltration or ultrafiltration alone.

The simplified process flow diagram is presented in Figure 73. Although the proprietary polymeric membranes form only a small part of the overall installation, they are of course critical to the performance of the system, and cannot be manufactured in any other material.

10.6 Cooling Towers and Cooling Tower Components

These are manufactured by:

- The Marley Cooling Tower Co
- The Baltimore Aircoil Co
- Hamon & Cie
- Addax Inc

Smaller units are also manufactured by a wide range of companies including:

- Polacel BV in the Netherlands, part of the GEA group; and
- CASE (Composite Aqua Systems and Equipment) in New Delhi, India.

A range of polyethylene towers is available from Delta Cooling Towers Inc.
The corrosion and associated problems in and around industrial cooling towers is well known, and can include one or more of the following:

- galvanic corrosion when different metallic materials are used;
- scaling due to the concentration of minerals; and
- presence of corrosive, saturated dissolved oxygen, and due to the warm, humid conditions, the potential proliferation of algae, fungi and bacteria, all of which may influence the type and speed of corrosion.

The Cooling Technology Institute (CTI) in Houston, Texas, USA, not only proposes a list of certified cooling towers, but also publishes a complete set of standard specifications and research reports, including specifications for the use of composite and thermoplastic materials. These include:

Std-131. Glass fibre reinforced plastic panels for applications on industrial water towers. This standard covers the classification of materials of construction, workmanship and methods of testing glass fibre reinforced plastic panels in various profiles intended for use as casing, louvers and similar applications on cooling towers.

Std-136. Polyvinyl chloride material used for film fill, splash fill, louvers and drift eliminators. This document covers the classification of rigid polyvinyl chloride (PVC); the physical properties, burning properties and recommended testing procedures employed to determine the defined values, whether processed from virgin or reground material.

Std-137: Glass fibre pultruded structural products for use in cooling towers. This specification offers recommendations for classification, materials of construction, tolerances, defects, workmanship, inspection, physical, mechanical and design properties of glass fibre reinforced pultruded structural shapes intended for use as construction items in cooling tower applications.

The CTI also publish a manual, which brings together in one document information pertaining to all aspects of cooling towers. The manual has been developed in the form of individual chapters, each of which stands on its own merit, and which can be purchased separately. Two chapters may be of particular interest to readers of this handbook:

- Chapter 9. Materials of construction for cooling towers
- Chapter 12. Fire protection of cooling towers

Returning to the manufacturers of composite cooling towers, one can examine one particular and representative application.

The Marley Cooling Tower Co produces a very wide range of cooling towers, from large, concrete counterflow hyperbolic towers handling flow rates of up to 500 000 gpm (1900 m^3/minute) or more essentially for large power stations and chemical plant down to small units with flow rates as low as 200 gpm (0.75 m^3/minute). Composite materials may be specified for components throughout the total range, but are of special interest for units such as the Sigma F series Fiberglass Composite Cooling Tower, the latest addition to the Marley series of cross-flow cooling towers. This product can be designed as a single or multi-cell unit, with individual units capable of handling 380 to 11 664 gallons/minute (1.4 to 44 m^3/minute). The F series tower relies on both pultruded sections and SMC mouldings to produce a corrosion-resistant structure, non-slip fan deck made from interlocking panels and end wall casings. Both internal and external piping systems can be made from standard composite pipe systems, and the internal spray system, Marley Spiral Target Nozzles, uses injection moulded polypropylene. The nozzles are equipped with a diffuser and snap-in orifice cap which offer a wide range of adjustment in flow rates and an even distribution over the fill area. Marley type MX film fill in PVC and integrally moulded drift eliminators and louvers completes the range of thermoplastic and composite materials used in these units.

The structure is impervious to a broad range of corrosive materials, is immune to rot and decay, and has a satisfactory fire resistance with a low flame spread rating of Class 1, self-extinguishing, to ASTM E 84. The stable

Non-skid Fan Deck
of interlocking pultruded panels, with fasteners concealed by the adjacent panel, offers a clean surface free of trip hazards.

Corrosion Resistant
pultruded glass fibre structural components and fan deck panels, GRP endwall casing and stainless steel hot water basins are all impervious to a broad range of corrosive materials.

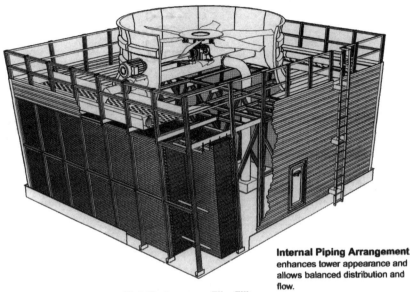

Internal Piping Arrangement
enhances tower appearance and allows balanced distribution and flow.

High-Performance Film Fill
Marley type MX film fill removes process heat efficiently, predictably. Integrally molded drift eliminators and louvers virtually eliminate drift spotting nuisance and help operation even in extremely cold weather.

Marley Spiral Target Nozzles
Injection molded polypropylene with a diffuser and snap-in orifice cap offer a wide range of adjustment in flow rates and basin water levels; even distribution over the fill area.

Figure 74. Sigma F series cooling tower from Marley. Courtesy of Marley Cooling Tower Co.

high-strength composite structure is claimed by Marley to be less costly than stainless steel, more durable than galvanised steel, and more consistent than wood. The clean aesthetic and durable appearance of the composite materials adds a final advantage.

The Marley organisation rely on the high strength and corrosion-resistant properties of composite materials for the construction of a critical element in all forced draft cooling towers, namely fan blades. The Marley HP7000 fan line has been designed specifically for industrial cooling tower applications and Marley's Flare Tip blade enhancement provides increased performance, overcoming tip clearance losses, and aiding the movement of air next to the fan cylinder.

A nylon leading edge erosion barrier is moulded into each blade to ensure a long service life. The hollow blade, a one-piece glass fibre reinforced epoxy laminate, is designed with a deep section airfoil, giving strength and low weight. All blades have an identical and consistent weight, removing the requirement to rebalance the fan should a blade need replacing. Each blade is gripped between two machined epoxy-coated cast iron clamps for attachment between dual hub plates made from heavy galvanised steel. All hub assembly hardware is series 300 stainless steel, with 316 stainless and Monel available for more corrosive environments. By loosening this hardware, the blades can be easily rotated in order to achieve the desired pitch.

The Marley HP7000 series of fans are all designed for continuous operation and are available in diameters from 240 to 394 inches (6.1 to 10.0 m).

The company Addax Inc, probably the world's leading supplier of composite mechanical power transmission equipment, adds an important link in composite materials in the transfer of power to fans such as those indicated above. Addax is specialised in resolving power transmission problems in a wide range of industries, by designing and manufacturing composite drive shafts. Composite materials allow drive shafts to be lighter in weight, span longer distances, be less prone to imbalance than metallic equivalents, whilst offering the well-known corrosion resistance associated with these materials. The materials used in these applications are made by a filament winding process using carbon and/or glass fibre reinforcement and an epoxy or vinyl ester resin.

Addax describe the development and use of its composite drive shafts in cooling tower applications as follows:

> One industrial application that had unending problems was the cooling tower fan coupling. The biggest issue concerning this application was that the cooling tower is an extremely corrosive environment. Chlorine, sulphides and other chemicals were present in the towers. The industry was required to use stainless steel shafts because that was the only material that did not present a corrosion problem. Because the spans were so long, many times two shafts were required with an intermediary support bearing and a flexible unit. This support bearing required grease and periodic maintenance. Another problem was that the sun would heat the top of the shaft when idle causing distortion. When the shaft began rotating there was a high vibration from the imbalance created by the thermal expansion of one side.
>
> With the use of carbon fibre, the entire distance could be spanned in one section of shaft. This lightweight shaft eliminates a centre support bearing and also eliminates the need for stainless steel shafting in a very corrosive environment.
>
> In addition to a composite tube, a composite flexible disc element was implemented to take up the misalignment between the motor and the

gearbox. The flexible element is also filament wound. It can accommodate up to 1 degree of angular misalignment as well as some axial misalignment. Rather than being designed as several shims subject to fretting, the composite pack is a unitised pack with no fretting surfaces.

Although the flexible element has high misalignment and torque capacity, it imposes relatively low bending loads on the bearings of the connected equipment. Before composite couplings were offered for this application, the most commonly used flexing unit in the industry was the universal joint or stainless steel shim pack elements which fretted and required frequent replacement.

The ease of installation and maintenance has been greatly improved. Shaft alignment is much easier with only one versus a two piece drive line. Composite shafts have been in service for over 10 years in cooling towers. It is estimated that there are over 12 000 of these shafts in service today.

Returning to the actual cooling tower, Hamon & Cie offer a complete range of towers from 500 m^3/hour with practically no limitation on maximum capacity, using composite materials as required or specified for particular applications. Hamon offers its own range of thermoplastic internal components, including film fill systems in PVC, under the trademark COOLFILM (SNCS 40) as a standard fill, and CLEANFLOW (AFNCS 40) as a low fouling fill, together with a splash fill system in polypropylene under the trade name COOLDROP (GRIDS).

The Baltimore Aircoil Company offer three different products based upon significant usage of composite materials:

- The PCS series designed for service in a broad range of air conditioning and industrial applications.
- The UNILITE cooling tower specifically designed around the properties of pultruded composite materials, destined for industrial, process and power generation applications.
- The Ultralite 100 cooling tower from Ceramic Cooling Tower Co, designed with a composite structure and corrosion-resistant components, for cell capacities from 47 to 232 nominal tons per cell.

Many smaller companies manufacturing cooling towers with high usage of composite and thermoplastic materials exist, and we take an example from each of Europe, India and the USA.

Polacel BV, a member of the GEA group, specialises in counterflow towers with capacities of 4 to 6000 m^3/hour, and cross-flow towers from 5 to 6000 m^3/hour, using both composite and stainless steel materials.

CASE (Composite Aqua Systems and Equipment) offers a range of small-capacity induced draft counterflow-type cooling towers with capacities from 7 to 360 m^3/hour, using a composite shell structure and a PVC honeycomb fill.

Delta Cooling Towers Inc has been manufacturing polyethylene cooling towers since 1971, and has also developed air stripping systems for VOC removal from process water and during groundwater remediation schemes. It offers a range of three models of cooling towers:

- Pioneer, a forced draft tower with a capacity of 15 to 400 gpm (3.4 to 91 m^3/hour).
- Paragon, an induced draft tower with a capacity of 185 to 1075 gpm (42 to 245 m^3/hour).
- Premier, an induced draft tower with a capacity of 475 to 2500 gpm (108 to 569 m^3/hour).

10.7 Certa-Lok C900/RJ Restrained Joint PVC Pipe

Certa-Lok C900/RJ restrained joint PVC pipe, manufactured by CertainTeed, is a first in thermoplastic pipe systems, offering a joint restraining system for use in municipal water and fire protection schemes, although the product is derived from a much older product line from CertainTeed, designated Certa-Lok Yelomine and destined originally for mining and industrial applications and more recently for use in association with micro-tunnelling.

The inability of certain soil conditions to support thrust blocks associated with non-restrained buried pipe systems, coupled with the increased demands placed on utility corridors, resulted in the development of this restrained joint PVC pipe concept, which meets the dimensional requirements on outer diameters of cast iron pipe and the performance requirements of the AWWA standard C900.

The Certa-Lok C900/RJ system is designed to eliminate costly and space-consuming concrete thrust blocks in a properly engineered water pipe system. The system provides the restrained joint via the utilisation of precision machined grooves on the pipe and inside the coupling. When the grooves on both components are aligned, a nylon spline is inserted through an opening in the coupling, resulting in a full circumferential restrained joint that locks pipe and coupling together. A flexible elastomeric O-ring joint located inside the coupling provides a hydraulic pressure seal.

There are no metallic components within the system, so no additional corrosion protection is required around the coupling, and a further feature of the system allows field fabrication of the groove on the pipe if required, through the use of a power router and grooving machine supplied by CertainTeed.

Certa-Lok C900/RJ Restrained Joint PVC Pipe is available with class 150 pressure rating (SDR18) in sizes of 4 to 12 inch, and a class 200 rating (SDR18 and 14) in sizes 4 to 12 inch. The system is FM approved, complies with NSF standard 61 for potable water service, and is listed by U/L.

The fact that the joint can be readily disassembled suggests that the system could be used in applications requiring frequent displacement or modifications, such as in irrigation systems, waste and effluent disposal.

CLASS 150 (DR 18) 4"-12" AND CLASS 200 (DR 14) 4"-12"
Includes Coupling, Gaskets, Splines, and Lubricant

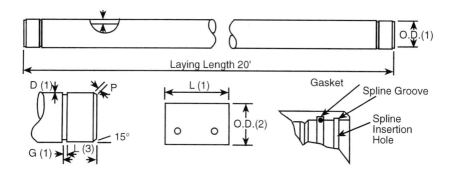

Pipe Dimensions											
Size	OD/(In)	T In/(In) Class150	T Mi /(In) Class 200	P/(In) M(In)	Max	L/(In)	G/(In)	L/(In)	D/(In)	D/(In) M(In)	Max
4	4.80	.267	.343	1/8	3/16	3.000	.375	8.25	5.964	.125	.130
6	6.90	.383	.493	3/16	1/4	3.000	.375	8.25	8.366	.125	.130
8	9.05	.503	.646	3/16	1/4	3.163	.500	10.50	10.947	.130	.135
10	11.10	.617	.793	5/8	11/16	3.500	.500	11.125	13.361	.200	.210
12	13.20	.733	.943	5/8	11/16	3.500	.500	12.00	15.836	.200	.210

Note: All demensions are subject to manufacturing tolerances.
CertainTeed C900/RJ Couplings are boxed and shipped with gaskets (o-rings) factory installed in coupling

Figure 75. Restrained joint pipe configuration and dimensions. Courtesy of CertainTeed Corp.

The system is completed with restrained joint PVC pipe sweeps, which again feature cast iron outside diameter dimensions, and are fabricated from tested Certa-Lok pipe, and are available in 4 to 12 inch, in 90°, 45° and 22.5° angles. Sweeps should be installed in the same fashion as the Certa-Lok pipe, using proper bedding and backfill. Correctly installed the restrained joint sweep does not require thrust blocking.

10.8 Aeration and Oxidation — Aerators and Fine Bubble Air Diffusers

Activated sludge treatment works on the basis of two different but associated processes, firstly the supply of oxygen (air) to the aerobic microorganisms, and secondly the intimate and uniform mixing of the live medium and the effluent. The first is obtained by the diffusion of air, via fine bubble air diffusers; the second by aerators, often low-speed aerators, both of which are supplied by the French company Europelec, and both of which call upon the properties of composite or thermoplastic materials.

The fine bubble air diffuser, marketed by Europelec under the name Aquadisc, uses the corrosion-resistant properties of polypropylene, together with the stability to handle the high temperatures of the compressed air, maintaining the structural stability necessary to ensure a regular distribution of the air bubbles.

Each complete diffuser, equipped with an EPDM membrane and stainless steel collar, has a diameter of 248 mm, is 57 mm high and weighs only 424 g,

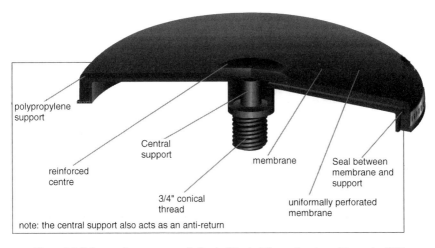

Figure 76. Polypropylene structure of a fine bubble air diffuser. Courtesy of Europelec/SFA.

296 *Aeration and Oxidation — Aerators and Fine Bubble Air Diffusers*

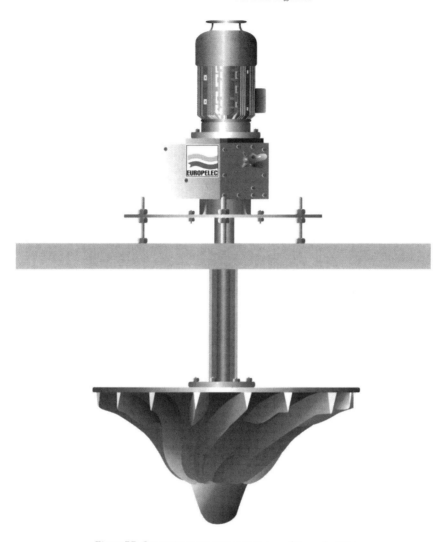

Figure 77. Composite aerator rotor. Courtesy of Europelec/SFA.

making installation and any subsequent removal for maintenance work in the basin extremely light work. The normal service life of the EPDM membrane is generally considered to be between 5 and 10 years, whereas the polypropylene support has a practically unlimited service life.

Diffusion rates can vary between 1 and 6 m^3/hour, and the membrane is designed to maintain an even distribution of very small bubbles, and to limit its speed of ascension, in order to ensure the longest contact time possible between the bubble and the effluent.

Europelec use the mechanical and corrosion-resistant properties of composite materials for its LTF range of low-speed aerators.

The rotor is made from glass fibre reinforced polyester resin and the centre is filled with expanded closed cell polyurethane foam. A circular steel reinforcing ring is imbedded into the composite material and facilitates assembly of the rotor onto a stainless steel drive shaft. The composite material also makes possible the construction of the hydrodynamic profile, designed to optimise mixing and aeration. This same profile discourages any attachment of solids to the rotor, and also helps to avoid any build-up of ice at negative ambient temperatures. The combined composite/foam structure results in a density of less than 1, which not only helps during installation, but also diminishes axial stress on the gearbox.

The aerator can be installed in both fixed and floating configurations, and for the latter, installed on a three-pontoon configuration, again manufactured in composite materials. The floating configuration avoids all problems associated with fixed rotors and variable water levels.

A full range of capacities is available, from 1.5 to 100 kW.

10.9 Dynamic Screening with the Discreen Family

Compared to the tradition fixed or articulated screens, the Discreen range of dynamic screens, manufactured by Mono Pumps Ltd, brings new ideas into the field of screening, the first stage of treatment for both surface water and wastewater, separating out and removing large particles which could damage or reduce the efficiency of further stages of the treatment process, without actually blocking up. This new equipment is self-cleaning, based on continuous movement of the filtering device, itself using corrosion-resistant materials. It requires no routine maintenance, and because of the modular construction, different channel widths can be easily accommodated.

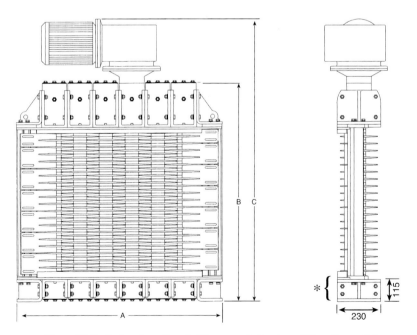

Figure 78. Typical configuration of a Discreen unit. Courtesy of Mono Pumps Ltd.

The Discreen equipment consists of a compact module, comprising a chassis and a set of vertical parallel shafts, onto which are fitted discs, which interlock with each other, with the distance between the discs being determined by the degree of screening required. As the discs are rotating, up to a maximum speed of 65 rotations per minute, they pass any entrapped solids along the face of the screen and evacuate them off the end of the screen. The interactive rotation of the discs guarantees the dynamic, continuous and automatic cleaning of the total surface of the screen.

Depending upon the size of the installation and the characteristics of the application, the discs and spacers, as well as side combs, are supplied in polypropylene, glass fibre reinforced polypropylene or stainless steel. The screening dimension can be set at 2.5, 5.0, 9.0, 13 and 18 mm.

Typical applications, apart from the inlet channel on water and wastewater treatment plants, are to be found on storm basins and storm overflow outlets, and they can also be associated with a grinder, with automatic feed from the Discreen, with a reintroduction of the ground-up solids into the feed channel.

10.10 Oblated FRP Tanks

Tankinetics developed and patented Oblation Technology, an innovative modular construction and shipping system for very large filament wound composite vessels. The oblation process takes advantage of the elastic properties of glass fibre reinforced plastics. The cylindrical shop-fabricated tank section is laterally compressed and restrained within a steel framework for transport to site.

Oblation technology is proposed for two basic situations:

- shipment of very large diameter shop-fabricated tanks which could not be constructed economically on site; and
- assembly of a tank within an existing structure into which full diameter sections could not be moved.

Figure 79. An oblated tank being assembled on site. Courtesy of Tankinetics Inc.

Tankinetics consider the oblated ring module as being similar to a spring, rather like a composite leaf spring in the automobile industry, and calculate the degree of oblation possible without subjecting the module to any excessive stress which could damage it.

Tankinetics consider that there are several advantages when using its system. First, the economic advantages over on-site filament winding. On-site filament winding of large vessels is generally only economic when several vessels have to be made and the set up costs can be spread over a significant workload. Oblation is possible when only one vessel is required.

Tankinetics also considers that shop fabrication is both cheaper and generally of higher quality than construction work carried out on site, and it is much easier to control.

The reduced time on site for assembling an oblated tank, as compared to the total time required to set up a site winding complex, plus the actual winding and assembly time, could be of interest to the client, as the period of disruption is also reduced.

Returning to the actual oblation system, it is evident that the displacement of the ring modules must be kept within the elastic limits of the material, and that consideration must be given to both the resin-rich inner liner and the structural laminate. Each tank is subjected to specific calculations in order to determine the appropriate degree of oblation to be obtained.

Large-diameter composite tanks are virtually all built with stepped wall thicknesses, that is each successive ring module is thicker, going from top to bottom, due to the increased hydrostatic head, with the bottom section being the thickest. This ring is oblated to the allowable width in a straight-sided slot shape. Upper sections, which are thinner and can therefore be oblated to a greater degree, are forced into an 'hour-glass' shape, which allows nesting into the lower ring.

Top and bottom head components are prefabricated for assembly on site, and as such are inserted into the oblated sections for transport.

When the total package arrives on site, the rings are removed from the framework and allowed to return to its normal size and shape, just as a leaf spring returns to its initial form when the loading is removed. In order to avoid working at high elevations, the tanks are normally assembled from the top down. The top head is attached to the top cylindrical section and jointed. This assembly is then placed onto the next cylindrical section and so on, until the entire shell is assembled onto the bottom. Final assembly involves the installation of nozzles, manways and holding down lugs. Once completed the vessel is filled with water and tested for leakage.

Since 1975, Tankinetics has fabricated numerous large-diameter composite tanks using this method. These vessels have performed well, with very limited maintenance, none of it being related to the oblation process itself. Oblation, like any other manufacturing process, must be carefully controlled and must be based upon detailed and appropriate analysis and design.

10.11 Gravity Filter Floors in Composite Sandwich Construction

CP Composites is a small family-owned company situated in the west of France, dedicated to the supply of various equipment in composite materials for the treatment of potable and wastewater, often working in close cooperation with the larger water treatment engineering and contracting companies. This application, as well as the following are typical of the results which can be obtained by a manufacturing company working in close cooperation with a specialised engineering organisation.

Gravity filters, generally rectangular concrete basins, are use in practically all potable water plants and some wastewater and effluent treatment plants. The floor of these filters are equipped with thermoplastic nozzles used to inject air into the filter media during the regular backwash of the filter media. The

Figure 80. Composite sandwich gravity filter floor. Courtesy of CP Composites.

filter floors must support the weight of the filter media and at the same time allow the distribution of air under the floor, and the passage through the washing nozzles.

The CPC composite floor is made using a sandwich construction in order to obtain the required level of strength and rigidity using an 8 mm hexagonal polypropylene honeycomb structure between two composite skins. These panels can be made in all sizes up to $4 \text{ m} \times 8 \text{ m}$, with a total thickness varying between 50 and 100 mm depending upon the overall dimension and the layout of concrete supports or beams in the filter bed. Filter floors have been supplied as original equipment or as replacement material during refits of existing plant.

The design basis of the standard panels is:

- maximum loading on to the panel — 2500 daN/m^2
- maximum loading under the panel (air pressure) — 4000 daN/m^2
- maximum deflection between support beams — $1/200$ of the span;
- number of nozzles per m^2 — 50 to 55.

Mechanical properties of the panels.

- Polypropylene honeycomb structure; density — 80 kg/m^3; compressive strength at rupture — 1.4 MPa; shear strength — 0.5 MPa.
- Composite skin; density — 1.5 kg/dm^3; flexural modulus — 8000 N/m^2; flexural strength — 190 N/mm^2.

The composite skins are generally manufactured with an isophthalic polyester resin, although vinyl ester formulations are available for aggressive industrial effluent. CPC supply all necessary fittings and fixtures required to assemble and install the panels.

The most significant advantages of composite floors over the concrete alternative are:

- light weight which facilitates installation;
- larger panels, meaning less joints and reduced problems of sealing between panels;
- a guaranteed long-term corrosion resistance; and
- facility to tailor make to fit any existing form and configuration.

10.12 Articulated Lamellae Filters for Easy Cleaning

Many different forms of lamellae filters are available, and various equipment and thermoplastic materials developed for this application have been examined elsewhere. CP Composites have developed a concept for lamellae decanters which bring some extra beneficial effects both to operation and maintenance.

Figure 81. Articulated composite lamella filter. Courtesy of CP Composites.

Optimum performance of a decanter can only be obtained if the total throughput is distributed in an equitable manner over the total surface of the decanter. The Equiflux lamellae decanter as manufactured by CPC uses a calibrated and adjustable orifice to introduce the water into the zone of decantation, with a controlled pressure drop. The lamellae can be rotated, using a hydraulic piston, during service to obtain a filtration angle of between $45°$ and $60°$, to suit the quality of the water being treated.

This same system of regulation can be used to place the lamellae in a vertical position for cleaning purposes.

This system has been used to treat both potable water and wastewater after an initial coagulation and flocculation.

The significant advantages of the system are:

- an equitable distribution of the fluid to be treated;
- a possibility to adjust the angle of the lamellae to suit actual conditions;
- assistance in cleaning the lamellae;
- a total corrosion resistance via the utilisation of an isophthalic resin based laminate; and
- easily adapted to existing conventional filters, with an increased throughput.

10.13 FlowGRiP Composite Planks

Pultruded and moulded gratings and other types of floor panels are available from a wide range of manufacturers, and information on principal suppliers are included in Chapter 14. The FlowGRiP composite planks manufactured and marketed by Redman Fisher Engineering Ltd in the UK offer some distinctive advantages over other standard products.

Redman Fisher, a member of the CI Group, located in the Birmingham area of the UK, is a company dedicated totally to the supply of open grid industrial flooring, essentially in mild steel, stainless steel and aluminium, together with associated hand railing, stairs and ladders.

Drawing upon its experience with aluminium extruded flooring, under the trade name Flowgrip its composites division produces its unique and patented composite version FlowGRiP.

The standard plank, 6000×500 mm is a pultruded one-piece section, supplied as either a solid or open grating panel, which can be used not only as flooring but also, for example, in the solid form, for covers on conduits and channels in water treatment plants linked to an odour control unit.

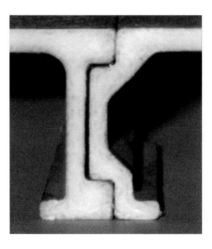

Figure 82. Tongue and groove joint. Courtesy of Redman Fisher Engineering Ltd.

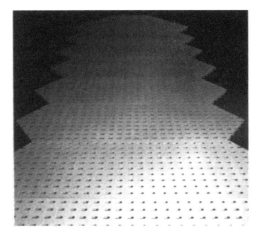

Figure 83. Flowgrip makes it difficult to see lower levels. Courtesy of Redman Fisher Engineering Ltd.

FlowGRiP is claimed to be the only one-piece interlocking FRP flooring system available, offering both solid or open grating panels, and which can be fabricated to any size.

The panels can be installed with spans up to 1500 mm with only a 1/200 deflection, supported on only two sides, and needs no joint clips. The high level of rigidity means less bounce than with conventional FRP grating, leading to increased operator confidence and safety. The unique tongue and grove joint interlocks the panels in such away that there are fewer holding down clips, no joint clips, no trip hazard due to differential deflection between panels, together with an improved load transfer between the panels. At the same time removable panels can be easily accommodated.

FlowGRiP panels weigh 18.5 kg/m^2, are only 40 mm deep, and can easily be cut with conventional tools on site. The grating form has relatively small open areas, designed to prevent a 15 mm diameter ball from passing through, and provides a more secure feeling for personnel as view of lower levels is inhibited.

In line with most other pultruded materials FlowGRiP is also available with fire-retardant resins which meet the requirements of such standards as BS476 Part 7 Class 1 and ASTM E-84 on flame spread ratings. The fire-retardant resins used are classed as self-extinguishing to ASTM D-635 and the low level of smoke emission obtained has allowed the material to receive the approval of the London Underground, the New York subway and the US Navy.

Redman Fisher offer three anti-slip finishes varying from a fine-grade finish which offers good anti-slip properties whilst minimising abrasion in case of contact with the surface, up to the coarse grade for use where the slip hazard or traffic is heavy and where it is necessary to improve the level of operator security.

The safety loads and associated deflections are indicated in Table 10.

Table 10. Loading and deflection of Flowgrip panels

Clear span (mm)	Flow GRIP* Solid (40 mm deep)				Flow GRIP* Grating (40 mm deep)			
	UDL		CL		UDL		CL	
	kN/m^2	Deflection (mm)	kN/m^2	Deflection (mm)	kN/m^2	Deflection (mm)	kN/m^2	Deflection (mm)
500	104.00	2.23	35.00	2.50	104.00	2.44	32.10	2.50
750	37.80	3.75	17.39	3.75	34.29	3.75	15.80	3.50
1000	16.50	5.00	10.20	5.00	14.29	5.00	9.20	5.00
1250	8.60	6.25	6.66	6.25	7.75	6.25	6.00	6.25
1500	5.02	7.50	4.68	7.50	4.52	7.50	4.20	7.50
1750	3.18	8.75	3.46	8.75	2.86	8.75	3.10	8.75
2000	2.13	10.00	2.66	10.00	1.92	10.00	2.40	10.00
2250	1.34	10.00	1.87	10.00	1.20	10.00	1.30	10.00

Loads based on 1/200 deflection.
UDL: uniform distributed load; CL: safe concentrated line load in kN/m width at midspan.
Light duty UDL 3.0 kN/m^2: access limited to one person; general duty UDL 5.0 kN/m^2: regular two-way pedestrian traffic; heavy duty UDL 7.5 kN/m^2: high-density pedestrian traffic.
Flowgrip panels conform to BS4592 Part 4 which states that deflections shall be 1/200 of the effective span, or 10 mm, whichever is the lesser.
Courtesy of Redman Fisher Engineering Ltd.

CHAPTER 11

Case Histories

11.1 Nanofiltration Plant at Méry sur Oise, Paris, France

Owner — SEDIF (Syndicat des Eaux d'Île de France)
Operator and main contractor — Vivendi Water (formerly Générale des Eaux)
Subcontractor for the nanofiltration equipment — OTV

The largest nanofiltration plant in the world producing potable water from river water, and in this case a river with a relatively high level of pollution from organic matter, was officially inaugurated in 1999 by the SEDIF. This unit, an extension of an existing unit using conventional filtration technology, will add 140 000 m^3/day of potable water to the existing 200 000 m^3/day supply. The use of water from the river Oise imposes an unusually high level of constraints on the design of the plant, due to possible extremes in temperature (1 to 25°C), and the variable level of solids in suspension and other pollutants. Mery sur Oise is situated to the northwest of Paris and the plant supplies a population of 800 000 inhabitants in the northern suburbs.

Although conventional materials and filtration processes abound within the new unit, the heart of the new plant, the actual nanofiltration plant, relies heavily on plastic materials for both the filtration membranes and the pressure vessels containing the membranes.

The sequence of treatment in the plant to the nanofiltration unit is as follows:

- Water is pumped from the river Oise, and allowed to decant naturally during 48 hours in storage basins.
- A sequence of pre-treatments is applied to the water prior to it reaching the nanofiltration plant, including:
    - Coagulation-flotation,
    - a passage through lamella decanters,
    - an ozone treatment using ozone produced from oxygen,
    - a passage though dual layer gravity filters (10 basins) using sand and anthracite, with an injection of a coagulant at the head of the filter in order to achieve optimum retention.

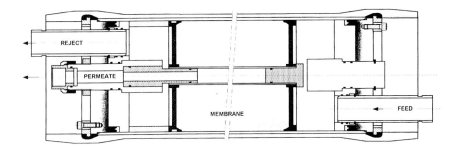

Figure 84. Composite pressure vessel for reverse osmosis. Courtesy of EWWA Sarl.

At the end of this pre-treatment phase, the clarified water is held in a storage tank, prior to entering the nanofiltration plant.

Before actually entering the nanofiltration vessels the water is pumped through a final pre-filter which will remove any remaining large particles which may result from any dysfunction in the previous stages of filtration. The pressure is then increased (from 8 to 15 bar depending upon the temperature of the water) and enters the filtration vessels.

The nanofiltration technology permits the production of a very high quality of potable water, by filtering water under pressure through membranes which have openings 10 000 times smaller than a human hair (1 nanometre=1 thousandth of a micrometre). The nanofiltration membranes ensure the elimination of organic materials and dissolved organic compounds, and in particular their biodegradable constituents which are the nutritive source for bacteria.

The nanofiltration unit is built up from eight separate trains, with each train comprised of three sequential stages, each stage being dimensioned to filter 50% of the individual input. Thirteen percent of the initial feed is rejected from the third stage, which means a total conversion level of 87%. Each stage of each train contains respectively 128, 54 and 28 pressure vessels for a total of 190 vessels per train and 1520 vessels for the total of eight trains. Each pressure vessel contains six spiral wound membranes 1 m long, that is 1140 membranes per train, and 9120 membranes for the total plant.

The pressure vessels are fabricated in dual laminate form, with a U-PVC inner liner and a filament wound structural laminate, with end caps manufactured from polypropylene. They are designed for a maximum service pressure of 20 bar at 25°C, and were qualified by a destructive hydraulic testing programme at 120 bar. Each membrane is built up with 29 spirally wound double sheets which results in an effective filtration surface of 37 m^2, or 42 000 m^2 per train, or 340 000 m^2 for the total plant (roughly equivalent to the total area of 70 football pitches) all included in a building of 3600 m^2. The membranes were developed specifically for the Mery sur Oise project by the American company FilmTec Corporation in Minneapolis, a subsidiary of the Dow Chemical Co, and had been tested on a prototype plant since 1993.

These membranes are made by laying up of two materials on a polyester structure, a polysulphone which ensures the quality of the state of the surface, and a polyamide (polypiperazine) which forms the actual nanofiltration layers. Ever since the beginning of this project, these membranes have constantly been improved in order to treat the river Oise water in the most efficient manner, whilst limiting the retention of calcium which would have made the water too corrosive and unsuitable for drinking. The membrane actually installed is designated NF 200B-400 (NF for nanofiltration, 200 symbolising the year 2000 and the actual industrial start-up date, and 400 corresponding to the effective filtration surface of 400 ft^2 (37.2 m^2).

The filtered water finally undergoes a CO_2 degassing, to eliminate CO_2 build-up due to the calcium retention, via eight degassers again built in composite materials and installed on the roof of the nanofiltration plant. Finally, and by measure of security, the water is subject to a disinfection by UV radiation, and at the same time is treated with sodium hydroxide in order to re-equilibrate a slight acidity resulting from an acidification (antiscalant) at the beginning of the process.

Finally, the installation of this new material will have no effect on the cost of supplying potable water to the Paris area. An extra cost of FF0.85/m^3 (US$0.50 per 1000 US gallons) due to the nanofiltration will be offset by economies in other parts of the plant.

11.2 Water Treatment Plant — Odour Control at Howdon, UK

Main consultant — WSP
Client — Northumbrian Water

One of the largest, and assuredly the most innovative, odour abatement scheme ever built in the UK, including large covers and ducting constructed from an association of composite and thermoplastic materials was successfully commissioned in 1999 at the Howdon sewage treatment works for Northumbrian Water. The UK specialist in odour abatement schemes, Aform Projects was responsible for both the conceptual and detailed design, tender preparation and approval of tenders, contract and installation management as well as the quality control of subcontractors.

The project, total value £8.5 million, required covering and ventilation of eight rectangular tanks, 25 m wide, with a total area of 13 000 m^2. Each tank, 65 m long, larger than six tennis courts placed side by side, has been successfully covered using a combination of structural composite beams and tensioned fabrics.

The tent-like covers, which incorporate ducting systems within the composite beams, were jointly developed by Consulting Engineers WSP and Aform Projects. The covers themselves are the largest of their kind in the UK, and this new concept means that now even the largest tanks can be completely covered, with the additional advantage that their aesthetic forms considerably improve the appearance of the treatment plant.

Computer-assisted structural analysis was necessary to develop the beam concept, based on box section fabrication using advanced adhesive technology supplied by Permabond, and once evolved, a prototype beam was subjected to a wide range of tests to prove the load-bearing capacities and associated deflections. The beams are linked by flexible covers made from a specialist PVC-coated polyester fabric.

Panels are assembled onto the beams with Keder beading, an industrial version of that used to attach an awning to a caravan, and held down by stainless steel cables. The fabric and fittings are designed to resist an uplift of

Figure 85. Tank cover FRP beams prior to installation of the polyester fabric. Courtesy of Aform Projects Ltd.

around 8 tonnes per panel, but can be removed to allow access for cleaning and maintenance within the tanks.

Composite materials were also used for the FlowGRIP flooring systems which provide both non-slip access walkways and covers on channels carrying effluent to the tanks. The FlowGRIP system supplied by Redman Fisher Engineering combines the traditional corrosion-resistant benefits of composite materials with a unique torque and groove jointing system, minimising deflection between adjacent panels and eliminating trip hazards.

11.3 Geothermal-Sourced District Heating

The use of low-temperature thermal and geothermal water in many sites in Europe, essentially in France, Italy, Germany and Austria has called for the use of epoxy-based composite pipe systems for a wide range of applications, due in some cases to the corrosive nature of the geothermal water itself, or because of the installation in a generally aggressive environment.

Geothermal water, at temperatures around 70 to 90°C, is available in several areas in Europe, including the Paris basin in France, where more than 60 geothermal wells have been drilled. The water is generally located at a depth of around 1800 m, and, during periods of high energy costs, many attempts have been made to replace or assist fossil fuelled heating plants with this interesting source.

The initial technique involved drilling a production well down to 1800 m, from which the hot water was drawn, coupled to a heat exchanger to transfer the heat from the highly aggressive geothermal water to normal process or potable water. A pipeline approximately 2500 m long was then used to take the cool, geothermal water to a re-injection well, again around 1800 m deep, in order to re-inject the water back into the same geological structure, but at a sufficient distance to avoid any cooling of the production source.

More recent techniques have benefited from technology developed in the oilfield, resulting in drilling two deviated wells from the same site, with a distance of around 1000 to 1500 m between the bottom of each well. This technology eradicates the requirement for the pipeline between the two well heads.

We summarise below three different uses of composite materials in the field of geothermal heating.

11.3.1 Jonzac, France

    Client — Jonzac town council
    Manufacturer — Sipap Pipe Systems
    Insulation and installation — Wanner Isofi
    Installed — 1979–82

Figure 86. Installation of insulated epoxy pipe.

Over a three-year period in the early 1980s a geothermal low-energy district heating scheme was installed throughout the town of Jonzac, located 60 km north of Bordeaux in France. The total length of pipe supplied by Sipap and installed in the system is approximately 20 km, with diameters of 50 to 200 mm.

The concept of this piping system, as used for many other geothermal heating schemes installed in France in the late 1970s and early 1980s, relies on three plastic materials:

- The carrier pipe is an amine-cured epoxy-based filament wound FRP pipe
- The insulation in polyurethane foam
- An outer casing, into which the polyethylene foam is injected, in HdPe.

In some other similar but later applications expanded epoxy foam has also been used as insulation, especially for on-site applications, at junctions between pipes and between pipes and fittings.

The operating conditions of the Jonzac installation are maximum 7 bar at maximum 90°C. After an initial period of 10 years, with no incident on the actual pipe system, a section was uncovered and a sample cut-out of one of the pipes. Once a thin soft deposit had been removed from the inner surface of the cut out pipe section, further examination showed absolutely no sign of corrosion, abrasion or any other sign of wear. The inner surface was smooth and shiny, and an examination of the structural laminate showed no sign of any deterioration. The installation has continued to function, lately with some minor incidents, probably related to installation practices which existed at that time and which have been revised since, especially concerning the design and fabrication of thrust blocks. In fact, some pipe weeping and leakage was experienced at the point of penetration into the concrete thrust block, probably caused by a displacement of the thrust block, and the absence of any cushioning material between the pipe wall and the concrete. This problem was overcome on later installations by wrapping an elastomeric band around the pipe, at the zone of penetration of the pipe into the thrust block, prior to pouring the concrete. The overall result is practically 20 years of service with only the minor incidents mentioned above being experienced during the second 10-year period.

11.3.2 Le Mée sur Seine, France

Client — CGCU (Compagnie Geothermique de Chauffage Urbain)
Supply and installation — Sipap Pipe Systems
Installed — 1989

One of the first major French geothermal heating schemes was installed in the town of Le Mée sur Seine, approximately 50 km to the south-east of Paris, and involved the drilling of two separate wells, one production, the other re-injection, as described above. The transfer line between the two wells was initially installed in coated steel. Operating conditions were 20 to 30 bar, with a temperature of between 30 and 40°C, handling geothermal water (a hot saline fluid, with a slightly acidic pH, with dissolved gas rich in H_2S and CO_2, together with a sulphate-reducing microbiological activity) and the pipe began to fail within two years. Eventually, with more than 30 major leakages requiring plant shut down, and, during later years, occurring on an almost monthly basis, the decision was taken to replace the line with composite materials. This decision was taken based on experience in other plants using FRP as a short loop between two well heads on the same site, as well as actual experience in the Le Mée sur Seine plant with composite pipe installed around the heat exchangers.

An underground adhesive bonded amine-cured epoxy system, diameter 200 mm, rated 30 bar, was installed and tested successfully at 45 bar for 24 hours, after an initial failure to hold pressure. This failure, a microscopic weeping from a single point, causing less than 1 bar pressure drop over a 12 hour period was finally traced by helium detection, and was found to be caused by vandalism, with damage caused by striking the pipe with steel bars before the trench had been backfilled. The pipe has performed satisfactorily ever since, without even the slightest incident during a period of more than 10 years of continuous operation.

11.3.3 Down hole tubing and casing

Because of the nature of geothermal water, as described above, composite materials are an automatic choice as far as corrosion resistance is required. Down hole tubing and casing requires a high level of mechanical strength, which in turn means relatively high wall thickness. As far as pure mechanical strength is concerned composites are comparatively more expensive than steel, and in spite of early successes with composite casing, due essentially to initial price levels, coated steel has dominated the low energy geothermal market.

An early and initial success for composites in this field can be found in the suburbs of Paris, where, in 1975, the French oil company Total drilled two wells as part of its research programme into new energy resources. The two wells, drilled approximately 10 m apart, were deviated by about 30° at a depth of 350 m, which resulted in a separation of approximately 1000 m at the final vertical depth of around 1750 m.

The geothermal water is high in salt content, approximately 23 000 ppm of NaCl, which presents a significant corrosion problem, solved in this case by using 7 inch composite casing supplied by Fibre Glass Systems Inc, steel casing with an epoxy coating, special connections and titanium heat exchangers.

Total obtained the following conclusions from this experience:

- This first installation of composite production casing in a deviated well presented no problems and the results have proven satisfactory.
- At the time this project was completed, Total concluded that composite casing could be used to a depth of 2000 m, but that new designs and fabrication techniques would enable their use in deeper deviated wells where corrosion was a problem.
- A composite casing of $13\frac{3}{8}$ inch was not available, and the upper portion of the hole (down to 350 m) was cased with $13\frac{3}{8}$ inch steel casing lined with epoxy resin and connected to the 7 inch composite casing by means of a corrosion-resistant Uranus B tie-back assembly.

Because of high costs related to the project mentioned above, many of them perhaps related to the somewhat experimental aspect of a first-off, the French geothermal industry turned to metallic alternatives for all the subsequent well casings, up to the invention of a new well concept as detailed below. The corrosion experienced with surface pipe, as demonstrated with the Le Mée sur Seine project, was ongoing, but less visible, in all underground applications. The down hole injection of corrosion and scaling inhibitors, associated in many cases with biocides, has been developed for treating many of the wells. Both composite and thermoplastic materials have been called on to handle this injection, either using down hole treatment tubing (DHTT), or auxiliary injection tubing (AIT).

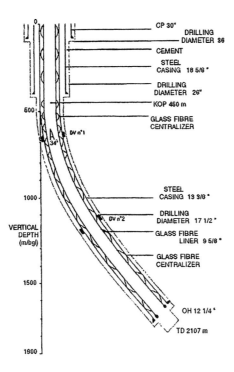

Figure 87. Combined steel/composite casing lined well. Courtesy of GPC.

The DHTT is a string of glass fibre threaded tubing, as manufactured and used in the oil production process, whereas development of AIT has in some cases called upon even more highly developed material properties and engineering technologies.

The new geothermal well concept mentioned above, developed and patented by Geoproduction Consultants (GPC), was, because of its innovative design, granted financial support from both the European Union and the French Ministry of the Environment. The new well is located on the existing geothermal site of Melun l'Almont, 50 km south of Paris, and employs 2000 m of $9\frac{5}{8}$ inch Star Fibreglass casing manufactured by Fibre Glass Systems Inc, reference series 2000-DHC-T&C.

This new well is the fourth to be drilled on this site, where in 1969 two initial boreholes were made, production and injection. The third well was drilled in 1989 to replace the original injection well which was abandoned after 20 years of service. The fourth well was conceived in order to reduce corrosion and scaling phenomena that severely affect the integrity and service life of wells in the mid-Jurassic (Dogger) carbonate formations of the Paris basin used as a source for district heating. Under this new concept the wells

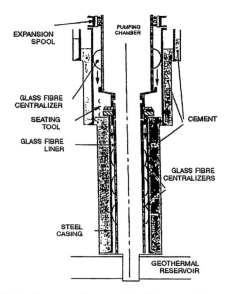

Figure 88. Configuration of the new Melun l'Almont well. Courtesy of GPC.

are completed by combining cemented steel casings and composite liners, whilst the annulus is kept free.

The steel casing provides mechanical strength (propping function) whereas the composite liners furnish chemical resistance (corrosion and scaling protection). The free annular space allows the circulation of corrosion/scaling inhibitors and/or biocides, which would otherwise be circulated through a down hole chemical injection line. This method also allows for the removal of the composite column for inspection and eventual repair or replacement.

It is worth noting that this design can accommodate a submersible pump set, and as such, the upper composite lining is placed under compression, and the lower part, diameter $9\frac{5}{8}$ inch, under tension, that is it is freely suspended and supports its own weight. Vertical displacement of the composite lining is eased by an upper expansion spool and composite centralisers. At Melun, due to exceptional reservoir performance, artificial lift is not required, and self-flowing at high production rates prevails, a fact which led to the simplified design shown in Figure 88.

The actual drilling called for a large diameter (44.5 cm), 2000 m long deviated (35°) section, in which was run a 24.5 cm diameter, 2000 m long centralised composite string. Well drilling, completion and testing operations were completed in 52 days starting in February 1995 and the well was on-line on March 21 of the same year, producing 200 m^3/hour at 70°C, with a wellhead overpressure of 2.5 bar. It is connected to the two existing wells, with two production and one injection wells.

The concept of using wells with steel casings and removable composite linings is being seriously considered in order to extend the lifetime and improve reliability of existing installations (over 30 double-well installations are currently being operated in the Paris area). Should these projects go ahead, a new production well would be drilled, and the two existing wells reconditioned and modified into injection wells. Total cost per project is estimated at around US$ 3 million.

In the meantime the operation at Melun l'Almont is now set for at least the next 20 years, and will supply 35 000 MWh per year to a group of about 3000 dwellings. This new generation well opens up prospects not only for the continuation of existing operations but also for new ones, as they will no longer be subject to corrosion and scaling.

Composite surface piping manufactured by several of the suppliers of epoxy-based systems, as well as down hole casing supplied by Fibre Glass Systems Inc have been installed in both geothermal and thermal water installations in France, Germany, Austria and Italy, handling a wide range of corrosion problems, and at the same time meeting the health and safety requirements and specifications for handling thermal water.

11.4 Vertical Composite Sewer Pipe — Ipswich, UK

Client — Anglia Water
Contractor — AMEC Tunnelling
Manufacturer — Johnston Pipes Ltd

In 1999 Johnston Pipes completed the supply of 250 m of 1400 mm diameter pipe for an unusual application. These pipes have been used in a vertical position to meet a major challenge on a large sewer relief tunnel project constructed for Anglia Water on the river Orwell in Ipswich. The pipes have been used to line five vertical drop shafts of up to 48 m deep, and which connect the existing sewer network to the new relief tunnel.

The pipes were moved from the horizontal to the vertical position in segments, using purpose-built lifting and support clamps. Specially designed trestles were also used at both sides of the vertical shaft entry, in order to support the base, intermediate and upper pipe sections throughout the lifting and positioning operations.

Mechanical harness rings were used to secure and joint as they were lowered down the shaft, and tie rods were positioned to the rings and torqued to achieve joint make up. This jointing sequence was repeated by adding 6 m lengths of pipe, as the assembled length was lowered into the shaft, and until the bottom was attained. Johnson Pipe and Amec Tunnelling co-operated to design and provide custom-designed, space-sensitive lifting equipment in order to resolve problems caused by the extremely limited space available on site.

Once the pipes were in place and secured the annulus between the steel casing and the composite shaft liner was filled with cementatious grout to which non-shrink additives had been added.

The main reasons for retaining composite materials were two-fold:

- The light weight of the pipe sections facilitated all lifting and positioning, of even greater significance when one considers the restrictions on space and manoeuvring.
- The long-term corrosion guarantee inherent with all composite materials, especially important in this application due to the aggressive

Case Histories 323

nature of the sewage, and the practically impossibility to repair or replace the material once installed.

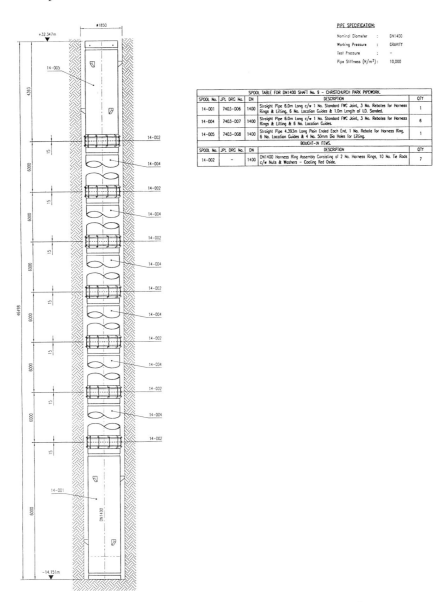

Figure 89. Layout of vertical piping. Courtesy of Johnston Pipes Ltd.

11.5 A 2100 Metre FRP Pipeline Insertion — Portland, USA

Supplier — Smith Fiberglass Products Company
Installed — 1983

The rocky coast of Maine is the location of a very early, and now well-proven application of relining an existing steel water main with a composite liner. Over 2100 m of 12 inch wrought iron main lying between Mackworth and Great Diamond Islands off Portland was completely relined with 83 Red Thread pipe from Smith Fiberglass Products Company.

Savings of several hundred thousand dollars were achieved by the Portland Water District in push–pulling the 8 inch diameter Red Thread pipe through the deteriorating 12 inch main rather than replacing it. The original main, installed in 1943, forms a long curve 12 m beneath the surface of Portland Harbor.

A thorough cleaning of the main was accomplished before insertion began. A total of 28 Girard Polly pigs of increasing size from 9 to 12 inch were forced through the main. A 10 mm rope was attached to the final pig and sent through the 2100 m line to complete this first phase of the project.

The next stage involved pulling a heavier 25 mm rope through, followed by a 16 mm steel cable. Two 35 tonne cranes, one at each end of the pipe, one for pulling and the other for pushing were major pieces of equipment for the project.

The Red Thread pipe was assembled into seven lengths of 350 m, plus one shorter section to make up the total required length. Each pre-assembled segment was pressurised a number of times to 20 bar and inspected for leakage. A special nose plug was bonded to the first section of the pipe. The 16 mm cable was attached and the pipe was placed inside the original main.

The second section was bonded to the first section, and this new joint was pressure tested, and this procedure continued as each subsequent section was joined.

Except for one obstruction, which was overcome by pulling the pipe back 4 m and then pulling it forward under tension, the project continued smoothly and the insertion was completed in seven days. The pipe appeared on Mackworth Island in excellent condition, with only minor surface scratches to show for its underwater trip, and has performed continuously ever since.

Figure 90. 1000 ft (305 metres) long sections of pre-assembled pipe for insertion. Courtesy of Smith Fibreglass Products Company.

11.6 FRP Folding Covers — Farmoor, UK

Supplier — Aform Projects Ltd
Installed — 1995

A new design of folding FRP covers was installed over the GAC Contact Tanks at the Farmoor reservoir. The particular requirements of the application necessitated a fresh approach to the design of the covers.

The GAC Contactors at Farmoor are contained in eight tanks each 6 m wide by 13 m long, and are installed in a single row standing 8 m high.

The particular requirements of this project were that the covers should:

- Exclude light in order to minimise growth of algae.
- Exclude wind-blown debris and insects.
- Resist all internal and external environmental conditions.
- Comply with Thames Water specification for FRP covers to tanks and chambers, especially as regards safe operating conditions and access.
- Comply with local planning requirements to restrict overall height to a strict minimum.

Figure 91. Folding FRP covers. Courtesy of Aform Projects Ltd.

- Comply with the requirements of the Health and Safety Executive.
- Expose 90% of the tank area when the covers are opened.
- Not encroach on the surrounding walkways and access area when in the full open condition.
- Be opened and closed by a single operator.
- Minimise solar gain.

The design which evolved consisted of 13 double curved FRP segments for each of the eight tanks. Each segment is 1 m wide, spans the width of the tank, and is able to stack closely within each other.

These segments were attached to a series of lever mechanisms which allow them to rotate through 90° and then concertina into a stack at the end of each tank. The mechanisms are electrically driven via a gearbox and cable-winding drum. The cable draws the covers open with a simple press of a button, and can be stopped at any point between fully open and fully closed. Closing is a reversal of the opening process.

The operating mechanism was fitted with all required safety locks and guards, and particular attention was paid to making it infallible, and includes an in-depth appreciation of the condition combining a partly open cover and maximum design wind loadings.

11.7 FRP Storage Tanks — Massachusetts, USA

Supplier — Ershigs Inc, Bellingham, Washington, USA
12 off FRP storage tanks, diameter 14.6 m by 12.2 m high (48 ft diameter × 40 ft high)
Installed — 1988

In 1988 Boston Edison completed a major wastewater containment tank farm project at two of its generating plants. These field-manufactured large diameter filament wound vessels are storing over 19 000 m³ (5 million gallons) of wastewater that previously were in ponds on site. The wastewater contains small amounts of acid and caustic substances with a pH range of 2 to 13. Each vessel is 14.6 m in diameter by 12.2 m high and holds 1600 m³ (430 000 gallons) of effluent.

Figure 92. Site manufactured storage tanks, diameter 14.6 m (48 ft). Courtesy of Ershigs Inc.

Design conditions include:

- Maximum water temperature 100°C (212°F).
- Seismic loadings as per Zone 3.
- Snow loading of 146 kg/m^2 (30 lb/ft^2).
- Wind loading of 146 kg/m^2 (30 lb/ft^2).
- Personnel loading on the covers of 245 kg/m^2 (50 lb/ft^2).
- Ambient temperature of between $-15°C$ and $38°C$.

The vessels were fabricated from a highly reactive Novolac-modified vinyl ester resin, with a tensile elongation greater than 5% for the anti-corrosion barrier, and standard resilient vinyl ester for the structural laminate. The toro-spherical covers were shop-fabricated in 16 sections and assembled on site, whereas all sidewalls were filament wound without any vertical or horizontal seams.

These 12 vessels represent one of the world's largest installation of FRP field-manufactured vessels, using around 300 tonnes of resin and 150 tonnes of glass fibre reinforcement.

11.8 FRP Pipe in Wastewater Treatment Plant — Arizona, USA

Supplier — Smith Fiberglass Products Company
5000 m of 2 to 24 inch pipe
Installed — 1983

More than 5000 m of FRP piping had been installed many years ago to carry solid and liquid sewage wastes at the Pima County, Arizona, Ina Road Water Pollution Control plant just outside of Tucson.

Green Thread, Polythread and Big Thread pipes and fittings in 2 to 24 inch diameters were selected because their corrosion resistance offered low maintenance and long-life potential, and their lighter weight in comparison to steel or concrete piping systems made them easier to handle and install.

Figure 93. Lifting into place of a 24 inch (600 mm) diameter prefabricated pipe spool. Courtesy of Smith Fibreglass Products Company.

Other considerations were that the glass fibre pipe selected had similar flow characteristics than more expensive glass-lined steel, and could still withstand the steam cleaning laid down by county officials who wanted a pipe that could be cleaned periodically to prevent clogging of lines with sludge.

The pipe supplied by Smith Fiberglass is used mainly to transport and recirculate wastewater between various clarifying tanks and other treatment apparatus. Since many of these circulation lines had to be installed underground in 4 m by 4 m tunnels, the lighter weight of FRP became particularly appealing in comparison with similar diameter pipe made of more traditional materials.

The FRP pipe has been used in the following areas:

- 850 m of 4 inch diameter Green Thread pipe carries primary sludge from a primary clarifying basin to a sludge thickener and control building where sand and other gritty material is taken out by a cyclone degritter.
- From the degritter the 4 inch pipe takes the degritted primary sludge to four sludge thickeners, where more settling takes place.
- A total of 600 m of 6 inch diameter Green Thread pipe takes the thickened sludge in two lines to a digester control building where 125 m of 24 inch Big Thread pipe circulates it through two 90 ft diameter digesters. The digesters reduce the volume of organic material by enabling it to decompose anaerobically and thus be partially converted to methane gas which is burnt off in the facility's power building.
- Some 350 m of 6 inch Green Thread pipe takes digested sludge from the digester control building to a vacuum filter where 20 to 30% of sludge is recuperated in a solid state.
- A total of 750 m of 14 inch Poly Thread takes returned activated sludge in four lines from the secondary clarifiers back to activated sludge reactors for further processing.
- 100 m of 12 inch Green Thread and 550 m of 16 inch Poly Thread take supernatant (the lighter material in thickened sludge) back from the previously mentioned sludge thickeners to the primary clarifiers for further processing.
- Another 350 m of 4 inch Green Thread pipe carries scum (floating material on top of the waste water–mostly grease) from scum troughs at the end of the primary clarifier trains in a single line to the sludge thickener where the scum is collected.
- A further 500 m of 6 and 8 inch Green thread pipe carries overflow from the top of a 36 inch diameter concrete line between the activated sludge reactors and the secondary clarifiers to a mixing box in the sludge thickener where it is mixed with primary sludge, digested sludge and service water. The resultant mixture is then transported to sludge thickeners for further processing.

- In addition to the above, a further 1000 m of 2 inch Green Thread pipe has been installed for various sampling lines, and some 350 m of 6 inch Green Thread pipe for sump drain lines in the secondary clarifiers.

At the time of construction, the Ina Road Water Pollution Control facility was the second largest sewage treatment plant in Arizona. It had a peak design capacity of 53 million gallons per day (200 000 m^3/day) and was intended to handle future sewage treatment needs of a regional area containing up to 500 000 inhabitants.

11.9 FRP Pressure Pipe in Sewage Transfer Scheme in Grimmen, Germany

Supplier — Owens Corning Eternit Rohr
4600 m of PN10 400 and 500 mm diameter pipe
Installed — 1995

The old sewage treatment plant of the small town of Grimmen in the province of Mecklenburg, Germany was built in 1960, before the territory surrounding the River Trebel was declared a water protection area.

Because of increased purification requirements, the plant had to be upgraded with a biological treatment stage, but the works had to be constructed outside of the water protection area. Therefore, the sewage treatment company of Grimmen decided to build a new plant and chose a location at the opposite side of the town, well away from the river and the protected area.

It would not be possible to transport the sewage to the new location by gravity, so two parallel pressure pipelines had to be installed in order to connect the plant with the existing sewer system.

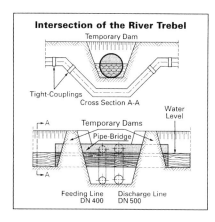 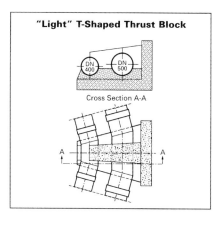

Figure 94. Details of pipe installation. Courtesy of Flowtite Technology A/S.

In the bidding process for the project, tenders were invited for both ductile cast iron and FRP pipes. The lowest price, 10% less than the lowest offer in ductile iron, was made by the contractor BAT with Owens Corning SN 10 000 pressure pipes and their Flowtite couplings.

The sewage system in Grimmen carries only a small volume of storm water, so the maximum sewage discharge rate of 467 m^3/hour was only raised by 40%. The line from the main pumping station to the plant and back to the river was consequently dimensioned for a total flow of 650 m^3/hour.

For the 1850 m feed line, the diameter was fixed at 400 mm, a service pressure of 3 bar, and a fluid velocity of 1 m/second, providing enough force for self-cleaning of the line.

For the 2750 m discharge line back to the river, the diameter was fixed at 500 mm because there is little risk of sedimentation in the clear discharge water. This diameter allows a lower service pressure of 2 bar and correspondingly lower operating costs.

A nominal design pressure of 6 bar was established for both lines, with a test pressure of 9 bar. The PN10 pressure class supplied by Owens Corning will allow for occasional short pressure surges, which will also provide an additional cleaning effect.

Slight bends in the lines were built up using the maximum permissible 3° deflection in the couplings, and for sharper turns specially built curved sections were used and constrained by thrust blocks. As the soil had low load-bearing capabilities the concrete thrust blocks were T-shaped, with broad bearing areas and relatively low weights.

A novel engineering concept was used to install a river crossing. To cross the river Trebel, a 6 m length of DN 1200 mm pipe was laid into and along the river bed, and temporary dams were built up at each end of the pipe, forcing the river to flow through the 1200 mm diameter pipe. The riverbed between the two dams was then excavated and the two pipelines installed below the riverbed without costly dewatering or sheet walls. Inverted elbows were prefabricated and were joined to the ends of the lines with bends and special couplings to prevent any leakage.

Some 500 m of pipeline runs beside a water-filled ditch, and the pipe bed level was below the water level. Drainage of the ditch was not necessary, as the 18 m long pipe sections could be laid, jointed and backfilled quickly, before the trench was flooded.

Not only was this solution the cheapest at the outset of the project, but additional economies developed. The pipes manufactured by Owens Corning Eternit Rohr were long, light in weight and easily handled, significant advantages in the wet, poor bearing soil, where rapid installation with minimum bedding preparation saved time and money.

11.10 Polyethylene Relining of 5500 Metres of 30 and 36 Inch Potable Water Pressure Mains

Supplier — Subterra
Client — Consolidated Edison Inc
Location — New York

Consolidated Edison Company of New York investigated various pipeline renewal processes that would help reduce operating costs, whilst minimizing the inconvenience of costly excavation, on two projects in New York. They retained Subterra Subline technology for two locations. In the Bronx a 30 inch diameter cast iron main was lined with 366 m of 760 mm SDR 32.5 high-density polyethylene pipe, and in Manhattan a 910 mm SDR 60 high-density polyethylene pipe was inserted into 183 m of 36 inch cast iron mains.

Figure 95. Preparation of a 'Subline' pipe liner for insertion. Courtesy of Subterra.

Both projects were carried out within urban areas with a minimum disruption of traffic, commercial activity and pedestrian access.

On both projects the mains pressures were tested and reconnected within 30 days of the mains being taken out of service.

In Manhattan, additional pipe lengths had to be butt welded on during the installation process due to the limited space available to string out the pipe on site. In the Bronx, the installation was stopped overnight after stray bullets from a shooting damaged the pipe. The damaged pipe was replaced and installation completed the following morning without further incident.

11.11 Renovation of 33 000 Metres of Sewer Pipe with 160–225 HdPe

Supplier — KWH Pipe
Contractor — Balfour Consulting Engineers
Client — Kuwait Ministry of Public Works
Location — Kuwait City

Kuwait is one of the richest countries in the world, the main source of income coming from the country's vast reserves of oil. Traditionally, the Kuwaiti government has always been concerned for the welfare of its people and has

Figure 96. VipLiner pipes being lowered into a manhole. Courtesy of KWH Pipe Ltd.

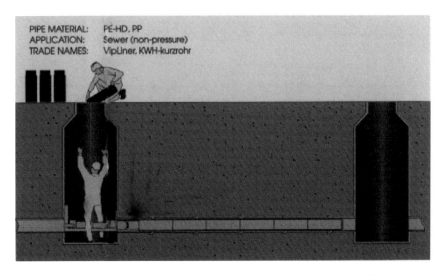

Figure 97. Lining with short length, in wall jointed VipLiner pipes. Courtesy of KWH Pipe Ltd.

been prepared to invest in the construction and maintenance of the country's infrastructure.

In the early 1980s the Kuwait Ministry of Public Works invited international consultants to tender for the design of, which was at the time, the largest sewer renovation project in the world. The contract was eventually awarded to Balfour Consulting Engineers who prepared a master plan for the Kuwait Sewer Renovation Project — Phase 1.

International contractors, in joint ventures with local companies were subsequently invited to bid for the construction and renovation work. In 1983 KWH Pipe was asked to bid for part of the rehabilitation project, which required relining of the existing asbestos cement pipes with polyethylene pipe. The following year KWH was awarded the sub-contract.

Over a contract period of three years KWH manufactured, delivered and installed 24 km of short-length HdPe in wall jointed relining pipes, VipLiner pipes.

The relining contract also required:

- Cleaning of pipes by high-velocity water jetting prior to relining.
- Pressure testing and joint sealing of pipes prior to relining.
- Grouting the annulus between the existing and relining pipe with a low-density foam concrete.
- Removal of the old house connection and replacement by new HdPe pipes.
- In addition, 8.5 km of standard wall, butt welded HdPe pipe was installed by slip lining.

See Chapter 7 for more details on the VipLiner system.

11.12 HdPe Submarine Pipeline in Malaysia

Main contractor — Antah Biwater JV
Client — Jabatan Kerja Raya/Jabatan Belakan AIR
Supplier — KWH Pipe

Pulau Ketam is an island off the coast of west Malaysia in the Indian Ocean. The only readily accessible water was rainwater, and during the dry season water had to be carried in by boat, over a distance of 35 km, to supply the 15 000 inhabitants of the two islands of Pulau Ketam and Pulau Lumut. A major water supply programme for that area of Malaysia included provision for a piped supply of potable water to the two islands.

The project required 24 500 m of HdPe pipe, diameter 200 to 400 mm, PN6, all of which was manufactured to DIN standards 8074 and 8075 in the KWH plant at Klang, Malaysia. A full range of tests was carried out on raw materials, pipe and fittings by both the KWH Pipe laboratory and by the Malaysian State laboratory SIRM.

Figure 98. Assembly of concrete collars onto the polyethylene pipe. Courtesy of KWH Pipe Ltd.

340 HdPe Submarine Pipeline in Malaysia

Figure 99. A 1000 metre long section of pipe being towed into position. Courtesy of KWH Pipe Ltd.

Some 18 km of pipe were installed offshore, and the remaining 6500 m as surface or buried pipe on essentially water-logged ground. The submarine pipe was assembled into lengths of 1000 m, by butt welding, and due to the low weight, each length was pulled manually into the sea. Each section was then towed into position and sunk onto the seabed where it was pressure tested, re-floated and flange jointed to adjacent pipe sections. The jointed pipeline was then sunk to its final position in the trench prior to being covered by 1 to 2 m of suitable fill material.

The deep seabed trench excavation was carried out using a combination of water jet and airlift systems. The shallower trenches were excavated using long-boom hydraulic excavators. The maximum water depth for excavation was 25 m, with a maximum water velocity of 5 to 6 knots. Visibility in the water was limited to 1 m.

The final testing and commissioning was completed in June 1991, some 21 months after the start of the contract.

11.13 Reverse Osmosis Disc Tube (DT) and Landfill Leachate Treatment

Supplier — Pall Rochem

Treating of leachate from landfill sites is a permanent problem which will remain with us for many years, well past the dates established in many countries to halt all new projects and close down existing ones. One way to treat landfill leachate, as developed by Pall Rochem is to reduce its volume by 75 to 80% using reverse osmosis, and then return the concentrate to the landfill by controlled re-injection. Conventional reverse osmosis and other membrane filtration technologies have been tried over the last few years, but most are hampered by rapid scaling and fouling, which results in poor membrane efficiency and reduced service life.

The reverse osmosis Disc Tube technology developed by Pall Rochem has shown a wide range of significant properties, including:

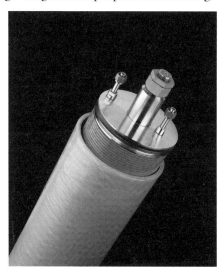

Figure 100. Exploded view of a Pall Rochem Disc tube module. Courtesy of Pall Corp.

- Minimisation of membrane scaling and fouling through the use of short flow paths, open channels and high-packing densities.
- Long membrane life, of five years or more.
- Low membrane replacement costs, as the DT system allows the replacement of individual sheets rather than complete elements as with spiral wound membranes, linked to easy access.

The Pall Rochem DT unit uses a filament wound composite housing and a range of polymeric membranes designed for each type of application.

Originally developed to produce drinking water from seawater, DT technology has now been successfully applied to other markets such as oil-contaminated water, industrial wastewater schemes and landfill leachate.

When handling leachate as feed water into a DT system, the suspended solids cannot easily settle out on the membrane surface due to the short, controlled and turbulent feed system. The hydraulics result in a high flow velocity across the membrane at minimum feed flow. The open channel over the membrane surface ensures optimal cleaning from fouling deposits.

When processing highly contaminated feed water extensive preconditioning is not required. A coarse sand filter is sufficient to pre-treat influent water containing up to 10 000 ppm dissolved solids and 5000 ppm suspended solids. As a basis for comparison, river water may contain 5 to 10 ppm suspended solids.

The system offers a large contaminant rejection rate combined with high recovery of water, being capable of recovering between 90 and 95% of the leachate volume as clean water (permeate).

The system's capabilities were well demonstrated by a 21-day trial to treat leachate from a contaminated section of the Central landfill in Johnson, Rhode Island. This demonstration was part of the EPA's Superfund Innovative Technology Evaluation programme. Approximately 33 000 gallons (125 m^3) of hazardous landfill leachate was processed using three DT units to maximise recovery and ensure the highest quality of permeate. The first unit operated at 1000 psi and the concentrate was temporarily in a holding tank before being fed into a unit operating at 2000 psi. The permeate from both the 1000 and 2000 psi units was then combined and processed by a third unit.

At the end of the test, the percentage of rejections were greater than the test criteria of 99% for total dissolved solids, 92% for total organic carbon and 95% for all target metals. In addition, the average percentage rejection for volatile organic compounds was greater than the test criteria of 90%. The average recovery rate during the demonstration was approximately 75%, which also met the test criteria.

The Pall Rochem DT system is now in service at 125 leachate sites in Europe and the USA. At the largest site in Ihlenberg, Germany, the system treats 750 000 gallons (2800 m^3) per day, which is about 80 times the flow of leachate generated at an average landfill.

The DT process relies on composite materials for strength and long-term corrosion resistance, especially valid for highly polluted and aggressive leachate, and on advanced polymers for the construction, corrosion resistance and performance of the membranes.

CHAPTER 12

Quality Systems and a Summary of Considerations Specific to the Use of Thermoplastic and Composite Materials

12.1 A Guide to the Design and Quality Control of Composite Vessels

Quality, and our appreciation of it, should be linked directly to the initial and long-term performance of the equipment purchased to a particular specification, and the quality and scope of this initial specification. It should not be uniquely established on the basis of an examination of a mass of documents, presented at completion of fabrication, although supporting relevant documentation in the exploitation and control of a quality system are of course of great importance.

Most major, and many of the smaller, manufacturers of thermoplastic and composite materials have put into place a quality system based upon or in conformity with ISO 9001 or ISO 9002. It is important to separate the suppliers into two groups. First, major national and more especially international suppliers of raw materials, that is thermoplastic and thermosetting resin systems, glass fibre reinforcements together with extruders and moulders of thermoplastic pipe systems. Second, manufacturers and fabricators of all types of equipment and protective coatings.

The first group of companies have set up in-depth comprehensive systems to manufacture and control what is essentially mass produced standard equipment and raw materials, in which, mainly because of the standardisation, production and control parameters are well established and controlled. This does not mean that the clients of such companies should not be attentive as to the quality of the material supplied, but from the author's own experience, it is most profitable to the user to understand the manufacturing and control tolerances employed by the supplier, and the influence of the extremes of these tolerances on subsequent manufacturing processes.

The second group of companies are generally involved in much shorter production runs, a high level of non-standard custom-designed equipment, and a much higher labour input, all of which suggest that a greater vigilance is required.

The ISO 9000 set of standards are guideline documents, adaptable to all forms of design, production, installation and back-up administrative activities, and the actual quality plan, and the depth and content of each individual plan, is left to the discretion of the manufacturing company and the controlling body. Each company may make an individual approach to the task of setting up their own quality plan within the overall framework of the standards, and it is not unusual to find substantially different content and scope of plans in different companies producing similar or even identical products.

The existence of validated quality plans should in no way discourage the client from investigating the content of a supplier's quality plan, and if required request clarification or modification to suit the client's particular requirements.

In some specialised areas, alternative audited quality plans are available which may be used instead of or alongside ISO standards. Typical examples are:

- ASME Boiler and Pressure Vessel Code, Section X, Fiber-Reinforced Plastic Pressure Vessels.
- API Specification Q1. Specification for Quality Programs.
- Most of the approval or listings concerning conformity of materials to specific specifications such as Factory Mutual, Underwriters Laboratories or NSF impose specific quality control patterns to ensure that the material conforms to the requirements of the specification, and many of the major users of anti-corrosion composite and thermoplastic equipment have edited their own specifications which include specific requirements as to the quality systems applied.
- National and international standards such as BS 4994: Specification for the Design and Construction of Vessels and Tank in Reinforced Plastics also contain specific instructions as to the quality control requirements of a particular design and production standard.

As some codes or standards require controls which are generally outside of the scope or level of controls written into an ISO 9001-type plan, it is important that the quality manual of the manufacturer reflects this possibility in some way, and that control systems are in place to deal with these exceptions to standard practice.

As mentioned in Chapter 8, in order to stress the importance of a quality system adapted to specific productions we include a reprint of a previous article called *Tank Failure* written by Bryan Broadbent, Sales Director of Forbes Plastics. Although this document is directed at users of composite and dual laminate tanks and vessels, the underlying philosophy can be used for all composite and thermoplastic corrosion-resistant equipment. In spite of the title, this article addresses the consideration of conceptual design and the various stages of inspection of the finished product, developing the notion of quality as long term, rather than a specific act related to the completion of a particular fabrication. We have, with the permission of the author, included a few small modifications to bring the original text up to date.

With the advent of the first edition of British standard, BS 4994: 1973, confidence was gained by engineers in specifying tanks of GRP (glass reinforced plastic) construction with or without thermoplastic linings. Prior to the British standard, the American specification PS 15.69 was used by some fabricators for the basis of their design.

Tanks and vessels designed and supplied to BS 4994: 1973 are now between 12 and 26 years old and should have been included in an inspection programme. BS 4994 was revised and re-issued in 1987, now as BS 4994: 1987.

Design specifications

The customer specification for a tank requirement should give a detailed list of the duty, i.e.

- contents and their specific gravity,
- temperatures (maximum and minimum),
- method of filling, if to be filled from the top or the bottom,
- the fill/empty cycle (that is the number of times the customer/user expects the tank to be filled and emptied per week).

It is important the customer/user is aware of the meaning of BS 4994: 1987 cyclic loading factor k_4 (see Chapter 8). Unless specified, the manufacturer/designer will take the figure as 1000 cycles over the tank design life, this is to say the tank has been designed for complete emptying and filling once per week over a design life of 20 years. From this it can be seen that, should the fill/empty cycle increase to twice per week, the design life is reduced by 50%.

With this in mind, the customer/user should consider the future use of the tank and the possibility of increasing the throughput. The marginal increase in cost this makes to the tank price is outweighed by the advantages in flexibility and life expectancy.

There have been a number of instances where tanks supplied to the original standard, designed with the initial k_4 factor of 1000 cycles were still in service after the expiry of the design life, and were still being filled/emptied in excess of once per day on a five-day working basis. These tanks were well past their design life and presented a potential hazard. A number of these tanks were being used with highly corrosive chemicals and filled by pressure delivery discharge road tankers. When at the end of a bulk delivery, the 'Pad Air' pressure from the tanker passes through the connecting pipework to the tank, and finally to the atmosphere via the receiving tank's venting system, it is adding to the potential hazard.

Because of the economic climate and the fact that the tanks were not giving problems it was difficult to persuade users to take the tanks out of service long enough for a full inspection to be carried out, not to mention giving consideration to replacing the tanks. There may still be many tanks in service supplied by various fabricators to different design standards, which are now geriatric. Operators may in all innocence continue to use these tanks, unaware of the design life or standards to which the tank has been supplied.

12.1.1 Vents and overflows

Still on the subject of design, it is important that thought is given to the fittings specified by the company for the duty the tank or vessel is to fulfil. The venting of a tank is important to the fill and empty rates. A normal guide is that the vent diameter should be a minimum of three times the fill pipe diameter, e.g. a 2 inch NB diameter fill pipe requires a 6 inch NB vent. Along with the tank vent, consideration should be given to the provision of a suitably sized overflow. Again as a guide, the overflow diameter may be a minimum of twice the inlet diameter, but this rule of thumb may be varied according to the rate and method of filling.

If for whatever reason it is not practical to fit an overflow, it is important that some form of high-level detection be fitted. Ideally, the vent should be positioned at the highest point in the tank top and the overflow at a point in the sidewall of the tank, to give the required working capacity of the tank and a minimum distance of 350 mm below the vent. This will ensure that the tank is not momentarily pressurised or vacuumed by the 'bow wave' that could occur from the 'air pad' at the end of a pressurised bulk delivery.

The use of combined vent/overflow should not be entertained in plastic tanks/vessels. The design and positioning of such combined fittings would inevitably be in the tank top. This means that should the tank become overfilled, the tank top would become pressurised before overflowing. Unless the tank was designed for the overflow situation, the tank top could fail, normally around the circumferential joint with the sidewall.

12.1.2 Fitting design

Consideration should be given to the design of, for example, fittings and branches specified by the users to suit ancillary equipment. Pads and threaded bosses below liquid level should be avoided. The excessive increases in wall thickness to create a pad could cause localised thermal expansion problems in many applications with warm fluids, leading to cracking and leakage. Also, the replacement of a threaded boss is very costly.

Flanged branches provide a much more effective way of creating tank inlets and outlets, and reinforcing gussets can be used for extra mechanical strength. Threaded connections can be provided by fitting adaptor plates. The smallest flange branch which should be fitted to a GRP tank is 50 mm (2 inch) as smaller sizes are vulnerable to mechanical damage which could also damage the integrity of the tank wall. The connecting pipework can be reduced as appropriate.

The selection of mating valves and pipework to flanged branches should be given careful consideration. Raised face flanges must not be used against full face flanges. Stub flanges with loose backing rings should not be over tightened.

12.1.3 Handling and installation

Great care is necessary when offloading and positioning plastic tanks. Soft nylon slings should be used wherever possible and the use of a 'tailing' crane is essential for large-capacity tanks. The use of chains should be restricted to use with lifting lugs, which would not allow direct contact of the chains with the tank or vessel body in any way. When the tank or vessel is in the upright position, care should be taken when lowering the tank onto the prepared plinth area, to avoid any impact damage.

It is essential that suitable support is provided and foundations or support structures should comply with the recommendations of BS 4994: 1987, paragraph 22 and appendices H and J. Discussions with the tank manufacturer may also be worthwhile. In all instances the proposed support structure must be capable of taking the total weight or the tank and contents, and for a vertical tank, the plinth should be of sufficient size to withstand the wind loadings and the overturning moments.

When installing a tank it is essential that all connecting pipework is adequately supported and that the total loads on branch fittings do not exceed design values. The use of bunding may be considered in relation to a tank's content and its proposed positioning. Bund walls must be designed to withstand the physical forces associated with the total release of tank contents on failure. The tank or vessel should be fixed to the bund to prevent floating when the tank is empty. Consideration should also be given to the effects and removal of rainwater from the bund and bund area.

12.1.4 Inspection

In section 5 of BS 4994: 1987, guidance is given on the inspection and testing procedures to be carried out after completion of the fabrication and before the tanks or vessels are commissioned. For tanks or vessels designed for use with hazardous contents, or in critical applications, it may be worth negotiating and incorporating in the contract with the supplier, an inspection programme which includes a full inspection of the tank or vessel after installation to verify integrity before commissioning. (This would eliminate any discussion at a later date as to whether the cause of a leak was in transportation, craneage or installation.) It is also recommended that a further full inspection be carried out during the following 12 to 24 months, and at planned intervals thereafter.

12.1.5 External inspection

All tanks and vessels should have periodic external examinations. These can be carried out while they are in use, and for outdoor tanks a fine, dry day should be chosen so any tell-tale leak signs such as wet or damp areas can be seen. The tank walls, the areas below fittings and branches and the base of the tank should all be checked. Areas behind branch flanges can be checked with the use of a mirror. This would show any cracking at the rear of the flanges, which could be the result of over tightening with mating manway covers, valves or pipework. Any defects, discoloration, blistering or cracking should be noted. If a leak is found or suspected while the tank/vessel is in service, it should be taken out of service and the cause investigated.

It is recommended that a competent person is called to carry out this examination. All inspections should be recorded, and a report issued. This should include repairs and confirmation of suitability for continued use. All dates should be recorded along with further inspections.

12.1.6 Internal inspections

Periodic internal inspections should be carried out on the tank's internal surfaces. Before such inspection is carried out, the tank should have been drained, washed through, made suitable for entry with necessary certification and 'isolated'. This means that *all* feed-in connection pipework is disconnected, along with the outlet pipework, particularly on manifolded tanks. The electrical supply to agitators, probes or other equipment should also be disconnected.

Care and thought should be taken before entering and carrying out any internal inspection. The choice of footwear should be considered together with the safety of entry and exit, particularly when only a top manway is available. Care should be taken with the use of ladders in order to prevent damage to the tank and possible injury to the inspector. Dependent on the size of the tank or vessel being inspected, safety harnesses should be considered. On tanks or vessels over 4 m in height, it will be necessary to erect scaffolding in order to carry out a full inspection of the inside surfaces.

Before erection of scaffolding, planks of timber should put down on the base to spread the loads. Great care should also be taken when erecting and dismantling scaffolding in confined spaces. At this point it is worth mentioning that thought should be given to side manways being incorporated in the design of tanks and vessels over 4 m in height, particularly to facilitate entry for inspection. The type of lining determines the inspection procedure to be carried out.

Tanks or vessels fabricated entirely of GRP would have a resin-rich internal coat to give the required properties for resistance to the contents being stored. Close and full visual inspection of such surfaces is important for continued use. Any breakdown in the surface, i.e. crazing, flaking or cracking, should be noted and the necessary remedial work carried out by a competent person.

For tanks or vessels with thermoplastic linings such as polypropylene, PVC-U, HdPe, PVDF and FEP, the entire surface of the lining should be inspected for bulging, cracking or discoloration. Bulging could indicate debonding of the lining, and cracking or cracks could indicate stress or impact damage.

The welds of the thermoplastic linings should be spark tested and any pinhole or hairline cracks found in the welds noted and repaired. However, great caution must be exercised should any potential leakage be discovered, as it is possible for the tank contents to affect the integrity of the GRP reinforcement behind the liner, and lead to failure. To aid the spark testing all welds in the thermoplastic lining should have a built-in conductive material or carbon-rich resin behind the weld to indicate the presence of pinholes or cracks. It is important that this feature is incorporated in the design and fabrication of the tank or vessel and reference made to it in the supplier's quoted specification. Before spark testing is carried out, all welds should be dry and free from moisture to avoid false readings.

The nature, position or type of any defect found governs the remedial work to be considered and carried out. If there is also evidence of leakage on the outside of the tank or vessel, it may be worth considering a non-destructive test, such as ultrasonic or acoustic emission testing to determine the magnitude of the defect. It is important that all repairs, including modifications which affect the integrity of any tank or vessel, are given careful consideration. The procedures adopted should be agreed by the user and the competent person carrying out the repairs.

The work should comply with the appropriate standard to which the tank or vessel has been fabricated and the necessary documentation drawn up to confirm the suitability for continued use in the application for which it was supplied.

12.1.7 Conclusion

When tanks fail they not only disrupt production and cause damage to plant, more importantly they can cause serious injuries to personnel, and all specifiers must be aware of their continuing responsibilities, their legal 'duty of care' in this regard. Tank failure can be generally summed up under the following major headings:

1. Inadequate specification and design.
2. Faulty manufacture.
3. Improper installation.
4. Use outside of design parameters.

It is obviously important to get the specification and design right at the outset and this is best accomplished by working with a reputable fabricator who can give guidance and justify the design with calculations.

The use of second-hand tanks is bound to be very risky as there may be no record of the tank's usage compared with the original specification. When re-use is being considered, a tank's history should be investigated and any proposed new use discussed with the original manufacturer.

Confidence in the use of GRP tanks and vessels in industry is widespread and the demand growing, but users should not become complacent because they are made with 'maintenance-free' materials. They should treat them with respect and include them in their inspection programmes along with other plant and equipment and continue to gain the maximum benefits from their investment.

A complementary document concerning guidance for specifying and purchasing of tanks called *Guide for Purchasing Reinforced Plastic Tanks* is available from the SPI (Society of the Plastics Industry) and guidance on all aspects of design, installation and use of composite pipe systems can be found in the *Fibreglass Pipe Handbook* published by the Fibreglass Pipe Institute in the USA.

CHAPTER 13

Raw Materials

13.1 Trade and Registered Names of Raw Materials and the Relevant Producers

Trade name	Material*	Producer
Advantex	Glass fibre reinforcement	Owens Corning
Amercoat	Coating	Ameron International
Altuglas	Polymethylmethacrylate	Elf Atochem
Aqualine	Coating	ITW Irathane International
Aquapon	Coating	PPG Industries Inc
Araldite	Epoxy curing agent	Ciba Speciality Chemicals
Aropol	Polyester resin	Ashland Chemical Co
Atlac	Polyester resins	Reichhold Chemicals Inc/ DSM Composite Resins
Beetle	Polyester resins	BIP Chemicals Ltd
Bidimat	Glass fibre reinforcement	Syncoglas NV
Caroguard	Coating	Carboline Co
Cellobond	Phenolic resin	Blagden Cellobond Ltd
Copon Hycote	Coating	E Wood Ltd
Corlar	Coating	EI Du Pont de Nemours and Co
Corroshield	Coating	Corroless International
Crestomer	Polyester resins	Scott Bader Co Ltd
Crystic	Polyester resins	Scott Bader Co Ltd
DEH	Curing agent	The Dow Chemical Co
DER	Epoxy resin	The Dow Chemical Co
Daron	Vinyl ester/urethane resin	DSM Composite Resins
Derakane	Vinyl ester resins	The Dow Chemical Co
Derakane Momentum	High reactive vinyl ester resins	The Dow Chemical Co

Trade name	Material*	Producer
Dion	Polyester resin	Reichhold Chemicals Inc
DION VER	Coating	Reichhold Chemicals Inc
ELC	Coating	Subterra Ltd
Epikure	Epoxy curing agent	Shell Chemicals
Epikote	Epoxy resin	Shell Chemicals
Flakeglass	Coating	ITW Irathane International
Geopox	Coating	Mercol Products Ltd
Halar	E-CTFE	Ausimont
Hetron	Polyester resins	Ashland Specialty Chemical Co
Hetron 800	Furane resin	Ashland Specialty Chemical Co
Hunting Waterline	Coating	Hunting Industrial Coatings Ltd
Hybon	Glass fibre reinforcement	PPG Industries Inc
Hylar	PVDF	Ausimont
Interline	Coating	International Paint Inc
Interpon	Coating	H B Fuller Coatings Ltd
Irathane	Coating	ITW Irathane International
Kevlar	Aramid fibre	EI Du Pont de Nemours Company Inc
Kynar	PVDF	Elf Atochem
Leigh's Waterline Blue	Coating	W&J Leigh & Co
Nexus	Surface veils	PFG Inc
Nitoline	Coating	Fosroc Ltd
Norpol	Polyester resins	Reichhold Chemicals Inc
Norsophen	Phenolic resins	Cray Valley
Nylon	Polyamide	EI Du Pont de Nemours and Co
Online	Coating	ITW Irathane International
Palatal	Polyester resins	DSM Composite Resins
Plascoat	Coating	Elf Atochem
Plasguard	Coating	Plastite Protective Coatings
Plexiglas	Polymethylmethacrylate	AtoHaas Americas Inc
Norsodyne	Polyester resins	Cray Valley Ltd
Polylite	Polyester resin	Reichhold Chemicals Inc

Trade name	Material*	Producer
Powercron	Coating	PPG Industries Inc
Resicoat	Coating	Akzo Nobel
Rilsan	Coating	Elf Atochem
Scotchkote	Coating	3M
Sikagard	Coating	Sika Ltd
Syncokomplex	Glass fibre reinforcement	Syncoglas NV
Syncoloop	Glass fibre reinforcement	Syncoglas NV
Synolite	Polyester resins	DSM Composite Resins
Teflon		EI Du Pont de Nemours and Co
Thortex Chemi-Tec PW	Coating	E Wood Ltd
Udimat	Glass fibre reinforcement	Syncoglas NV
Vipel	Polyester resins	AOC
Vyguard	Coating	Ameron International

*The material description is limited to the generic description only. Polyester resins indicate the potential supply of the full range of isophthalic, orthophthalic, terephthalic, bisphenol, vinyl ester and HET resins. Coating is usually an epoxy based material, but vinyl ester, urethane and other resin formulations are also included, and examples used are essentially approved or suitable for contact with potable water.

Trademarks and registered and non-registered trade names

Trade name	Material	Producer
Red Thread	Amine-cured epoxy piping system	Smith Fiberglass Products Company
Blue Streak	Anhydride-cured epoxy piping system	Smith Fiberglass Products Company
Green Thread	Amine-cured epoxy piping system	Smith Fiberglass Products Company
Pronto-Lock	Jointing technology	Ameron International
TMJ	Jointing technology	Smith Fiberglass Products Company
TAB	Jointing technology	Smith Fiberglass Products Company
Bondstrand	Composite pipe system	Ameron International

Trade name	Material	Producer
Thermothan	Pre-insulated epoxy piping system	Fiberdur-Vanck Group
Navicon	Electrically conductive epoxy piping system	Fiberdur-Vanck Group
Wavistrong	Amine-based epoxy pipe systems	Future Pipe Industries
Coil-Lock	Jointing technology	Ameron International
Key Lock	Jointing technology	Ameron International
Pronto-Lock	Jointing technology	Ameron International
F-Chem		Fiberdur-Vanck Group
Red Thread 11 Performance Plus	Amine-cured epoxy piping system	Smith Fiberglass Products Company
Flowtite	Continuous filament wound pipe	Owens Corning Engineered Pipe Systems
Big Thread	Reciprocal filament wound pipe	Smith Fiberglass Products Company
Fibaflow	Reciprocal filament wound pipe	Fibaflow Reinforced Plastics
Sarplast	Reciprocal filament wound pipe	Sarplast Iniziative Industriali
Centricast	Centrifugally cast composite pipe	Fibercast Co
Novacast	Vinyl ester-based composite pipe	Fibercast Co
FM	Factory Mutual-approved epoxy piping system	Fibercast Co
Dualcast	Double containment pipe system	Fibercast Co
Apollo Bi-axial	Bi-oriented PVC pipe	Wavin
BiOroc	Bi-oriented PVC pipe	DYKA Plastic Pipe Systems
BO	Bi-oriented PVC pipe	Alphacan
Aquaforce	Enhanced performance PVC pipe	Uniplas Ltd
SYGEF	PVDF pipe	Georg Fischer
PUMA	FRP tank	Forbes Plastics
BioDisc	Effluent treatment unit	Klargester Environmental Engineering Ltd
Minibulk	Bunded safe storage system	Forbes Plastics
TIF	Modular tank system	Brimar Plastic Fabrications
FilmTec	Spiral wound membranes	The Dow Chemical Co

Trade name	Material	Producer
Aquasource	Membrane filtration unit	Aquasource
Composit	Composite pressure vessel	The Structural Group
Polyglass	Composite pressure vessel	The Structural Group
Bajonet	Internal distribution system	The Structural Group
Certa-Lok	PVC pipe and well casing	CertainTeed
Rolldown	Pipe lining process	Subterra
Subline	Pipe lining process	Subterra
Compact Pipe	Pipe lining process	Wavin
NuPipe	Pipe lining process	Insituform Technologies
Omega-Liner	Pipe lining process	Uponor Anger
Safeguard	Pipe lining process	Durapipe S&LP
Remotel and Remoteline Patch	Pipe lining process	Subterra
Insituform	Pipe lining process	Insituform Technologies
Insituform Point Repair	Pipe lining process	Insituform Technologies
Trolining	Pipe lining process	TROLINING GmbH
VipLiner	Pipe lining process	KWH Pipe Ltd
Neofit	Pipe lining process	Wavin
Bekaplast	Lining system	Steuler Surface Protective Systems
Korlok	Pultruded sheet pile	International Grating Inc
SCRIMP	Seemann Composite Resin Infusion Process	Seemann
Seapile	Fender system	Seaward International Inc
Comby	Settling tank	FZ Fantoni
Fäkalex	Prefabricated pumping stations	Feluwa Pumpen GmbH
ACCU-PAC	Modular biomedia	Brentwood Industries Inc
BIOdek	Modular biomedia	Munters Euroform GmbH
Biolox	Modular biomedia	Rauschert Verfahrenstechnik
Bioflow	Modular Biomedia	Rauschert Verfahrenstechnik
MARPAK	Cooling tower fill and modular biomedia	The Marley Cooling Tower Co
TUBEdek	Plastic lamellae	Munters Euroform
VF	Cooling tower fill media	Brentwood Industries Inc
TURBOsplash	Cooling tower fill media	Brentwood Industries Inc
ZeeWeed Process	Membrane filtration process	Zenon Environmental Systems

Trade name	Material*	Producer
Marley HP7000	Cooling tower fan	The Marley Cooling Tower Co
Flare Tip	Cooling tower fan blade	The Marley Cooling Tower Co
COOLFILM	Cooling tower fill media	Hamon & Cie
CLEANFLOW	Cooling tower fill media	Hamon & Cie
COOLDROP	Cooling tower fill media	Hamon & Cie
UNILITE	Cooling tower	Baltimore Aircoil Company
Ultralite	Cooling tower	Ceramic Cooling Tower Co
Pioneer	Cooling tower	Delta Cooling Towers Inc
Paragon	Cooling tower	Delta Cooling Towers Inc
Premier	Cooling tower	Delta Cooling Towers Inc
Discreen	Dynamic screening device	Monopumps Ltd
FlowGRIP	Composite flooring	Redman Fisher Engineering Ltd
RO DiscTube	Membrane filtration system	Pall Rochem
Vee-wire	PVC screens	USF Filtration and Separations

13.2 Suppliers of Raw Materials (for Composite Products)

Please note that, in general, for international or multinational companies, only the location of the corporate headquarters or the principal location handling the materials in question are indicated. In some cases, where applicable, a website address is indicated rather than a fax number.

Superscripts: WRc indicates that the company manufactures materials approved by the WRc; DWI indicates that the company has both WRc and DWI approval for contact with potable water, as the DWI approval overrides a previous WRc approval; and NSF indicates that the company manufactures material approved by NSF for contact with potable water.

13.2.1 Polyester, vinyl ester, bisphenol and HET resins

AOC
950 Highway 57 East, Collierville, TN 38017, USA
Tel: +1 901 854 2800
Fax: +1 901 854 7277

Ashland Specialty Chemical Co
5200 Blazer Parkway, Dublin, OH 43017, USA
Tel: +1 614 889 4191
Fax: +1 614 889 3735

BIP Ltd[WRc]
PO Box 3180, Oldbury, Warley, West Midlands, B69 4PG, UK
Tel: +44 121 544 2333
Fax: +44 121 5447193

CERSA Produtos Quimicos Ltd
Av. Luiz Henrique de Oliveira, 600 Vila Quitauna, CEP 06186-130-Osasco SP, Brazil
Tel: +55 11 7208 3366
Fax: +55 11 7208 3396

Cray Valley[WRc]
Structural Resins Dept, Laporte Road, Stallingborough, DN37 8DR, UK
Tel: +44 1469 572464
Fax: +44 1469 572988

Cray Valley, a subsidiary of the French group Total, now includes Cook Composites & Polymers and Sartomer in North America, and has recently acquired the structural resin activities of South Korea's DAE SANG group. This situation, together with the purchase of Resisa, the Cepsa group's resin subsidiary in Spain, and other smaller acquisitions in France and Germany, now makes the group the world's second largest resin producer.

Dow Chemical Co
2040 Dow Center, Midland, MI 48674, USA
Tel: North America: 1 800 441 4369; Europe: +31 20 691 6268; Japan: +81 35 4602114

DSM Composite Resins[WRc]
PO Box 615, 8000AP Zwolle, The Netherlands
Tel: +31 38 4569660
Fax: +31 384569502

Reichhold Chemicals Inc[WRc, NSF]
PO Box 13582, Research Triangle Park, NC 27709, USA
Tel: +1 919 990 7500

Scott Bader Co Ltd[WRc]
Wollaston, Wellingborough, NN29 7RL, UK
Tel: +44 1933 663100
Fax: +44 1933 664592

13.2.2 Epoxy resins

Bakelite AG
Postfach 12 05 52, D-47125 Duisburg, Germany
Tel: +49 203 429602
Fax: +49 203 425454

Ciba Speciality Chemicals[WRc]
Kliybeckstrasse 141,m CH-4002 Basle, Switzerland
Tel : +41 61696 11 1
Fax: +41 61696 4474

At the time of going to press the acquisition of Ciba Speciality Chemicals by Morgen Grenfell Private Equity was in the process of finalisation, and the name of the company and ownership of Araldite trademarks may be modified.

Dow Chemical Co
(see above)

Shell Chemicals
Shell Centre, London SE1 7NA, UK
Tel: +44 71 934 1234
Fax: +44 71 934 8060

13.2.3 Phenolic and/or furane resins

Bakelite AG (see above)

Borden Chemicals Ltd
Sully, South Glamorgan, UK
Tel: +44 1446 731293
Fax: +44 1446 731450

Georgia Pacific Resins Inc
2883 Miller Road, Decatur, GA 30035, USA
Tel: +1 404 652 8617
Fax: +1 404 749 2682

13.2.4 Glass fibre reinforcements

Ahlstrom Paper Group
PO Box 275, 2130 AG Hoofddorp, The Netherlands
Tel: +31 23 554 7000
Fax: +31 23 554 7100

Cam Elyaf Sanayii AS
Cayirova 41401 Gebze-Kocaeli, Turkey
Tel: +90 262 6534741
Fax: +90 262 6532657

Chomarat (Les Fils d'Auguste Chomarat & Cie)
07160 Le Cheylard, France
Tel: +33 475 29 70 00
Fax: +33 475 29 17 75

CPIC/Fiberglass — Chongqing Glassfiber Co Ltd
Dadukou District, Chongqing, China 400082
Tel: +86 23 68906144
Fax: +86 23 68833143

Fiberex Glass Corp
6602 45th Street, Leduc, Alberta, Canada
Tel: +1 780 980 1300
Fax: +1 780 980 1330

Lantor BV
PO Box 45, 3900 AA Veenendaal, The Netherlands
Tel: +31 318 537360
Fax: +31 318 537420

Owens Corning
1 Owens Corning Parkway, Toledo, OH 43659, USA
Tel: +1 419 248 8000
Fax: +1 419 248 7420

Parabeam
PO Box 134, NL 5700 AC Helmond, The Netherlands
Tel: +31 492 570625
Fax: +31 492 570733

PPG
1 PPG Place, Pittsburgh, PA 15272, USA
Tel: +1 412 454 5131

Syncoglas NV
Drukkerijstraat 9, 9240 Zele, Belgium
Tel: +32 52 45 76 11
Fax: +32 52 44 95 02

Vetrotex International
767 quai des Allobroges, 73009 Chambery Cedex, France
Tel: +33 479 755300
Fax: +33 479 755399

13.2.5 Glass fibre and synthetic surfacing veils

Freudenburg (Carl)
Postfach 180, D-6940 Weinheim, Germany
Tel: +49 6201 801
Fax: +49 6201 69300

Lantor BV
PO Box 45, 3900 AA Veenendaal, The Netherlands
Tel: +31 853 7360
Fax: +31 853 7420

Precision Fabrics Group Inc (PFG)
Suite 600, 301 N Elm Street, Greensboro, NC 27401, USA
Tel: +1 901 279 8000
Fax: +1 910 279 8002

13.2.6 Coating materials and formulations

3M/Corrosion Protection Products[DWI, NSF]
6801 River Place Boulevard, Building 130-3N-5Y, Austin, TX 78726-9000, USA
Tel: +1 512 984 5320

Akzo Nobel Powder Coatings GmbH[WRc]
Postfach 2433, D 72714 Reutlingen, Germany
Tel: +49 7121 519100
Fax: +49 7121 519199

Ameron Inc[NSF, WRc]
245 South Los Robles, Pasadena, CA 91101, USA
Tel: +1 714 529 1951

Carboline Co[NSF]
350 Hanley Industrial Court, St Louis, MO 63144, USA
Tel: +1 314 644 1000
Fax: +1 314 644 4617

Corroless International — Joseph Mason plc[DWI]
Nottingham Road, Derby, DE21 6AR, UK
Tel: +44 1332 295959
Fax: +44 1332 295252

E Wood Ltd[DWI, NSF]
Standard Way, North Allerton, DL6 2XA, UK
Tel: +44 1609 780170
Fax: +44 1609 780438

Elf Atochem SA[DWI, NSF]
Cours Michelet, La Defense 10, 92091 Paris La Defense Cedex, France
Tel: +33 1 49 00 80 80
Fax: +33 1 49 00 83 96

Fosroc Ltd[DWI, NSF]
Coleshill Road, Tamworth, Staffordshire, B78 3TL, UK
Tel: +44 1827 262222
Fax: +44 1827 262444

H B Fuller Coatings Ltd[WRc, NSF]
95 Aston Church Road, Nechells, Birmingham, B7 5RQ, UK
Tel: +44 121 322 6900
Fax: +44 121 322 6902

Hunting Industrial Coatings Ltd[DWI, NSF]
Derby Road, Widnes, Cheshire, WA8 9ND, UK
Tel: +44 151 495 3505
Fax: +44 151 495 3522

International Paint Inc[NSF]
6001 Antoine Drive, Houston, TX 77210-4806, USA
Tel: +1 713 682 1711
Fax: +1 713 682 0065

ITW Irathane International[DWI]
Brunel Close, Park Farm Industrial Estate, Wellingborough, NN8 6QX, UK
Tel: +44 1933 675099
Fax: +44 1933 674643

Mercol Products Ltd[DWI, NSF]
Carr Vale, Bolsover, Chesterfield, S44 6JD, UK
Tel: +44 1246 822521
Fax: +44 1246 240021

Sika Ltd[DWI]
Watchmead, Welwyn Garden City, Hertfordshire, AL7, 1BQ, UK
Tel: +44 1707 394444
Fax: +44 1707 372431

Subterra Ltd[DWI]
Dullar Lane, Sturminster Marshall, Wimborne, Dorset BH21 4DA, UK
Tel: +44 1258 857556
Fax: +44 1258 857960

W&J Leigh & Co[DWI]
Tower Works, Kestor Street, Bolton, Lancashire BL2 2Al, UK
Tel: +44 1204 521771
Fax: +44 1204 382115

13.2.7 Thermoplastic sheet and pipe designed specifically for incorporation into dual laminates

Agru/Alois Gruber+Sohn OHG
A-4540 Bads Hall, Austria
Tel: +43 7258 2172
Fax: +43 7258 3863

Simona AG
Postfach 133, D-6570 Kirn, Germany
Tel: +49 6752 140
Fax: +49 6752 14211

Symalit AG
Lenzhard, 5600 Lernzburg, Switzerland
Tel: +41 62 8858150
Fax: +41 62 8858181

CHAPTER 14

Trademarks, Equipment Suppliers and Manufacturers

Many manufacturers now have plant and/or distributors in several countries. It is impossible to include all addresses, telephone, fax and e-mail details in this handbook. We have in general limited our indications to the main plant or corporate headquarters, except for companies having significant manufacturing facilities on more than one continent. In all other cases, an asterisk indicates that other significant manufacturing and/or sales centres exist in different countries.

Superscripts: WRc indicates that the company manufactures materials or equipment approved by the WRc; DWI that the company has both WRc and DWI approval for contact with potable water, as the DWI approval overrides a previous WRc approval; and NSF indicates that the company manufactures material approved by NSF for contact with potable water.

We have not attempted to identify the specific products which are effectively approved, and which may in certain cases only concern a small percentage of the total range of products manufactured by a certain company.

Brief comments on the company, or groups of companies are included where the author believes it appropriate, and in order that the reader may achieve a better understanding of the structure of the anti-corrosion composites industry.

The author has assumed that most readers are aware of the principal manufacturers in their own country, and has indicated all telephone and fax numbers with the most used international dialling code prefix + followed by the applicable country code.

We have included website identification for companies supplying products and materials for the water-related markets where such sites have sufficient and relevant information readily available, especially technical data rather than basic commercial information, and where the website connection is more practical than a fax number.

14.1 Composite Pipe Systems

14.1.1 Filament wound epoxy pipes and fittings

Amiantit Fiberglass Industries
PO Box 589, Damman 31421, Saudi Arabia
Tel: +966 3 8571160
Fax: +966 3 8579397
Amiantit Fiberglass Industries is part of the Amiantit group of companies, reputed to be the largest pipe manufacturer in the Middle East. **Amiantit Fiberglass Industries** is the licensed manufacturer of the Bondstrand epoxy range from Ameron in Saudi Arabia

Ameron — Fibreglass Pipe Division*NSF
Trademarks — Bondstrand, Key-Lock, Quick-Lock, Pronto-Lock, Dualoy
www.ameron.com

PO Box 878, Burkburnett, TX 76354, USA
Tel: +1 817 569 1471
Fax: 1 817 569 4012

J F Kennedylaan 7, NL 4191MZ Geldermalsen, The Netherlands
Tel: +31 3455 72241
Fax: +31 3455 75254

7A, Tuas Avenue 3, Singapore 2263
Tel: +65 862 1301
Fax: +65 862 1302

Fiberdur-Vanck GmbH
Postfach 1410, Feldenend, 52234 Eschweiler, Germany*
Tel: +49 2403 7020
Fax: +49 2403 702151
The Fiberdur-Vanck group is a subsidiary of the Beteiligungsgesellschaft Aachener Region mbH (BGA), having separate pipe and tank plants in

Germany. The group also includes the Vetroresina plant in Italy, participation in Fiberdur France and Budaplast in Hungary, and also licensed operations in Indonesia (MECO) and in Australia (3D-Composites).

Fibre Glass Systems
Trademark — STAR
PO Box 37389, San Antonio, TX 78237, USA
Tel: +1 210 434 5043
Fax: +1 210 434 7543

Future Pipe Industries BV[*, WRc]
Trademark — Wavistrong
J C Kellerlaan 3, 7770 AG Hardenberg, The Netherlands
Tel: +31 523 288 811
Fax: +31 523 288 441

The Future Pipe Group also includes three large-diameter pipe plants in the Middle East (see below). The epoxy plant in The Netherlands was acquired from the previous owner Wavin Repox.

Smith Fiberglass Products Company[*, NSF]
Trademarks — Red Thread, Green Thread, Blue Streak
2700 West 65th Street, Little Rock, AR 72209, USA

Also operates epoxy plants in Wichita, Kansas and Harbin, China.

14.1.2 Filament wound polyester and vinyl ester pipes and fittings

Engineered Pipe Systems, a wholly owned subsidiary of Owens Corning[*, WRc]
Trademark — Flowtite
Owens Corning Science and Technology Center, 2790 Columbus Road, Route 16, Granville, OH 43023-1200, USA
Tel: +1 740 321 5530
Fax: +1 740 321 7433
Owens Corning operate directly, or indirectly via licensees, with plants located in many countries. For example: Flowtite Botswana (Pty) Ltd, Botswana; Owens-Corning Eternit Rohre GmbH in Germany; Flowtite Pipe and Tank A/S, Norway; Flowtite Iberica S.A. in Spain; Owens Corning Yapi Merkezi Sanayive Ticaret A.S in Turkey; Amiantit Fiberglass Industries in Saudi Arabia; Gulf Eternit Industries Co. in Dubai; Flowtite Argentina SA in Argentina; and Owens Corning Andercol Tuberias SA in Colombia.

Ershigs
742 Marine Drive, Bellingham, WA 98227, USA
Tel: +1 360 733 2620
Fax: +1 360 733 2628

Ershigs is part of the Denali group. The group owns many of the composite manufacturers referred to in this chapter, in both Europe and North America. The ownership in Europe is composed of the Plasticon group of companies, that is Plasticon and Plasticon Kialite in The Netherlands, Plasticon KTD in Germany, Plasticon Garlway in the UK, Plasticon Sovap in France, as well as participation in Plasticon Kialite in Thailand and Metalchem in Poland. The organisation in North America includes Ershigs, Fibercast, Containment Solutions Inc, Plastifab and Belco, all of which are involved in composite anti-corrosion equipment for water and wastewater treatment, as well as a wide range of other industrial activities.

Fibaflo Reinforced Plastics Ltd
Holton Heath Trading Park, Holton Heath, Poole, Dorset, BH16 6LP, UK
Tel: +44 1202 624141
Fax: +44 1202 631589

Fiberdur-Vanck GmbH*
(see epoxy pipe systems above)

Kialite Plasticon BV
Jellinghausstraat 32, 5048 AZ Tilburg, The Netherlands*
Tel: +31 134 631592
Fax: +31 134 671201

Ollearis SA
PO Box 127, 08107 Martorelles, Spain
Tel: +34 93 5796520
Fax: +34 93 5933616

Plasticon BV*
Parallelstraat 52, 7570 AH Oldenzaal, The Netherlands
Tel: +31 541 858 500
Fax: +31 541 858 501

Plastillon Oy
Muovikuja 7, FIN 55120 Imatra, Finland
Tel: +358 5 68231
Fax: +358 5 6823236

Protesa
Carretera Nacional 11, Km 599, 08780 Palleja (Barcelona), Spain
Tel: +34 93 6630002
Fax: +34 93 6630901

Reinforced Plastic Systems Inc
PO Box 299, Mahone Bay, Nova Scotia, Canada, B0J 2E
Tel: +1 800 343 9355
Fax: +1 902 624 6395

Sarplast Iniziative Industriali
Via Rosignanina 14, 56040 Santa Luce, Pisa, Italy
Tel: +39 0506 84502
Fax: +39 0506 84506

Sguassero SpA
Via Pasubio 36, I 20037 Paderno Dugnano (MI), Italy
Tel: +39 0291 05510
Fax: +39 0291 83293

Smith Fiberglass Products Co* (see epoxy pipe systems above)
Trademarks — Poly Thread, Big Thread

Speciality Plastics Inc
15915 Perkins Road, Baton Rouge, LA 70879-7011, USA
Tel: +1 504 752 2705
Fax: +1 504 752 2757

Vidropol
Apdo 2001, Castelo da Maia, 4474 Maia Codex, Portugal
Tel: +351 2981 2017
Fax: +351 2981 2476

Zinkal Group — Fiberglass Pipes Ltd
PO Box 3131, Petach Tikva 49131, Israel
Tel: +972 3 9223551
Fax: +972 3 9223374

14.1.3 Centrifugal cast and filament wound epoxy and vinyl ester pipes and fittings

Fibercast
Trademarks — Centricast, Novacast, F-Chem
P O Box 968, Sand Springs, OK 74063, USA
Tel: +1 918 245 6651
Fax: +1 918 241 1143

14.1.4 Centrifugal cast polyester pipes and fittings

Hobas Engineering GmbH*
Pischeldorfer Str 128, 9020 Klagenfurt, Austria
Tel: +43 463 482424
Fax: +43 463 482121

Hobas operates its own pipe plants in Austria, Germany, Switzerland, the USA, and the Czech Republic and has appointed licensees in the UK, Italy, Belgium, Slovenia, Japan, Lebanon, Australia, the Philippines, South Africa, China, the United Arab Emirates and Spain. For details on some of these companies see below. For further information the reader should contact Hobas directly.

Eternit (Licensee of Hobas)
Kuiermanstraat 1, B 1880 Kapelle op den Bos, Belgium
Tel: +32 15 71 80 80
Fax: +32 15 71 80 09

Future Pipe Industries Co SAL (Licensee of Hobas)
Al Kalaa, PO Box 13-5009, Beirut, Lebanon
Tel: +961 1 869757

Hobas Pipe USA Inc[NSF]
1413 Richey Road, Houston, TX 77073, USA
Tel: +1 281 821 2200

Johnston Pipes Ltd[DWI] (Licensee of Hobas)
Dosely, Telford, TF4 3BX, UK
Tel: +44 1952 630300
Fax: +44 1952 501537

14.1.5 RTM epoxy, vinyl ester and polyester fittings

Rotec Composite Group BV
Bouwweg 25, 8243 PJ Lelystad, The Netherlands
Tel: +31 320 260933
Fax: +31 320 261283

14.1.6 High pressure filament wound pipe, tubing and casing

Centron (a subsidiary of Ameron, see above)

Fiberglass Systems (see above)

14.1.7 Low-pressure filament wound tubing and casing

Burgess Well Co[NSF]
742 East Highway 6, Minden, NE 68959-2569, USA
Tel: +1 308 832 1645
Fax: +1 308 832 0170

Fibaflo Reinforced Plastics Ltd[WRc]
Holton Heath Trading Park, Holton Heath, Poole, Dorset, BH16 6LP, UK
Tel: +44 1202 624141
Fax: +44 1202 631589

Sarplast Iniziative Industriali
(see above)

14.1.8 Dual laminate pipe systems

Garlway-Plasticon
Grovehill Industrial Estate, Beverley, East Yorkshire, HU17 0JT, UK
Tel: +44 1482 862194
Fax: +44 1482 871398

Kialite-Plasticon (see above)

Keram Chemie
PO Box 1163, D 5433 Siershahn, Germany
Tel: +49 2623 600-0
Fax: +49 2623 600513

MB Plastics
Reinforced Division, Forward Works, Woolston, Warrington, WA1 4BA, UK
Tel: +44 1925 822811
Fax: +44 1925 818907

Ollearis (see above)

Plastilon Oy (see above)

14.1.9 Thermoplastic lined steel pipe

3P
Rue Etoile de Langres, 52200 Langres, France
Tel: +33 3 25 87 23 24
Fax: +33 3 25 87 66 27

Crane Resistoflex
PO Box 1449, Marion, NC 28752, USA
Tel: +1 828 724 4000
Fax: +1 828 724 9469

MB Plastics
Fluoropolymer Division, Bridge Lane, Woolston, WA1 4BA, UK
Tel: +44 161 877 1963
Fax: +44 161 877 1964

14.2 Thermoplastic Pipe Systems

14.2.1 Extruded thermoplastic pipes

Agru Kunststofftechnik GmbH
Ing Pesendorfer Strasse 31, 4540 Bad Hall, Austria
Tel: +43 725 8790
Fax: +43 72583863

Alphacan
12–18 Avenue de la Jonchère, 78170 La Celle St Cloud, France
Tel: +33 1 30 82 58 00
Fax: +33 1 39 69 19 99

CertainTeed Corp[NSF]
Pipe and Plastics Group, 1400 Union Meeting Road, Blue Bee, PA 19422, USA
Tel: +1 610 341 6637

CertainTeed, a subsidiary of the French Saint-Gobain group, specialises in a wide range of PVC pipe systems, including pipes for potable and wastewater, well equipment and irrigation systems.

CSR Poly Pipe[NSF]
16701 Greenspoint Drive, PO Box 60297, Houston, TX 77205-0297, USA
Tel: +1 940 665 1721
Fax: +1 281 874 0962

Diamond Plastics Corp
1212 Johnstone Road, Grand Island, NE 68803, USA
Tel: +1 308 384 4400

Durapipe-S&LP[*, DWI]
Norton Canes, Cannock, Staffordshire, WS11 3NS, UK
Tel: +44 1543 279909
Fax: +44 1543 279450
www.durapipe-slp.co.uk

Durapipe-S&LP is the result of bringing together the companies Durapipe, Stewarts and Lloyds Plastics, Vulcathene and Calder Equipment. Durapipe-S&LP is part of the Glynwed International plc Group, with Glynwed Pipe Systems Ltd being one of the world's largest producer of thermoplastic pipes. The organisation has manufacturing and distributing companies in the UK, France, Germany, The Netherlands, Austria, Spain, Singapore, Australia, South Africa and the USA. Other pertinent companies within the group include: FIP in Italy — pipe fittings and valves; Friatec in Germany — electrofusion fittings; Philmac in Australia — Pipe fittings and irrigation systems. The recent acquisition of IPEX Inc (see below) in Canada now probably makes Glynwed International the largest thermoplastic pipe producer in the world.

Dyka Plastic Pipe Systems[*, DWI]
Prodktieweg 7, 8331 LJ Steenwijk, The Netherlands
Tel: +31 521 534911
Fax: +31 521 534889

Dyka, a subsidiary of the Tessenderloo Chemie group of companies, has production/sales outlets in the UK, Netherlands, Belgium, France and the Czech Republic.

Eagle Pacific Industries Inc[NSF]
2430 Metropolitan Centre, 333 South Seventh Street, Minneapolis, MI, USA
Tel: +1 612 371 9650

Recently merged with PWPipe (see below).

Elson Thermoplastics[NSF]
PO Box 240696, Charlotte, NC 28224, USA
Tel: +1 704 889 2431

Hepworth Building Products International Ltd[*, DWI]
Hazlehead, Stocksbridge, Sheffield, S30 5HG, UK
Tel: +44 1226 763561
Fax: +44 1226 765110

Industrias Vassallo Inc[NSF]
PO Box 52, Coto Laurel 00780, Puerto Rico
Tel: +1 787 848 1515

IPEX Inc[NSF]
Port of Montreal Building (Wing 3), Montreal, Quebec, Canada, H3C 3R5
Tel: +1 514 861 7221
www.ipexinc.com

IPEX was formed via the fusion of Scepter Manufacturing Company Ltd and Cannon Ltd, and has now been acquired by the Glynwed organisation in the UK.

J-M Manufacturing[NSF]
9 Peach Tree Hill Road, Livingstone, NJ 07039, USA
Tel: +1 800 621 4404
Fax: +1 800 451 4170

KWH Pipe Ltd[*, NSF]
PO Box 21, FIN-65101 Vaasa, Finland
Tel: +358 632 65511
www.kwhpipe.com

Produces large-diameter Pe pipes, 1600 mm by extrusion and 3000 mm by spiral welding. Has plants in Finland, Sweden, Denmark, Czech Republic, Poland, Portugal, Canada, USA, India, Malaysia and Thailand.

Mahaveer Industries
SP-1, First phase, industrial area, Barmer 344001, India
Tel: +91 2982 21339
Fax: +91 2982 20471

North American Pipe Corp[NSF]
The Westlake Center, 2801 Post Oak Boulevard, Suite 300, Houston, TX 77056, USA
Tel: +1 713 585 2639

Philips Driscopipe[NSF]
PO Box 833866, Richardson, TX 75083, USA
Tel: +1 800 527 0662
www.driscopipe.com

Philips Driscopipe, a subsidiary of Phillips Petroleum Company, manufactures polyethylene pipe and fittings in diameter $\frac{1}{2}$ to 72 inch, in six different locations in the USA.

Plexco Performance Pipe[NSF]
1050 IL Route 83, Suite 200, Bensenville, IL 60106-1048, USA
Tel: +1 630 350 3700
Fax: +1 630 350 2704

Founded as Pipeline Service Company, it changed name to Plexco, and was acquired by Chevron. It merged with another Chevron Company, Spirolite.

Plexco has plants throughout the USA and Mexico, specialising in a full range of Pe pipes and fittings.

PWPipe[NSF]
1550 Valley River Drive, Eugene, OR 97401, USA
Tel: +1 541 343 0200
www.pwpipe.com

Recently merged with Eagle Pacific Industries (see above).

Saudi Plastic Product Co Ltd[WRc]
PO Box 4916, Damman 31412, Saudi Arabia
Tel: +966 3 857 5784
Fax: +966 3 857 1969

Uniplas Ltd
Seagoe, Portadown, Co Armagh, BT63 5HU, UK
Tel: +44 2838 333311
Fax: +44 2838 350806

Uponor Oyj[*, DWI]
Kimmeltie 3, FIN-02110, Espoo, Finland
Tel: +358 947 8962
Fax: +358 947 896400
www.uponor.com

Uponor manufactures a wide range of thermoplastic pipe systems for potable and wastewater — PVC and Pe — and hot water distribution via production units in 15 countries. The Finish oil and petrochemical group Neste Oy is a major share holder in the Uponor organisation. Companies within the Uponor group include Wirsbo, Wirsbo Bruk and Ecoflex Polytherm.

Wavin BV[*, DWI]
PO Box 173, 8000 AD Zwolle, The Netherlands
Tel: +31 38 4294911
Fax: +31 38 4294238
www.wavin.com

The Wavin group, via more than 40 sites in 23 European countries, offers a wide range of both polyethylene and PVC pipe systems for all aspects of the water business.

14.2.2 Spiral wound thermoplastic pipes

ADS — Advanced Drainage Systems
3300 Riverside Drive, Columbus, OH 43221, USA
Tel: +1 614 457 3051
Fax: +1 614 538 5204

ADS claims to be the largest manufacturer in the world for polyethylene drainage products, with 21 production units in North America, supplying pipe diameter 3 to 60 inch, having supplied more than 6 billion feet of ADS pipe worldwide (1.8 million kilometres).

Bauku-Troisdorfer Bau- und Kunststoff GmbH*
Industriestrasse 9, 51674 Wiehl, Germany
Tel: +49 2262 72070
Fax: +49 2262 720726

Bauku also manufactures in Canada, the USA, Mexico, India, Iran, South Korea, Norway and Yugoslavia and has licensed operators in Japan (Dainippon Plastics Co Ltd, Osaka), Italy (Societa del Gres, Petosino), Australia (James Hardies Pipelines, Gladesville), Malaysia (Spirolite, Kuala Lumpur) and Taiwan (UDE Co Ltd, Taipei).

Hancor Inc
PO Box 1047, Findlay, OH 45839, USA
Tel: +1 888 367 7473
Fax: +1 888 329 7473

KWH Pipe (see above)

Polypipe Civils Ltd
Union Works, Bishop Meadow Road, Loughborough, LE11 5RE, UK
Tel: +44 1509 615100
Fax: +44 1509 610215

14.2.3 Thermoplastic pipe fittings

Fusion Group[DWI]
Fusion House, Chesterfield Trading Estate, Chesterfield, S41 9PZ, UK
Tel: +44 1246 260111
Fax: +44 1246 450472
www.fusiongroup.co.uk

Georg Fischer
CH 8201 Schaffhausen, Switzerland*, WRc
Tel: +41 52 631 11 11
Fax: +41 52 631 28 00
www.georgfischer.com

Georg Fisher also controls or has a significant share holding in, Tecno Plastics Spa, Busalla, Italy; and Plasson Group, Israel (see below).

GIRPI
Rue Robert Ancel, 76700 Harfleur, France
Tel: +33 2 32 79 60 00
Fax: +33 2 32 79 60 27

Glynwed Pipe Systems — Glynwed International plc*, WRc
Headland House, New Coventry Road, Sheldon, Birmingham, B26 3A2, UK
Tel: +44 121 742 2366
Fax: +44 121 722 7607
www.glynwedpipesys.com

Glynwed Pipe Systems offer products from the following subsidiaries: FIP Spa — PVC, PP, Pe, CPVC and PVDF fittings; AVF Astore Valves and Fittings — PVC compression fittings; Innoge/PE industries — PE electrofusion systems; Friatec — Pe electrofusion systems; Harrington Industrial Plastics Inc — pipe fittings; Philmac Pty Ltd — compression fittings; and GPI Industries Pty (JV) — compression fittings.

Hydrodif — Hydro Difusion SL
Calle, Industria s/n 08295 Sant-Viccenc de Castellet (Barcelona), Spain
Tel: +34 93 833 3960
Fax: +34 93 833 3961

Plasson Ltd
Maagan Michael D.N. Menashe 37805 Israel
Tel: +972 66394711
Fax: +972 66390887
www.plasson.com

Raufoss Water and Gas Division
PO Box 21, N2831 Raufoss, Norway
Tel: +47 61 15 20 00
Fax: +47 61 15 17 20

STP
Ctra de Montcada 608, 08223 Terrassa, Spain
Tel: +34 937 31 27 00
Fax: +34 93784 21 66

14.2.4 Large diameter fabricated fittings and access chambers for water and effluent mains

ADS — Advanced Drainage Systems (see above)

Bauku (see above)

B&M Pipeline Services Ltd
Dock Road South, Bromborough, Wirral, CH62 4SQ, UK
Tel: +44 151 334 0122
Fax: +44 151 334 5067

KWH Pipe (see above)

MB Plastics Ltd
Thermoplastics Division, Mackenzie Industrial Park, Birdhall Lane, Stockport, SK3 0SB, UK
Tel: +44 161 428 0131
Fax: +44 161 428 0134

Nyloplast Europe BV
Mijlweg 45, 3295 KG 's-Gravendeel, The Netherlands
Tel: +31 78 6 73 20 44
Fax: +31 78 6 73 48 99

14.2.5 Thermoplastic well casings and screens

CertainTeed Corp (see above)[NSF]

National Plastics & Building Materials Industry LLC
PO Box 1943, Sharjah, UAE
Tel: +971 6 331 830
Fax: +971 6 335 629

US Filter/Johnson Screens[NSF]
1950 Old Highway 8, PO Box 64118, New Brighton, MN 55112, USA
Tel: +1 651 636 3900
Fax: +1 651 638 3132

14.2.6 Thermoplastic valves

EFFAST
Loc. Pianmercato 5C-5D, 16044 Monleone di Cicagna, Italy
Tel: +39 01 8592399
Fax: +39 01 8592699

Georg Fischer (see above)

Glynwed Pipe Systems Ltd (see above)

SAFI
Route de Montélimar, 26770 Taulignan, France
Tel: +33 4 75 53 56 29
Fax: +33 4 75 53 62 78

14.3 Composite Vessels

14.3.1 Standard composite storage tanks, vessels for storage and treatment of water and silos

Many of these companies will also supply custom-built vessels, although not all of them are listed under this second category (see Section 14.3.2).

Brimar Plastic Fabrications Ltd[WRc]
North Rd, Yates, Bristol, BS17 5PR, UK
Tel: +44 1454 322111
Fax: +44 1454 316955

Campas Technology Ltd
Heol Ffaldau, Brackla Industrial Estate, Bridgend, CF31 2XB, UK
Tel: +44 1656 657482
Fax: +44 1656 767127

Carlier
BP 8, 62470 Calonne Ricouart, France
Tel: +33 3 21 65 54 54
Fax: +33 3 21 65 69 95

Casals Cardona
Moragas 14–20, 08240 Manresa, Spain
Tel: +34 938727000
Fax: +34 938725734

Conder Tanks
Pullman House, Baron Park, Chicken House Lane, Eastleigh, Hampshire, SO50 6RR, UK
Tel: +44 2380 687 100
Fax: +44 2380 687 101

Eternit*
Kuiermanstraat 1, B 1880 Kapelle op den Bos, Belgium
Tel: +32 15 71 80 80
Fax: +32 15 71 80 09

Fiberdur-Vanck GmbH
D 5520 Bitburg-Staffelstein, Germany
Tel: +49 6563 510
Fax: +49 6563 5128

Forbes^{WRc}
Ryston Road, Denver, Downham Market, PE38 0DR, UK
Tel: +44 1366 388941
Fax: +44 1366 385274

FRP
2 Whitehouse Way, South West Industrial Estate, Peterlee, County Durham, SR8 2HS, UK
Tel: +44 1915 865311
Fax: +44 1958 61274

F Z Fantoni
Via Benaco 1, 25081 Bedizzole (BS), Italy
Tel: +39 030 674057
Fax: +39 030 675338

Hermex
Z I, 45270 Bellegarde, France
Tel: +33 2 38 95 02 30
Fax: +33 2 38 90 47 33

Klargester Enviromental Engineering Ltd
College Road, Aston Clinton, Aylesbury, HP22 5EW, UK
Tel: +44 1296 630190
Fax: +44 1296 630263

LF Manufacturing Inc
PO Box 578, Giddings, TX 78942, USA
Tel: +1 409 5428027
Fax: +1 409 5420911

MFG Justin Tanks
8 Cedar Creek Ave, Georgetown,
DE 19947, USA

Tel: +1 800 245 5821
Fax: +1 302 856 3527

Pactech
2032 Foothill Drive, Fullerton, CA 92833, USA
Tel: +1 714 992 1336
Fax: +1 714 738 4582

Standard composite septic and wastewater treatment tanks manufactured in Korea.

Plasticon — Sovap
ZA de l'Eraudiére, F 85170 Dompierre sur Yon, France
Tel: +33 2 51 34 17 18
Fax: +33 2 51 34 14 02

Plavisa
Aptdo 3323, 09007 Burgon, Spain
Tel: +34 947 484658
Fax: +34 947 481733

Polem
PO Box 65, 5004 AB Lemmer, The Netherlands
Tel: +31 514 562447
Fax: +31 514 564214

Reposa
Ctra Mayorga km 1, 24234 Villamanan, Spain
Tel: +34 987 767324
Fax: +34 987 767375

Rousseau SA[*]
F-79160 Fenioux, France
Tel: +33 5 49 75 22 06
Fax: +33 5 49 75 24 44

Selip
Via Provinciale 36, 43012 Fontanellato, Italy
Tel: +39 05 21822306
Fax: +39 05 21821944

SOCAP
La Vega 23, Apdo 35, 48300 Guernica, Viscaya,
Spain
Tel: +34 94 6250326
Fax: +34 94 6255365

Tankinetics Inc
230 Industrial Park Road, Harrison, AK 72601, USA
Tel: +1 870 741 3636
Fax: +1 870 741 0911

Tung Hsing FRP Corp*
12F-5, 270 Chung Hsiao E Road, SEC 4, Taipei, Taiwan
Tel: +886 2 2788 1122
Fax: +866 2 2781 1182

VERA Klippan
PO Box 9, S 26421 Klippan, Sweden
Tel: +46 435 140 90
Fax: +46 435 149 44

14.3.2 Custom-fabricated vessels

AC Plastiques
1395 Montée Chènier, Les Cédres (Quebec) Canada, J7T 1L9
Tel: +1 450 455 3311
Fax: +1 450 452 2037

ACS Plastiques
PO Box 1190, B 7170 Manage, Belgium
Tel: +32 64556301
Fax: +32 64555252

An-Cor Industrial Plastics Inc
Melody Lane, North Tonawada, NY 14120, USA
Tel: +1 716 6953141
Fax: +1 716 6950465

Carlier (see above)

Chemposite Inc*
1758 Webster Avenue, Delta, British Columbia, Canada, V4G 1E4
Tel: +1 604 946 7688
Fax: +1 604 946 7038

Chemposite Inc has its engineering and sales offices in Canada, and two production plants in China.

Forbes Plastics (see above)

Loyal Impact M S B
298A Jalan Gopeng, Gunung Rapat, Ipoh Perak, Malaysia 31350
Tel: +60 5 4668888
Fax: +60 5 3132828

MB Plastics
Reinforced Division, Forward Works, Woolston, Warrington, Cheshire, WA1 4BA, UK
Tel: +44 1925 822811
Fax: +44 1925 818907

MFG Justin Tanks (see above)

Plastic Engineering Works
19 B.T. Compound, Malad (W), Mumbai — 40064, India
Tel: +91 22 88 20 847
Fax: +91 22 88 20 663

Plasticon BV
Parallelstraat 52, 7575 AN Oldenzaal, The Netherlands
Tel: +31 54 1858585
Fax: +31 54 1858586

Plasticon Sovap (see above)

Poly Plast C.P(I) Pvt Ltd
Thakore Estate, Vidyavihar (W), Mumbai Maharashtra, 400 086, India
Tel: +91 22 5116659
Fax: +91 22 5117698

Precision Poly Products
A/17/2/4, TTC MIDC, Khairne, Navi Mumbai Maharashtra, 400 705, India
Tel: +91 22 7672269
Fax: +91 22 7686418

Ruia Chemicals Pvt Ltd
105/2B Ultadanga Main Road, Calcutta — 700 067, West Bengal, India
Tel: +91 33 3562656
Fax: +91 33 3523103

T3 Tanks Manufacturing
80 Geneva Drive, Camps Bay, Cape Town, Western Cape,
South Africa 8005
Tel: +27 21 8423200
Fax: +27 21 8423200

Troy Manufacturing
PO Box 1269, Brockville, Ontario, Canada, K6V 5W2
Tel: +1 613 345 1306
Fax: +1 613 345 1257

Uribe Fiberglass
Carretera Sanchez, Km 19, Barsequillo, Haina, Santa Domingo D.N., Dominican Republic
Tel: +1 809 542 1420
Fax: +1 809 542 1116

14.3.3 Standard high and low pressure composite storage tanks/filter bodies

K&M Plastics Inc
1601 Pratt Boulevard, Elk Grove Village, IL 60007, USA
Tel: +1 847 439 3311
Fax: +1 847 439 1053

Park International Corp
1401 Freeman Avenue, Long Beach, CA 90804, USA
Tel: +1 562 494 7002
Fax: +1 562 494 4809

Structural Group[*, NSF, DWI]
200 Industrial Parkway, Chardon, OH 44024, USA
Tel: +1 440 286 4116
Fax: +1 440 286 6759

14.3.4 Modular composite storage tanks

AC Plastic Industries Ltd
Armstrong Road, Daneshill East, Basingstoke, RG24 8NU, UK
Tel: +44 1256 329334
Fax: +44 1256 817862

Anchor — link Sdn Bhd
15-B, Jalan SS 21/1A, Damansara Utama, 47400 Petaling Jaya, Selangor Darul Ehsan, Malaysia
Tel: +603 7196907
Fax: +603 7177293

Balmoral Composites^{WRc}
Balmoral Park, Loirston, Aberdeen, AB9 2BY, UK
Tel: +44 1224 859000
Fax: +44 1224 859059
Product designation — Balmoral M100.

Bridgestone Corp^{WRc}
10-1 Kyobashi 1 Chome, Chuo-ku, Tokyo 104, Japan
Tel: +81 3 575 23 4111
Fax: +81 3 3535 2581
Product designation — Bridgestone.

Brimar Plastic Fabrications Ltd^{WRc}
North Road, Yates, Bristol, BS17 5PR, UK
Tel: +44 1454 322111
Fax: +44 1454 316955
Product designation — Brimar.

Decca Plastics Ltd^{WRc}
Victoria Mill, Lincoln Street, Preston, Lancashire, PR1 6RE, UK
Tel: +44 1772 825757
Fax: +44 1772 204967
Product designation — DECCAsectional.

Dewey Waters and Co Ltd^{WRc}
Cox's Green, Wrington, Bristol, BS18 7QS, UK
Tel: +44 1934 862601
Fax: +44 1934 862602
Product designation — DWC-B range.

Mitsubishi Plastics Inc^{WRc}
5-2 Marunouchi 2 chome, Chiyoda-Ku, Tokyo 100, Japan
Tel: +81 3 3283 4154
Fax: +81 3 3213 4089
Product designation — Hishi.

Nicholson Plastics Ltd[WRc]
Riverside Road, Kirkfieldbank, Lanark, ML11 9JS, UK
Tel: +44 1555 664316
Fax: +44 1555 663056
Product designation — Longlife.

Permali UK Ltd[WRc]
Bristol Road, Gloucester, GL1 5TT, UK
Tel: +44 1452 528671
Fax: +44 1452 304215
Product designation — Hydroglas.

Robust Tanks Ltd[WRc]
Fountain House, Brindley Close, Holly Lane Industrial Estate, Atherstone, Warwickshire, CV9 2GA, UK
Tel: +44 1827 714343
Fax: +44 1827 714534
Product designation — Robust Tank.

Sekisui Plant Systems Co Ltd[WRc]
21st floor, Tower West, Umeda Sky Building, 1-30, 1-chome, Oyodo-Naka, Kita-Ku, Osaka 531, Japan
Tel: +81 64402508
Fax: +81 64402518
Product designation — Sekisui.

14.3.5 Very large filament wound composite storage tanks (on site manufacture, diameter above 8 m/25 ft)

Ershigs
742 Marina Drive, Bellingham, WA 98227-1717, USA
Tel: +1 360 7332620
Fax: +1 360 7332628

Ollearis SA
PO Box 127, 08107 Martorelles, Spain
Tel: +34 93 5796520
Fax: +34 93 5933616

RPS[*]
740 Feaubeaux Road, Mahone Bay, Nova Scotia, Canada, B0J 2E0
Tel: +1 902 6248383
Fax: +1 902 624 6395
Subsidiary in Germany.

14.4 Thermoplastic Vessels

14.4.1 Rotationally moulded thermoplastic storage tanks

Assman Corp
300 North Taylor Road, Garnerr, IN 46738, USA
Tel: +1 219 357 3181
Fax: +1 219 357 3738

Balmoral Tanks[WRc]
Balmoral Park, Loirston, Aberdeen, AB9 2BY, UK
Tel: +44 1224 859000
Fax: +44 1224 859059

Linpac G.P.G[WRc]
Luton Road, Dunstable, Bedfordshire, LU5 4LN, UK
Tel: +44 1582 664255

14.4.2 Other standard thermoplastic storage tanks

Allibert Manutention
2 rue de l'Egalité, 92748 Nanterre Cedex, France
Tel: +33 1 41 20 09 95
Fax: +33 1 41 20 45 80

Cadiou Industrie
BP 4, 29180 Locronan, France
Tel: +33 2 98 91 73 01
Fax: +33 2 98 91 79 50

Forbes Plastics (see above)

14.5 Other Equipment

14.5.1 High-pressure filters

Park International Corp
1401 Freeman Avenue, Long Beach, CA 90804, USA
Tel: +1 562 494 7002
Fax: +1 562 494 4809

Structural Group[*, WRc]
200 Industrial Parkway, Chardon, OH 44024, USA
Tel: +1 440 286 4116
Fax: +1 440 286 6759

14.5.2 Reverse osmosis pressure vessels

Bekaert Composites[*]
Barrio Atela 6 — Appatado 71, 48100 Munguia, Spain
Tel: +34 946 740316
Fax: +34 946 741369

Codeline — Structural Group[*, DWI]
200 Industrial Parkway, Chardon, OH 44024, USA
Tel: +1 440 286 4116
Fax: +1 440 286 6759

EWWA
2 rue des Quatres Vents, 27170 Le Plenis Sainte Opportune, France
Tel: +33 2 32 35 47 88
Fax: +33 2 32 67 78 50

Phoenix Vessels Ltd[WRc]
Unit 2, The Old Bakery, Lower Tuffley Lane, Gloucester, GL2 5DP, UK
Tel: +44 1452 311673
Fax: +44 1452 310295

Plasticon-Sovap
ZA de l'Eraudière, 85170 Dompierre sur Yon, France
Tel: +33 251 35 17 18
Fax: +33 251 34 14 02

Spaulding Composites Co
1 Monogram Place, Rochester, NH 03866-1748, USA
Tel: +1 800 964 0555
Fax: +1 603 332 5357

14.5.3 Spiral wound membranes

DuPont — Permasep Products
Willmington, USA
Tel: +1 302 451 9681
Fax: +1 302 451 9686
www.dupont.com

FilmTec Corp,
The Dow Chemical Co
Liquid Separations, PO Box 1206, Midland, MI 48642-1206, USA
Tel: +1 800 447 4369
Fax: +1 517 832 1465
www.dow.com

Hydranautics
401 Jones Road, Oceanside, CA 92054, USA
Tel: +1 760 901 2500
Fax: +1 760 901 2578
www.membranes.com

Koch Membrane Systems — Fluid Systems[DWI]
850 Main Street, Wilmington, MA 01887-3388, USA
Tel: +1 619 695 3840
Fax: +1 619 695 2176
www.kochmembrane.com

Osmonics
5951 Clearwater Drive, Minnetonka, MN 55343-8995, USA
Tel: +1 612 933 2277
Fax: +1 612 933 0141
www.osmonics.com

Ropur AG
Grabenackerstrasse 8, CH-4142 Munchenstein 1, Switzerland
Tel: +41 61 415 87 10
Fax: +41 61 415 87 20

Toray Plastic Films Co Ltd
5 Sakura-machi 1-chome, Takatsuki, Osaka 569-0807, Japan
Tel: +81 726 82 1101
Fax: +81 726 85 7700

Trisep Corp
93 S, La Patera Lane, Goleta, CA 93117, USA
Tel: +1 805 964 8003
Fax: +1 805 964 1235
www.trisep.com

14.5.3.1 Other membrane filtration technologies referred to in this handbook

Aquasource
20 Avenue Didier Daurat, 31000 Toulouse, France
Tel: +33 561 36 30 36
Fax: +33 561 54 14 13

Pall Corp
25 Harbor Park Drive, Port Washington, NY 11050, USA
Tel: +1 516 484 3600
Fax: +1 516 434 3651

Zenon Environmental Inc
845 Harrington Court, Burlington, Ontario, Canada, L7N 3P3
Tel: +1 905 639 6320
Fax: +1 905 639 1812

14.5.4 Tank covers and odour control equipment

AFORM Projects
Denford Manor, Hungerford, Berkshire, RG17 0UN, UK
Tel: +44 1488 686433
Fax: +44 1488 686434

AKS Enviromental Inc
4510 E. Tazarv Street, Tucson, AZ 85706, USA
Tel: +1 520 574 3844
Fax: +1 520 574 3811

Ardep Ltd
Greenforge Way, Cwmbran, Gwent, NP44 3UZ, UK
Tel: +44 1633 675600
Fax: +44 1633 876466

Armfibre Ltd
Drove Road, Everton, Sandy, Bedfordshire, SG19 2HX, UK
Tel: +44 1767 651811
Fax: +44 1767 651901

Corporate Engineering Ltd
Culham Mill, Little London Road, Silchester, Reading, RG7 2PP, UK
Tel: +44 118 9701366
Fax: +44 118 9701566

Cor-Pro Inc
PO Box 49067, Jacksonville Beach, FL 32240, USA
Tel: +1 904 246 7550
Fax: +1 904 246 8347

Delsystems
La Croix Rouge, 35530 Brécé, France
Tel: +33 2 99 00 17 72
Fax: +33 2 99 00 21 79

Delta Fibreglass Structures
1235 S. Pioneer Rd, Salt Lake City, UT 84104, USA
Tel: +1 801 977 0091
Fax: +1 801 977 0095

Hygrade Industrial Plastics Ltd
Hunters Lane, Rugby, CV21 1EA, UK
Tel: +44 1788 571316
Fax: +44 1788 541184

Trioplast
BP 556, 62411 Bethune, France
Tel: +33 3 21 63 23 73
Fax: +33 3 21 56 63 69

14.5.5 Biological carrier media

Brentwood Industries Inc
PO Box 605, Reading, PA 19603, USA
Tel: +1 610 374 5109
Fax: +1 610 376 6022
www.brentw.com

2H Kunststoff GmbH
Dieselweg 5, D-48493 Wettringen, Germany
Tel: +49 2557 93900
Fax: +49 2557 939049
www.2h-kunststoff.de

Marley Cooling Tower Co
7401 West 129 Street, Overland Park, KS 66213, USA
Tel: +1 913 664 7400
Fax: +1 913 664 7439
www.marleyct.com

Munters Euroform GmbH^{DWI}
PO Box 1089, D-52011 Aachen, Germany
Tel: +49 241 89000
Fax: +49 241 8900199
www.munters.com

Rauschert Verfahrenstechnik
Paul-Strasse 6, D 96349 Steinwiesen, Germany
Tel: +49 9 262 770
Fax: +49 9 262 97151
www.rauschert.com

14.5.6 Cooling towers (and components)

Addax Inc
PO Box 81467, Lincoln, NE 68501, USA
Tel: +1 402 435 5253
Fax: +1 402 435 0566
www.addax.com

Baltimore Aircoil Co
PO Box 7322, Baltimore, ME 21227, USA
Tel: +1 410 799 6200
www.baltaircoil.com

CASE (Composite Aqua Systems and Equipment) Pvt Ltd
39 Shahpur Jat, Opp. Asian Village Complex, New Delhi-110049, India
www.compositeaqua.com

Delta Cooling Towers Inc
134 Clinton Road, Fairfield, NJ 07004, USA
Tel: +1 800 289 3358
Fax: +1 973 227 0458
www.deltacooling.com

Hamon & Cie[*]
Rue Capouillet 50–58, 1060 Brussels, Belgium
Tel: +32 2 5351239
Fax: +32 2 5370039
www.hamon-industrie.com

Marley Cooling Tower Co (see above)

Polacel BV
PO Box 296, 7000 AG Doetinchem, The Netherlands
Tel: +31 314 37 14 14
Fax: +31 314 34 48 84
www.polacel.nl

Tower Tech
PO Box 891810, Oklahoma City, OK 73189, USA
Tel: +1 405 290 7788
Fax: +1 405 979 2131
www.towertechinc.com

14.6 Composite Structures and Gratings

14.6.1 Manufacturers of composite profiles and gratings

American Grating Inc
19138 Walnut Drive, Suite 204, Rowland Heights, CA 91748, USA
Tel: +1 626 913 5812
Fax: +1 626 913 3192

ATP Srl
Via Casa Pagano 31, 84012 Angri, Italy
Tel: +39 081 947777
Fax: +39 081 947740

Beckaert Composites
Industriepark De Bruwaan 2, 9700 Oudenaarde, Belgium
Tel: +32 55 33 30 11
Fax: +32 55 33 30 40

Bedford Reinforced Plastics Inc
264 Reynoldsdale Road, Bedford, PA 15522-7401, USA
Tel: +1 814 623 8125
Fax: +1 814 623 6032

Biji Profielen BV
Arbeidsweg 4, 4794 SZ Heijningen, The Netherlands
Tel: +31 1 67 52 43 43
Fax: +31 1 67 52 29 11

BSF (Beijing Sunrise Fiberglass)
PO Box 249, Beijing, 101117, China
Tel: +86 10 62279378
Fax: +86 10 62279379

BTM
PO Box 523, 8000 AM Zwolle, The Netherlands
Tel: +31 3 84 54 37 79
Fax: +31 3 84 53 21 87

Chemgrate Corp
4115 Keller Springs Road, Addison, TX 75244-2034, USA
Tel: +1 972 732 8045

Creative Pultrusions
Pleasantville Industrial Park, Alum Bank, PA 15521, USA
Tel: +1 814 839 4186
Fax: +1 814 839 4276

EBO Systems
Boulevard de l'Europe, 67211 Obernai Cedex, France
Tel: +33 3 88 49 51 82
Fax: +33 3 88 49 50 14

EXEL Oy Kivara
Muovilaaksontie 2, 82110 Heinavaara, Finland
Tel: +358 13 73711
Fax: +358 13 7371500

Fiberforce Composites Ltd
Fairoak Lane, Whitehouse Industrial Estate, Runcorn, WA7 3DU, UK
Tel: +44 1928 701515
Fax: +44 1928 713572

Fibergrate Composite Structures Inc
4115 Keller Springs Road, Addison, TX 75001, USA
Tel: +1 972 250 1633
Fax: +1 972 250 1530

Fiberkonst AB
Magasingatan 9, 21613 Malmö, Sweden
Tel: +46 40160905
Fax: +46 40151101

Fiberline Composites AS
Nr Bjertvej 88, 6000 Kolding, Denmark
Tel: +45 75 565333
Fax: +45 75 565281

Hardcore Dupont Composites
801 East Sixth Street, New Castle, DE 19720, USA
Tel: +1 302 326 0900
Fax: +1 302 326 0620

International Grating Inc
7625 Parkhurst, Houston TX, 77028, USA
Tel: +1 713 633 8614
Fax: +1 713 633 3210
www.igicomposites.com

Kuhne Kunststoffwerk
Hagenacker 6, 79761 Waldshut-Tiengen, Germany
Tel: +49 7751 9110
Fax: +49 7751 911130

Lancaster Composite Inc
1000 Houston Street, Columbia, PA, USA
Tel: +1 717 684 4440
Fax: +1 717 684 4445

Menzolit-Fibron GmbH
Alte Hünxer Strasse 139, 46562 Voerde 2, Germany
Tel: +49 2811 3231
Fax: +49 2811 3233

Nordic Supply Composites AS
6260 Skodje/Aalesund, Norway
Tel: +47 702 44500
Fax: +47 702 44549

PlaCo Srl
Via Bruno Buozzi 10, 20047 Brugherio (MI), Italy
Tel: +39 39289181
Fax: +39 392891828

Redman Fisher Engineering Ltd
Birmingham New Road, Tipton, West Midlands, DY4 9AA, UK
Tel: +44 1902 880880
Fax: +44 1902 880446

Rochling Haren KG
D-49724 Haren, Ems, Germany
Tel: +49 5934 701362
Fax: +49 5934701

Seaward International Inc
3470 Martinsburg Pike, Clear Brook, VA 22624-1548, USA
Tel: +1 540 667 5191
Fax: +1 540 667 7987

Top Glass SpA
Via Bergamo 15, 20096 Pioltello, Italy
Tel: +39 2 9291861
Fax: +39 2 92918620

Trival Kompoziti
Bakovnik 3, 1241 Kamnik, Slovenia
Tel: +386 61814394
Fax: +386 61812294

Vello AS
Hahjem, Naeringspark, 6260 Skodje, Norway
Tel: +47 70244770
Fax: +47 70244780

14.6.2 Companies specialising in the design and fabrication of composite structures from pultruded sections

DCN (Direction des Constructions Navales)
Arsenal Maritime, 56100 Lorient, France
Tel: +33 2 97 12 12 12
Fax: +33 2 97 21 12 14

Fibergrate BV
4530 AA Terneuzen, The Netherlands
Tel: +31 115 644550
Fax: +31 115 644599

Fibreglass Grating Ltd
Earls Colne Business Park, Earls Colne, Colchester, Essex, UK
Tel: +44 1787 223993
Fax: +44 1787 224330

Lionweld Kennedy
CED Division, Brignell Road, Middlesbrough, TS2 1PS, UK
Tel: +44 1642 220529
Fax: +44 1642 232978

IMCO Reinforced Plastics Inc
858 N Lenola Road, Moorestown, NJ 08057, USA
Tel: +1 609 235 7254
Fax: +1 609 234 3964

MM Srl
Zona Industriale Udinese, 33100 Udine, Italy
Tel: +39 432602218
Fax: +39 432522253

Motherwell Bridge Nordic
149 Glasgow Road, Wishaw, ML2 7QQ, UK
Tel: +44 1698 266111
Fax: +44 1698 269774

Trioplast
BP 556, 62411 Bethune, France
Tel: +33 321632373
Fax: +33 321566369

14.7 Coating and Linings

Names and addresses of suppliers, and in some cases applicators, of coating or lining formulations are contained in Chapter 13. We also include in this section some of the applicators approved and qualified by either the manufacturers or third parties such as the WRc and the DWI in the UK.

14.7.1 Pipe linings

CertainTeed Corp (see above)

CSR Pipeline Systems
16701 Greenspoint Park Drive, Suite 350, Houston, TX 77060, USA
Tel: +1 281 872 3500
Fax: +1 281 874 0962

Durapipe-S&LP (see above)

Hobas Rohre AG (see above)

KWH Pipe (see above)

Insituform Technologies Inc[*]
1770 Kirby Parkway, Suite 300, Memphis, TN 38138, USA
Tel: +1 901 759 7473
Fax: +1 901 759 7500

Ka-Te Systems Ltd
Leimbachstrasse 38, 8041 Zurich, Switzerland
Tel: +41 1 482 8888
Fax: +41 1 4828978

Pipeway Ltd
Pipeway House, Bishopstone Lane, Sayers Common, Hassoks, West Sussex, BN6 9HG, UK

Tel: +44 1444 230011
Fax: +44 1444 230055

Stanton plc
Lows Lane, Stanton by Dale, Ilkeston, Derbyshire, DE7 4QU, UK
Tel: +44 115 9305000
Fax: +44 115 9329513

Steuler Industriewerke GmbH
Georg Steuler Strasse, D-56203 Höhr — Grenzhausen, Germany
Tel: +49 2624 130
Fax: +49 2624 13350

Subterra Ltd[WRc]
Travel house, Rough Hay Road, Grimsargh, Preston, PR2 5AR, UK
Tel: +44 1772 653750
Fax: +44 1772 653740

Trolining GmbH
Kaiserstrasse, D-53840 Troisdorf, Germany
Tel: +49 2241 853125
Fax: +49 2241 853594

Uponor Anger (see above)

Wavin (see above)

14.7.2 In-situ tank and vessel linings — applicators

Colvic plc
Earls Colne Business Park, Earls Colne, Colchester, Essex, CO6 2NS, UK
Tel: +44 1787 223993
Fax: +44 1787 224330

Kennedy Utility Management Ltd[DWI]
Nash Road, Trafford Park, M16 1JJ, UK
Tel: +44 161 848 7666
Fax: +44 161 872 6887

Mercol Products[DWI] (see above)

Pipeway[DWI] (see above)

Subterra Ltd[DWI] (see above)

14.8 Mini/modular Wastewater Treatment Plants

Bio-plus Environmental Systems Ltd
St Cuthberts House, Durham Way North, Ayecliffe Industrial Park, Newton Aycliffe, DL5 6YS, UK
Tel: +44 1325 313141
Fax: +44 1325 307700

CPI
35 Chemin de servette, 74380 Cranves-Sales, France
Tel: +33 4 50 39 38 43
Fax: +33 4 50 36 76 83

POW[WRc]
Conitor House, Denbury Road, Newton Abbot, Devon, TQ12 6AD, UK
Tel: +44 1626 333385
Fax: +44 1626 333359

WPL Ltd
14/15 Bridge Industries, Broadcut, Fareham, Hampshire, PO16 8SX, UK
Tel: +44 1329 239951
Fax: +44 1329 823111

14.9 Thermoplastic Pipe and Sheet Welding Equipment

Durapipe — S&LP (see above)

Fusion Group Manufacturing
Chesterfield, UK
Tel: +44 1246 260111
Fax: +44 1246 450472

Georg Fisher (see above)

KWH Tech Ltd
PO Box 383, FIN 65101 Vaasa, Finland
Tel: +358 632 65666
Fax: +358 632 65690

Leister Process Technologies
Riedstrasse, CH 6060 Sarnen, Switzerland
Tel: +41 6627474
Fax: +41 6627416

MAROK
S.S. Padana Superiore n. 317, 20090 Vimodrone (MI), Italy
Fax +39 2250 0941
www.marok.it

McElroy
PO Box 580550, Tulsa, OK 74158-0550, USA
Tel: +1 918 836 8611
Fax: +1 918 831 9256
www.mcelroymfg.com

Munsch Kunststoff-Schweiβtechnik GmbH
Postfach 142, D-56221 Ransbach-Baumbach, Germany
Tel: +49 2623 8980
Fax: +49 2623 89821

Omicron Srl
Padua, Italy
Tel: +39 49 8975721
Fax: +39 49 633324
Recently acquired by the Georg Fischer group.

14.10 Specialised Water Treatment Equipment as Described in this Handbook

CP Composites
ZA Le Vivier, 85430 Nieul le Dolent, France
Tel: +33 251 07 91 40
Fax: +33 251 07 94 16

Europelec
8 rue d'Aboukir, 75002 Paris, France
Tel: +33 1 44 82 39 50
Fax: +33 1 44 82 39 51

Feluwa Pumpen GmbH
Beulertweg, 54570 Mürlenbach, Germany
Tel: +49 6594 10-0
Fax: +49 6594 1640

FZ Fantoni (see above)

Mono Pumps Ltd
Martin Street, Audenshaw, Manchester, M34 5DQ, UK
Tel: +44 161 339 9000
Fax: +44 161 344 0727

14.11 Engineering and Contracting Companies Referred to in this Handbook

ACWa Services Ltd
ACWa House, Keighley Road, Skipton, North Yorkshire, BD23 2UE, UK
Tel: +44 1756 794794
Fax: +44 1756 790898

Degrémont
183 Avenue du 8 Juin 1940, 92508 Rueil-Malmaison Cedex, France
Tel: +33 1 46 25 60 00
Fax: +33 1 42 04 16 99

OTV
L'Aquarène, 1 Place Montgolfier, 94417 Saint-Maurice Cedex, France
Tel: +33 145 11 55 55
Fax: +33 145 11 55 00

14.12 Consultants

Al May, Rich Hickman and Associates
4710 Sam Peck Road, Unit 121, Little Rock, AK 72223-5005, USA
Tel: +1 501 227 7436
Specialises in design, manufacture and marketing of composite pipe systems.

GPC — Geoproduction Consultants (Pierre Ungemach and Roland Turon)
Paris Nord II, 14 rue de la Perdrix, 95946 Roissy CDG Cedex, France
Tel: +33 1 48 63 08 08
Fax: +33 1 48 63 08 89
Specialises in the application of coil tubing, composite tubing and casing in both geothermal and oil wells.

ICCR — International Composites Consultancy and Representation
(Derick Scott)
12 Rue Abbé Vacandard, 76000 Rouen, France
Tel: +33 2 35 89 48 43
Fax: +33 2 35 88 61 64
Specialises in design, manufacture, installation and marketing of composite and thermoplastic pipe systems and vessels.

CHAPTER 15

Governing Bodies, Authorities and Associations of Relevance to the Scope of This Handbook

15.1 The Field of Water Supply and Treatment

ADA, American Desalting Organisation
915 L Street, Ste. 1000, Sacramento, CA 95814, USA
Internet: www.desalting-ada.org

AWWA, American Water Works Association
6666 West Quincy Avenue, Denver, CO 80235, USA
Internet: www.awwa.org

DWI, Drinking Water Inspectorate
Floor 2/A1, Ashdown House, 123 Victoria Street, London SW1E 6DE, UK
Internet: www.dwi.detr.gov.uk

EDS, European Desalination Society
Abruzzo Science Park, Via Antica Arischia 1, 67100 L'Aquila, Italy
Internet: www.edsoc.com

EUREAU, European Union of National Associations of Water Suppliers and Wastewater Services
Chaussée de Waterloo 255-bte 6, B-1060 Brussels, Belgium

IDA, International Desalination Association
PO Box 387, Topsfield, MA 01983, USA
(The IDA has a number of affiliates including EDS and WSTA for which further details are included, as well as InDA in India, PakDa in Pakistan, JDA in Japan and ADA and SEDA in the USA.)
Internet: www.ida.bm

IWA, International Water Association
Alliance House, 12 Caxton Street, London SW1H 0QS, UK
(Formed in 1999 by the merger of IAWQ (International Association on Water Quality) and IWSA (International Water Services Association).
Internet: www.iawq.org.uk

IWRA, International Water Resources Association
4535 Faner Hall, Southern Illinois University, Carbondale, IL 62901-4516, USA
Internet: www.iwra.siu.edu

WQA, Water Quality Association
4151 Naperville Road, Lisle, IL 60532-1088, USA
Internet: www.wqa.org

WRc plc
Henley Road, Medmenham, Marlow SL7 2HD, UK
Internet: www.wrcplc.co.uk

WSTA, Water Science and Technology Association
PO Box 20018, Manama, Bahrain
Internet: www.wsta.org.bh

15.2 The Field of Composites and Thermoplastics and Products Manufactured from These Materials

API, American Petroleum Institute
1220 L Street, Northwest, Washington DC 20005-4070, USA
Internet: www.api.org

ASME, American Society of Mechanical Engineers
3 Park Avenue, New York NY 10016-5990, USA
Internet: www.asme.org

ASTM, American Society for the Testing of Materials
100 Barr Harbor Drive, West Conshohocken, PA 19428-2959, USA
Internet: www.astm.org

BPF, British Plastics Federation
6 Bath Place, Rivington Street, London EC2A 3JE, UK
Internet: www.bpf.co.uk

CFA, Composites Fabricators Association
1655 N. Fort Meyer Drive, Suite 510, Arlington, VA 22209, USA
Internet: www.cfa-hq.org

Cooling Technology Institute
530 Wells Fargo Drive, Houston, TX 77090, USA
Internet: www.cti.org

EPTA, European Pultrusion Technology Association
PO Box 344, 3840 AH Harderwijk, The Netherlands
Tel: +31 341 422424
Fax: +31 341 425614
Internet: www.pultruders.com

ISTT, International Society for Trenchless Technology
15 Belgrave Square, London SW1X 8PS, UK
(The ISTT has more than 20 affiliated societies throughout the world, e.g. NASTT (North American Society for Trenchless Technology): www.nastt.org; GSTT (German Society for Trenchless Technology): www.gstt.de; United Kingdom Society for Trenchless Technology: www.ukstt.org.uk)
Internet: www.istt.com

PPI, Plastics Pipe Institute Inc
1825 Connecticut Ave NW, Suite 680, Washington DC 20009, USA
Internet: www.plasticpipe.org

SPI, Society of the Plastics Industry
1801 K St, NW600, Washington DC 20006-1301, USA
Internet: www.plasticsindustry.org

Other major National Associations concerned with the composite and plastics industries are:

Australia — Australian Composite Structures Society
GPO Box 2476V, Melbourne, Victoria 3001, Australia

Canada — CPIA, Canadian Plastics Industry Association
5925 Airport Road, Suite 500, Mississauga, Ontario, Canada, L4V 1W1

Europe — EuPC, European Plastics Converters Association
Avenue de Cortenbergh 66, b4 -PO Box 4, B1000 Brussels, Belgium

Japan — JPIF, Japan Plastics Industry Federation
5-18-17 Roppongi, Minatu-ku, Tokyo 105, Japan

CHAPTER 16

Glossary of Abbreviations/Technical Terms — Bibliography

16.1 Summary of Abbreviations and Technical Terms Frequently used in the Water or Plastic Industries

Abbreviation/ technical term	Description
ABS	Acrylonitrile butadiene styrene
Accelerator	Accelerates the catalytic reaction required to cure a resin, an example is cobalt naphthanate
API 8RD or API 8 round	Standard thread configuration (oilfield specification)
Axial thrust	Axial force generated by the pressurisation of a pipeline, or by an increase in temperature should the expansion of the pipeline be restrained
Binder	Product to hold chopped fibres together, either as a mat or as a preform
BMC	Bulk moulding compound
BOD	Biochemical oxygen demand
Catalyst	Product required to cure a resin, for example methyl ethyl ketone peroxide
Circumferential winding	Filament winding process wherein the fibres are wound at almost 90° to pipe or vessel axis
COD	Chemical oxygen demand
Corrosion barrier	See liner
Cure	Change of state in a resin, from a liquid to a solid state
Cure time	The elapsed time from the moment of addition of a catalyst to a liquid resin up to the moment that the resin has attained a solid condition

Abbreviation/ technical term	Description
DMC	Dough moulding compound
Dual laminate	Combination of a thermoplastic liner with a composite structure
ECTFE	Ethylene-chlorotrifluoroethylene
Exothermic reaction	Heat given off during the curing cycle of a thermosetting resin
FEP	Fluorinated ethylene propylene
FRP	Glass fibre reinforced plastic (also known as GRP)
GAC	Granular activated carbon
Gel	Semi-solid state attained by a resin during a curing process
Gel time	Elapsed time for a resin to attain the state of gelification, once the catalytic process has been initiated
Glass/resin content	Percentage glass/resin content of a composite structure or laminate
GRP	Glass (fibre) reinforced plastic
HDB	Hydrostatic design basis
HdPe or HDPE	High-density polyethylene
HDS	Hydrostatic design stress
HDT	Heat distortion temperature
Helical winding	A filament winding process in which the fibres are wound along a helical reciprocal path
HMW-PE	High molecular weight polyethylene
Inhibitor	Product used to delay or slow up a curing process
Knuckle area	Area between the end or bottom of a tank and the cylindrical shell
LdPe	Low density polyethylene
Liner	Resin-rich or thermoplastic layer used as the internal surface of, for example, a pipe or tank
Mandrel	Cylindrical mould used for the production of pipes and tank walls
Matched taper joint	An identical conical/conical spigot/socket pipe joint
MdPe	Medium-density polyethylene
MEKP	Methyl ethyl ketone peroxide, a catalyst
Mitred fittings	Fitting, such as an elbow or tee-piece, manufactured by cutting and assembling sections of a pipe, as opposed to a moulded fitting
Non-woven fabric	Textile structure obtained by bonding or stitching fibres together, as opposed to a woven fabric
Pa	Polyamide

Abbreviation/ technical term	Description
PBT	Polybutylene terephtalate
PC	Polycarbonate
Pe	Polyethylene
PET	Polyethylene terephthalate
PFA	Perfluoro-alkoxy
PMMA	Polymethylmethacrylate
Polar winding	A filament winding process in which a virtually closed body can be wound in one piece, with both ends and the cylindrical shell being wound in one piece
POM	Polyacetal
Post-cure	A heating process used to obtain a maximum level of cure for a thermosetting resin having undergone an initial curing process
Pot life	Period during which a thermosetting resin may be used in the construction of a laminate, timed as from the moment that the curing system is initiated by the addition of curing agents, up to the time that the resin starts to solidify
PP	Polypropylene
PTFE	Polytetrafluoroethylene
PVC, PVC-U or U-PVC	Polyvinyl chloride
PVC-C	Chlorinated PVC
Regression curve	Method used, in association with a testing programme such as ASTM D2992, to record and then forecast changes in physical properties over specific time periods
RTMP	Reinforced thermosetting mortar pipe, that is a pipe in which aggregate materials have been added to the resin
RTRP	Reinforced thermosetting resin pipe
Shelf life	Storage life of a thermosetting resin or adhesive
Size	Product used to coat glass fibre materials in order to facilitate their adhesion to different resins
Spigot/socket joint	Simple push-in male/female pipe jointing technique and configuration
Surface mat	See veil
TBC	Total bacteria count
TDS	Total dissolved solids
Tex	Unit used to define the linear density of glass fibre, equal to the mass in grams of a length of 1 km, e.g. 2400 tex is equal to a mass of 2400 g/km of roving
T_g	Glass transition temperature

Abbreviation/ technical term	Description
Thermoplastic resin	A resin that can be repeatedly softened and reshaped
Thermosetting resin	A resin once cured by heat and/or chemical reaction or any other means cannot be reshaped
Thrust block	A block, generally in concrete, used to transfer an axial loading in a buried pipe system from the pipe to the surrounding soil
TOC	Total organic carbon
TSS	Total suspended solids
Ultraviolet deterioration	Degradation of a resin due to the exposure to ultraviolet radiation
Veil	Thin glass or synthetic material used on internal or external surfaces of composite pipes and vessels
Warp	The fibres running lengthwise in a fabric
Weeping	Phenomena of leakage due to transmission of a fluid under pressure through the wall of a composite structure
Weft	The fibres running crosswise in a fabric
Woven fabric	Fabric obtained by weaving of glass rovings
Yield	Refers to the linear density of glass fibre, and expressed as number of yards of roving per pound

16.2 Bibliography

16.2.1 The Water Industry

Asian Water. First Asia Publishing PTE Ltd, Singapore
D&WR — The International Desalination & Water Reuse Quarterly. International Desalination Association.
Dow Liquid Separations — FilmTec Membranes. Technical manual
European Approval Systems — effects of materials on water quality. WRc plc, UK
Filtration + Separation, a monthly magazine published by Elsevier Advanced Technology, UK
US Water News, a monthly magazine published by US Water News, USA
Water and Waste Treatment, a monthly magazine published by Faversham House Group Ltd, UK
Water Technology, a monthly magazine published by National Trade Publications Inc, USA
Water Treatment Handbook. Degrémont, France
World Water and Environmental Engineering, a monthly magazine published by Faversham House Group Ltd, UK

16.2.2 The composite and plastics industry

Composites Technology, a bimonthly magazine published by Ray Publishing Inc, USA
Fibreglass Pipe Handbook, published by SPI (Society of Plastics Industry), USA
Reinforced Plastics, a monthly magazine published by Elsevier Advanced Technology, UK
Reinforced Plastics Handbook. Elsevier Advanced Technology, UK
Many suppliers of both raw materials and finished goods publish excellent documents concerning both their own products and composite materials in general, including:
Derakane Epoxy Vinyl Ester Resins – Chemical Resistance and Engineering Guide. The Dow Chemical Co

Evaluating Isopolyester Resins. BP-Amoco

Handbook of FRP Products, Processes and Design Procedures. Vetrotex-CertainTeed, USA

Hobas GRP Pipeline Textbook. Hobas Engineering GmbH, Germany

Success by Design. Published in an interactive electronic form by Smith Fiberglass Products Company, USA

Editorial Index

A

abrasion resistance 148, 201, 238
ABS (acrylonitrile butadiene styrene) 44, 69, 157, 159, 169
absorption towers 220, 260
access chambers 172–3
acoustic emission testing 83, 352
acrylics 67
acrylonitrile butadiene styrene (ABS) 44, 69, 157, 159, 169
acrylonitrile styrene acrylic (ASA) 44
activated sludge treatment 21, 195, 295–7, 331
additives 36–7, 54
adhesive bonding 128, 129, 130, 143, 149, 150, 157–8
aeration 21, 295–7
  filters 21, 275, 277–8
  stripping towers 187
aliphatic polyamines 33
amine cured epoxy resins 33, 35, 54, 57, 133, 135, 205, 254
  see also polyamines
anhydride cured epoxy resins 32, 35, 57, 135, 230, 254
API (American Petroleum Institute) 75
  see also North American standards
Aquasource process 209
aromatic amines 33, 207
ASA (acrylonitrile styrene acrylic) 44
asbestos 112, 338
ASME (American Society of Mechanical Engineers) 75,
  see also North American standards
ASTM (American Society for Testing and Materials) 75,
  see also North American standards
Austria, potable water materials testing 121
auxiliary injection tubing (AIT) 318–19
AWWA (American Water Works Association) 75,
  see also North American standards

B

Bajonet systems 215–16
Bakelite 23
Belgium
  composite pipe systems standards 139
  potable water materials testing 121
Biodisc treatment plants 185
biological filters 281
biomass 277
biomedia, modular 281–4
biotechnology 206
bisphenol A epoxy resins 29, 35
bisphenol A vinyl ester resins 246
bisphenol fumarate resins 26, 29, 30–1, 35
BMC (bulk moulding compound) 43, 54
boiler work 71
bridges 258
brine saturators 186–7
brine tanks 210
British Standards
  composite pipe systems 85, 143
  composite tanks and other equipment 93, 94, 179–84, 188, 307, 347, 348
  thermoplastic pipe systems 96–7, 148, 162
  see also WIS standards
brominated vinyl ester 246
buried fire mains 264
buried pipelines 18, 159, 170–2, 253–4, 317, 333–4
  renovation and rehabilitation 232–44, 324–5, 335–6, 337–40
butt and strap joints 127, 128, 143
butt welded polyfusion 163–6, 168, 336
bypass interceptors 185, 188, 195

C

C-glass 38
catch pits 172
CCTV applications 232, 233
celluloid 23
cellulose acetate 70
cellulosics 67

cement-based mortar linings 5, 6, 233
CEN standards 73
centrifugal casting 55, 56, 57, 62–3
  pipes and fittings 138–41, 145
  vessels 176, 189
centrifugal pumps 280
centrifugal separation 21, 272
cesspools 184, 188, 189
chemical resistance
composite materials 56–7, 59, 65, 82, 83, 126, 134, 257
  dual laminates 154
  testing 82–3
chemicals
  membrane filtration 204
  positive lists 113, 117
  storage and preparation equipment 154, 186, 196, 231
  see also process vessels
chlorendic resins 34, 246
chlorides, and steel corrosion 6
chlorinated polyethylene (CPE) 150
chlorinated PVC *see* PVC-C
chlorine disinfection 16–17, 210, 233
chopped strand mats/chopped fibres 37, 38, 39, 40, 41, 257
  in moulding processes 56–7, 58, 59, 61, 64, 65
CHP (cyclohexanone peroxide) 48
coagulation 15, 310
coatings 231, 238, 244–52
  processes 72
  resin-based 33, 34, 37, 72
  see also liners
COMPOSIT process 210–12
composite materials 24–43
  gravity filter floors 302–3
  linings 239–40
  pipe systems 125–47, 254, 262–3, 264, 315–24
  planks 306–8, 314
  production technology 53–66
  pultruded profiles 65–6, 109, 257–9, 288, 291
  standards 73, 75–94, 109, 254
  tank covers and odour control 218–22, 313–14, 326–7
  tanks and vessels 175, 176–94, 272, 275–8, 313, 328–9, 347–53
  cylindrical 176–90, 210–16, 275, 328–9
  rectangular and square 191–4, 250, 275, 313
  tubing and well casings 224, 227–30
  see also laminates; thermosetting resins
compression joints 158, 159
compression moulding 55, 57–8, 64–5, 67, 178
  pipe fittings 147

concrete, lining 247, 248–52
concrete–based mortar linings 5, 6
constructions
  composite 191–4
  dual laminate 201–2
  protective coatings 244–52
contact moulding 35, 40, 55, 57, 58–9, 176, 178, 267
cooling systems 153, 254, 255
Cooling Technology Institute (CTI) 287–8
cooling towers 284, 287–92
copolymers 67
copper pipes 159
corrosion 5–6, 19–20
  cooling towers 290
corrosion resistance
  composite materials 29, 30, 54–8, 65, 145, 219, 303, 318, 343
  thermoplastic pipes 148, 154, 268
  vessels 201
cost considerations 5–8
CPE (chlorinated polyethylene) 150
crack propagation 152–3
cross-linked polyethylene pipe 153–4
culverts 157
curing 29, 32–3, 36, 54, 212
  agents 138
  compression moulding 64, 191
  continuous filament winding 62
  for dual laminates 48
  HOBAS process 63
cutting (thermoplastics) 71, 72
cyclohexanone peroxide (CHP) 48

D
D-glass 39
DDS (diamino diphenyl sulphone) 33
decantation 16, 272, 273, 304–5
degassers 187, 281, 283, 312
degritters 331
deionised water 11, 155, 255
demineralised water 11, 255
denitrificaton 283
Denmark
  potable water materials testing 121
  standards 113
desalination 203, 206, 207
design standards 73–110, 169–70, 179–84, 348–50
design trends 5–8
diamino diphenyl sulphone (DDS) 33
diffusers 21, 295–7
dip moulding 72
Disc Tube technology 341–3
district heating and cooling 253–4
  geothermal 315–24
domestic effluent 11
  package pumping stations 188, 279–80

package sewage treatment plants 275–8
pipes and other equipment 151, 261, 281–4
treatment processes 19–22
down hole treatment tubing (DHTT) 318–19
Drinking Water Inspectorate (DWI) 116–17, 233, 234
drive shafts 290, 297
drying beds 21
dual laminates 24, 26, 48–9, 56
  constructions 201–2, 231
  pipe systems 149, 154–5, 171
  tanks and vessels 182–3, 185, 189, 201–2, 231, 311, 347–53
ducts 220, 260, 313

E
E-glass 39, 63, 257
economical trends 5–8
ECR-glass 39
ECTFE (ethylene–chlorotrifluoroethylene) 44, 51
  dual laminates 48, 154
  liners 56, 211
  tanks and vessels 196
effluent
  composite vessel applications 184–90, 272–3
  corrosion-resistant resins for 29–32, 247
  pipes 151, 154, 157, 172, 293, 330–2
  treatment process 19–22, 275, 277–8
  see also domestic effluent; industrial effluent
elastomeric seals 140, 159
electrofusion couplers 237
engineering trends 5–8
EPDM 140, 295
epoxy adhesives 130, 132
epoxy foam 316
epoxy resins 24, 26, 31–4, 35, 54
  coatings and linings 232–4, 243, 244–5, 286
  pipes and fittings 57, 126–37, 145, 146, 147, 255, 262–3, 264, 315–21
  pressure vessels 205, 212
  pultrusion 65, 257
  tubing/well casing/screens 224, 228–30
  ethylene–chlorotrifluoroethylene see ECTFE
European regulations, potable water 12, 112–13, 114, 117
European standards (EN and ETS) 73, 113
  composite pipe systems 86–7
  composite vessels 188
  irrigation systems 269
  pultruded profiles 100
  thermoplastic pipe systems 105–7, 148, 162
  thermoplastic tanks and vessels 108,
    see also individual countries
extrusion 68–9, 148, 149–55, 266

F
fabrics, glass fibre 37, 39–41, 57
Factory Mutual (FM) 264, 293, 347
failure 7–8, 317, 318, 352–3
fan blades, cooling towers 289–90
fender piles 266
fenders 266
FEP (fluorinated ethylene propylene) 45, 51
  dual laminates 48, 154
  liners 56, 155, 352
  tanks and vessels 196
filament winding 35, 40, 41, 55, 56, 57, 58, 60–2, 140
  continuous 62, 139, 139–40, 178
filament wound pipes and fittings
  epoxy-based 126–37, 146, 315–24
  polyester 138–41, 142–4
  vinyl ester 138–41, 142–4, 146
filament wound tanks and other equipment 176–8, 189–90, 211, 212, 266, 300–1, 328–9
fill 284, 288, 291
filter screens 258, 298–9
filters
  anthracite 310
  biologically aerated 188, 277–8
  candle/frame 16
  carbon 210
  granular 16
  gravity 310
  flooring 302–3
  high-pressure 203–9
  lamellae 304–5, 310
  sand 210, 278, 310, 342
  thermoplastic 21
  trickling 21, 281, 283
filtration 16, 21, 275, 277–8
  membranes 203–9, 222, 285–6, 310–12
  see also microfiltration; nanofiltration; ultrafiltration
fine bubble air diffusers 295–7
Finland, potable water materials testing 121
fire/flame resistance 29, 83, 145, 154, 257, 262–3, 288–9, 293, 307
fittings
  epoxy-based pipe systems 127–30, 145, 146, 147
  inspection considerations 349–50, 351, 352
  polyester pipe systems 141, 147
  thermoplastic pipe systems 155, 159–61, 162, 167, 169, 172, 268
  vinyl ester pipe systems 141, 143, 145, 146, 147
  well casings 224–30

430 *Index*

flanged composite panel constructions 191–4
flanges 130, 160–1, 349, 350
flocculation 15–16, 20, 196
flooring
  composite planks 306–8, 314
  gravity filters 302–3
FlowGRIP composite planks 306–8, 314
fluoridation, potable water 17
fluorinated ethylene propylene *see* FEP
foaming agents 69
Food and Drug Administration (FDA) 117–18
food industry 204, 247, 264
France
  case studies
    geothermal-sourced district heating 315–21
    nanofiltration plant 310–12
  standards
    composite pipe systems 83–4
    composite tanks and vessels 94
    irrigation systems 269
    thermoplastic pipe systems 99–100
    testing of potable water materials 114, 121–2
FRP *see* glass fibre reinforcement
furane resins 34, 35
fusion welding 162–8, 197, 198–9

G

gas scrubbers 71
gases 22, 154
  geothermal wells 11, 223, 229, 315–21
Germany
  case study, FRP pressure pipeline 333–4
  standards
    composite pipe systems 84–5, 143
    composite tanks and vessels 94
    irrigation systems 269
    thermoplastic pipe systems 98–9
    thermoplastic tanks and vessels 108
    testing of potable water materials 115, 122
glass fibre reinforcement 24, 25, 37–42, 257
  forms of glass fibre 38–42
  linings 239–40, 243, 324–5, 352
  manufacture 37–8, 145
  moulding procedures 54–66, 69, 72
  vessels 185–90, 196, 212–15, 255, 275–8, 301, 351–2
  wastewater treatment pipes 330–32
glass flake linings 243, 245–7
gratings 258–9, 306–8
gravity flow drains 157
grease removal 20, 184, 188
grinders 299
grit removal 20
ground-probing radar (GRP) 232
groundwater stripping systems 187, 292
grouting 234, 241, 242, 250, 322, 338

H

hand lay up 57, 178
  see also contact moulding
handrails 258
hardening agents 36, 54
HdPe (high density polyethylene) 45, 51
  dual laminates 154
  hot gas welding 199
  linings 241–2, 248, 352
  pipe casing 316
  pipes 151–4, 155–7, 162, 167, 169, 255, 268, 335–40
  tanks and vessels 196, 197, 280
heat resistance 246
heating, district 253–4
helium detection 317
HET resins 34–6
high-density polyethylene *see* HdPe
high-frequency welding 70–1
high-performance pipe systems 31
high-performance vinyl esters 29
high-purity equipment 154, 256
high-temperature centrifugally cast epoxy and vinyl ester pipe systems 145
high-temperature vinyl esters 29
HOBAS process 62–3, 139, 140–1
hot gas extrusion welding 162, 200
hot gas welding 162, 199
hot water storage heaters 210
HPPe pipes 151, 167, 234
hygienic surfaces 247–8

I

in-pipe lining 239
industrial effluent 11, 19–22, 342
  pipes and other equipment 151, 261, 303
industrial plants 11, 153, 204, 255
injection moulding (resins) 55, 57–8, 63–4, 67, 212–14
injection moulding (thermoplastics) 67–8
Insituform process 240
inspection chambers 172
inspection considerations 347, 348–9, 350–2
interceptors 188, 195
international standards *see* ISO standards
iron piping 335–6
irrigation 151, 268–9, 280, 293
ISO standards 73–4
  composite pipe systems 87–9
  irrigation systems 269
  quality systems 346–7
  thermoplastic pipe systems 101–5, 169
isophthalic resins 24, 25, 26, 27, 28, 35, 65, 202, 246, 267, 303

Italy
  composite pipe systems standards 139
  potable water materials testing 121

J
Japanese standards
  composite pipe systems 89–90, 139
  composite profiles and associated
    structures 100
  composite tanks and vessels 94
  thermoplastic pipe systems 100
jetties 265–6
jointing and assembly
  composite pipes 127–30, 135–7, 140–1,
    143
  composite planks 307, 314
  thermoplastic pipes 149, 150, 157–69,
    254, 293–4
  well casings 224–5

K
Kuwait, sewer pipe renovation case study
  337–8

L
ladders 258
lake water 10–11
lamellar decanters 304–5, 310
laminates 24, 41
  composition 55–8
  production technology 54–66
  structural 247–8
  see also dual laminates; liners
landfill sites 151, 341–3
LdPe (low density polyethylene) 45, 151–3,
  268
leachate treatment 341–3
lead pipes 8, 112, 159, 242
leakages 317, 350–2
linear polyesters 67
liners
  production technology 55–7
  resin-based 33, 34, 37, 182–3
  thermoplastic 155, 182, 211, 234–9,
    248–52, 255, 256
  see also dual laminates
lining of pipe systems 155, 231–44
lip ring seals 129, 149, 159, 205
lock gates 267
locking systems, epoxy pipes 129–30, 135–7

M
machining, thermoplastics 71–2
maintenance platforms 258
Malaysia, HdPe submarine pipeline case
  study 339–40
manhole-to-manhole lining systems 239,
  250

manholes 157
MdPe (medium density polyethylene) 45,
  151–3, 167, 234, 268
mechanical joints 159
media, modular 281–4
membranes 203–9, 222, 285–6, 310–12,
  341–3
metal pipe systems 155, 159
  see also copper; iron; lead
metallic materials 3, 5, 6, 26, 112
metaphenylene diamine (MPDA) 33
microbiological growth 13–14, 112, 205
microfiltration (MF) 204, 206–7, 285–6
modular biomedia 281–4
modular composite storage tanks 190,
  191–4, 300–1
  components 215–16
modular settling tanks 272–3
modular sewage treatment packages 275–8
module slip–lining system 242
moulding *see* compression; contact; dip;
  injection; resin transfer; spray up
  moulding
moulding compounds 42–3, 54
MPDA (metaphenylene diamine) 33
MRS (minimum required strength) 151
municipal effluent *see* domestic effluent

N
NACE International (National Association of
  Corrosion Engineers), pipe lining
  standards 243–4
nanofiltration (NF) 204, 207, 310–12
National Sanitation Foundation (NSF)
  standards 117–20, 293, 347
Netherlands laws and regulations, potable
  water 113, 115–16, 122
nitrification 283, 284
North American laws and regulations,
  potable water 113, 117–20, 121
North American standards 73
  composite pipe systems 75–83, 139, 143
  composite tanks and vessels 75, 90–4,
    204, 347, 348
  cooling towers 287–8
  fire resistance 83, 307
  irrigation systems 268–9
  pipeline rehabilitation 240, 243–4
  potable water materials 117–20, 121
  pultruded profiles 109–10
  quality plans 347
  thermoplastic pipe systems 83, 95–6, 148,
    162, 169
  thermoplastic tanks and vessels 108
  thermoplastics lined metal pipe systems
    155
  well casings 223–4
Norway, potable water materials testing 121

432 *Index*

notch sensitivity 152
novolac-based epoxy resins 29, 35
novolac-based vinyl ester resins 246, 329
Nylon 45

O

O-ring seals 129, 130, 140, 149, 159, 170, 205, 228
oblation system 190, 300–1
odour control 21–2, 71, 218–22, 281, 283, 284, 306, 313–14
oil interceptors 184, 188, 195
oil pollution 184, 185, 342
oil removal, effluent treatment 20, 204, 284
oilfield-related applications 135, 227, 262–3
orthophthalic polyester resins 26, 27–8, 35
outfalls 153, 157
oxidation 15, 295
ozone disinfection 16–17, 310

P

Pa (polyamides) 45
package pumping stations 188, 275, 279–80
package sewage treatment plants 188, 275–8
Palatal A 410–11 isophthalic resin 24, 25, 48
PB (polybutylene) 154
PBT (polybutylene terephthalate) 45
PC (polycarbonate) 45
Pe (polyethylenes) 45, 69, 72
 access chambers 172–3
 cooling towers 287
 covers and odour control equipment 220, 222
 linings 234, 238–9, 240, 246
 pipes 151–4, 159, 172–3, 254, 268
 structured media 281–4
 tanks and vessels 195–7, 201–2, 211–12
 see also HdPe; LdPe; MdPe; PET
Pe-X 153–4
PET (polyethylene terephthalate) 45, 242
petrol interceptors 184, 188, 195
PFA (perfluoro-alkoxy) 45, 51
 dual laminates 48, 154
 liners 56, 155
 pipes and fittings 256
pharmaceuticals 206
phenolic resins 34, 35, 65, 257
piles 266
pipe bending radii 153
pipe bursting 237–8
pipe systems
 composite
  products and applications 125–47, 254, 330–2
  standards 75–90, 254
  lining 231–44, 245
 in situ 231–44, 324–5
 thermoplastic
  products and applications 125, 148–73, 293–4
  standards 95–107, 169–70
  miscellaneous uses 253–6, 262–4, 268–9
 see also buried pipelines
planks, composite 306–8, 314
PMMA (polymethylmethacrylate) 46
pollution control 20, 187, 188, 195, 209, 251, 260, 275, 285, 310–11, 341–3
polyacetal (POM) 46
polyamides (Pa) 45
polyamines 33, 35, 69
polybutylene (PB) 154
polybutylene terephthalate (PBT) 45
polycarbonates (PC) 45, 69
polyester resin concrete (PRC) 243
polyester resins 24, 26–8, 35, 54
 composite floors 303
 moulding 67
 pipes and fittings 138–41, 142–4, 147, 254
 structures and gratings 257–9, 312
 tubing/well casings/screens 224, 227–8
polyether ketone 46
polyethylene *see* Pe
POLYGLASS process 210, 212–15
polymethylmethacrylate (PMMA) 46
polyolefins 67
polyphenylenes 46, 67
polypiperazine 312
polypropylene *see* PP
polystyrene (PS) 47, 67
polysulphones 46, 67, 206, 207–8, 312
polytetrafluoroethylene (PTFE) 47, 72, 155
polyurethane (Pu) 24, 34, 233, 244, 247, 254, 297, 316
polyvinyl chloride *see* PVC
POM (polyacetal) 46
positive lists (chemicals) 113, 117
potable water 12–18
 consumption 12
 filters 209, 285–6, 302–5, 310–11
 laws and regulations 112–13
 portable production unit 209
 quality 12–14, 111–23, 205, 286
 storage tanks 191–4, 195–6
 testing requirements 12–14
 thermoplastic pipes 151, 293–4, 335–6
 treatment processes 14–18
 wastage 12
power stations 153, 290
PP (polypropylene) 46, 51
 cooling towers 288, 291
 covers and odour control equipment 220
 dual laminates 48
 gravity filter floors 303

Index   433

hot gas welding 199
linings and coatings 56, 72, 155, 211, 248, 352
pipes 154, 162, 169, 255
structured media 281–4
tanks and vessels 196, 197, 201–2, 216
PPE (polyphenylene ether) 46
PPO (modified polyphenylene oxide) 46
PPs (flame-retardant PP) 154
PPS (polyphenylene sulphone) 46
PRC (polyester resin concrete) 243
press moulding *see* compression moulding
pressure pipelines 333–4
pressure testing 79–82, 204–5
pressure vessels 31, 203–16, 255, 310–12
process vessels 71, 186, 196
process water 11, 204, 255, 292
production standards 7–8, 73–110, 346–50
production technology 7–8, 53–72
PS (polystyrene) 47, 67
PTFE (polytetrafluoroethylene) 47, 72, 155
Pu *see* polyurethane
pull winding and forming 66
pultrusion 39, 55, 58, 65–6
planks 306–8, 314
profiles and sections 65–6, 100, 109–10, 257–9, 288, 291
sheet pile 266
pump chambers 172
pumping stations 188, 279–80
pumps 280
PVC (polyvinyl chloride) 47
bi-axial 150, 169
extrusion 69
high-frequency welding 70
linings and coatings 56, 72, 202, 237, 243, 352
pipes 149–55, 159, 169, 255, 293–4
structured media 281–4
use in cooling towers 288, 291
well casings and screens 223–7
PVC-C (chlorinated PVC) 47, 48, 51
extrusion 69
liners 56, 202
pipes 149–50, 154, 157, 169
PVC-U (unplasticised PVC) 47, 48, 51, 154, 157, 196, 197, 237, 311
PVDC (polyvinylidene chloride) 155
PVDF (polyvinylidene fluoride) 47, 51
extrusion 69
linings and coatings 48, 56, 72, 155, 211, 248, 352
pipes 48, 154, 162, 166, 169, 255, 256
tanks and vessels 196, 197, 216, 256
PVDF-HP 154, 256

Q
quality control 7–8, 181–2
quality plans 346–7
quality systems 346–53

R
R-glass 39
radar surveys 232
radiation resistance 256
rainwater 10
reactors 71, 196
recuperation, effluent 19–20
remote lining systems 239–40
reservoirs 247
resin transfer moulding (RTM, resin injection) 55, 57–8, 63–4, 67, 212–14
composite pipe fittings 147
composite tank panels 191
resins, thermosetting 24–36
curing 29, 32–3, 36
retaining walls 266
reverse osmosis (RO) 203, 204, 206, 207, 210, 262, 285–6
Disc Tube technology 341–3
Rilsan 45, 72
river water 10–11, 310–12, 342
rolldown lining 234–6
rotational moulding 69–70, 195–6
rovings, continuous 37, 38, 39, 40, 41, 58
in moulding processes 58
rovings, woven 37, 39–41, 189, 257
in moulding processes 57, 58, 64, 65
rubber-modified resins 29, 30

S
S-glass 39
safety 351–3
salt saturators 186–7
SAN (styrene acrylonitrile) 44
saws 71
scrapers 222
screening 15, 20, 298–9
screens 223–30
SCRIMP process 266
scrubbers 71, 187, 220, 260, 281, 286
SDR (standard dimension ratio) 151
seals *see* elastomeric; lip ring; O-ring seals
seawater 11
desalination 203, 206, 207, 342
intakes 153, 157
and metallic materials 6
oilfield applications 262
salinity 11
separation technology 203–9
separators 184–5
septic tanks 184, 188, 189, 195
Sequential Batch reactor (SBR) 195
service life 7, 348, 349
service pressure 151–2, 153
service temperature 148, 153–4

settling 20, 21
  tanks 188, 210, 247, 260, 272–3, 275–8, 284
sewage treatment plants 195, 247
  package 188, 275–8
sewage treatment processes 19–22, 331
  see also effluent
sewers 11
  lining and relining 157, 231–44, 248–52, 322–3, 337–8
shafts, lining 322–3
silage effluent tanks 188, 189
silage tanks 184
silos 71, 175, 186, 188, 273
silt interceptors 184, 188
slip lining 234, 242
sludge treatment 21, 195, 222, 295–7, 331
slurry transport 154
SMC (sheet moulding compound) 43, 54
socket fusion welding 162–3
  electrically heated 166–8
solid epoxy resins 33–4
solvent cement bonding 149, 150, 157–8
solvent contamination 187
Spain, potable water materials testing 121
spark testing 352
spigot seals *see* O-ring seals
spiral wound filtration membranes 204, 311–12
spiral wound pipe 155–7
spiral wound tank shell construction 196
spray up moulding 55, 57, 58, 59–60, 176
standards 8, 26, 73–110
  contact laminates 58–9
  dual laminates 48–9
  irrigation systems 268–9
  metallic materials 26
  production 7–8, 73–110, 346–50
  well casings 223–4
  see also individual countries
steel 5–6, 26, 65, 71, 204, 258
  casings 318, 320
  coated/lined 72, 155, 231, 267, 286, 317, 324
  fittings 258
  strip laminate (SSL) 135–7
storage tanks 185–6, 189, 195, 231, 328–9
  lining 231, 244–52
storm sewers 157
stripping systems 187, 284, 292
structural laminates 247–8
structural profiles, pultruded 65–6, 100, 109–10, 257–9, 288
structured media, modular 281–4
styrene acrylonitrile (SAN) 44
submarine pipelines 324–5, 339–40
sulphates, and steel corrosion 6
sunlight/ultraviolet light 57, 171, 205, 283

surface veils 37, 38, 39, 41–2, 56
  moulding 56, 58, 65
Sweden
  composite pipe systems standards 139
  potable water materials testing 121

T

tanks
  composite
    production 58–60, 328–9
    products and applications 176–90, 191–4
    standards 73, 82–3, 91–4, 175, 186, 347–53
  covers 70, 218–22, 258, 306, 313–14, 326–7, 329
  dual laminate 195–7, 201–2
  failure 352–3
  inspection considerations 347, 348–9, 350–52
  linings 231, 244–52
  thermoplastic
    production 70, 71, 163
    products and applications 175, 195–7
    standards 108, 175
  see also storage tanks; vessels
terephthalic resins 26, 31
testing
  composite pipe systems 76–83, 127, 139, 264
  composite tanks and vessels 91–3, 205, 350, 352
  potable water 12–14, 114–23
  thermoplastic pipes 170–72
  see also standards
thermal expansion 153
thermoforming 70
thermoplastics 44–51
  linings 155, 182, 211, 233–9, 248–52, 255, 256, 352
  moulding 67–72
  pipe systems 148–73, 254, 255, 268
    standards 95–107, 148, 268
  sheets 70, 71, 163, 197, 198–200
  welding 70, 71, 163, 197
tank covers 222
tanks and vessels 195–7
  welding 70–71, 154, 161–9, 197, 198–200, 256
  well casing and screens 223–7
thermosetting resins 24–36
  glass fibre reinforcement 37–42
  moulding 53–66
  see also composite materials
threaded and bonded (TAB) joints 128, 129, 130, 134, 230, 349
threaded joints 161, 230
thrust blocks 170

Index   435

transfer moulding *see* resin transfer moulding (RTM/resin injection)
TROLINING systems 240–42
tubing 224, 227–30, 318–19
tubular sedimentation media 283–4

U
U-PVC *see* PVC-U
UK
  FRP folding covers case study 326–7
  odour control case study 313–14
  potable water legislation and testing 113, 116, 121
  standards *see* British Standards
  vertical composite sewer pipe case study 322–3
ultrafiltration (UF) 204, 206, 207, 209, 285–6
ultrapure water 11, 154, 155, 204, 256
ultrasonic testing 352
ultraviolet light/sunlight 57, 171, 205, 283
ultraviolet radiation curing 239
ultraviolet radiation disinfection 16–17, 312
underfloor heating schemes, pipes 154
underground fire mains 264
underground pipelines *see* buried pipelines
underground storage tanks 185–6
underground water sources 11
Underwriters' Laboratories 264, 347
UP (unsaturated polyester) resins 24
USA
  Disc Tube landfill leachate treatment trial 342
  FRP pipe in wastewater treatment plant case study 330–32
  FRP pipeline insertion case study 324–5
  FRP storage tanks case study 328–9
  HdPe relining of potable water main case study 335–6
  see also North America

V
vacuum filtration 21
vacuum forming 70
valve chambers 172
ventilation systems 154

vessels
  composite 184–90, 187, 204–5
  dual laminate 195–7, 201–2
  internal structures 196, 202
  lining 231, 244–52
  standards 73, 82–3, 91–4, 108, 204
  thermoplastic 195–6
  see also tanks
vinyl ester resins 26, 27, 28–30, 35, 54, 246, 329
  composite floors 303
  linings 30, 243
  pipes and fittings 57, 138–41, 142–4, 145, 146, 147, 237
  pultrusion 65, 257
  tanks 329
  tubing/well casings/screens 228
vinyls 67
VipLiner system 242, 338

W
walkways 258, 314
wastewater treatment
  odour control 21–2, 71, 218–22
  processes 19–22, 298–9, 331
  pumping stations 188, 279–80
  vessels and pipes 71, 215, 218–22, 281–4, 302–3, 304–5, 328–9, 330–2
  see also effluent
Water Byelaws Scheme (UK) 116
water industry overview 9–22
Water Industry Specifications *see* WIS standards
water injection 262
water mains 172, 324–5, 335–6
water makers 262
water quality, potable water 12–14, 114–20
Water Research Centre (WRc) 97–8, 113, 114, 116, 123, 234
water softeners 210
water towers 247
water treatment vessels 184–90, 196, 215–16, 260, 313–14;, *see also* tanks
welded shell construction 196
welding, thermoplastics 70–71, 154, 161–9, 197, 198–200, 256, 352
well and borehole equipment 135, 223–30, 315–21
WIS standards 97–8, 162, 169, 243